endgame

VOLUME II

RESISTANCE

Also by Derrick Jensen

Railroads and Clearcuts
Listening to the Land
A Language Older Than Words
Standup Tragedy (live CD)
The Culture of Make Believe
The Other Side of Darkness (live CD)
Strangely Like War
Walking on Water
Welcome to the Machine

endgame

VOLUME II

RESISTANCE

derrick jensen

SEVEN STORIES PRESS

new york • london • melbourne • toronto

A Seven Stories Press First Edition

Seven Stories Press
140 Watts Street
New York, NY 10013
www.sevenstories.com

In Canada:
Publishers Group Canada, 559 College Street, Suite 402, Toronto, ON M6G 1A9

In the UK:
Turnaround Publisher Services Ltd., Unit 3, Olympia Trading Estate, Coburg Road, Wood Green, London N22 6TZ

In Australia:
Palgrave Macmillan, 15–19 Claremont Street, South Yarra VIC 3141

Library of Congress Cataloging-in-Publication Data

Jensen, Derrick, 1960–
  Endgame / by Derrick Jensen.
     p. cm.
  Vol. 1 has subtitle: The end of civilization; vol. 2: Resistance.
  Includes bibliographical references.
  ISBN: 978-1-58322-730-5 (pbk. : alk. paper)
  ISBN: 978-1-58322-724-4 (pbk. : alk. paper)
  1. Nature--Effect of human beings on. 2. Human ecology. 3. Civilization. I. Title.

GF75.J45 2006
304.2--DC22
2006003645

College professors may order examination copies of Seven Stories Press titles for a free six-month trial period. To order, visit www.sevenstories.com/textbook or fax on school letter-head to (212) 226-1411.

Book design by Jon Gilbert

Printed in the USA

9 8 7 6 5

Contents

Volume I: The Problem of Civilization

Volume II: Resistance

Premises

PREMISE ONE: Civilization is not and can never be sustainable. This is especially true for industrial civilization.

PREMISE TWO: Traditional communities do not often voluntarily give up or sell the resources on which their communities are based until their communities have been destroyed. They also do not willingly allow their landbases to be damaged so that other resources—gold, oil, and so on—can be extracted. It follows that those who want the resources will do what they can to destroy traditional communities.

PREMISE THREE: Our way of living—industrial civilization—is based on, requires, and would collapse very quickly without persistent and widespread violence.

PREMISE FOUR: Civilization is based on a clearly defined and widely accepted yet often unarticulated hierarchy. Violence done by those higher on the hierarchy to those lower is nearly always invisible, that is, unnoticed. When it is noticed, it is fully rationalized. Violence done by those lower on the hierarchy to those higher is unthinkable, and when it does occur is regarded with shock, horror, and the fetishization of the victims.

PREMISE FIVE: The property of those higher on the hierarchy is more valuable than the lives of those below. It is acceptable for those above to increase the amount of property they control—in everyday language, to make money—by destroying or taking the lives of those below. This is called *production*. If those below damage the property of those above, those above may kill or otherwise destroy the lives of those below. This is called *justice*.

PREMISE SIX: Civilization is not redeemable. This culture will not undergo any sort of voluntary transformation to a sane and sustainable way of living. If we do not put a halt to it, civilization will continue to immiserate the vast majority of humans and to degrade the planet until it (civilization, and probably the planet) collapses. The effects of this degradation will continue to harm humans and nonhumans for a very long time.

PREMISE SEVEN: The longer we wait for civilization to crash—or the longer we wait before we ourselves bring it down—the messier the crash will be, and the worse things will be for those humans and nonhumans who live during it, and for those who come after.

PREMISE EIGHT: The needs of the natural world are more important than the needs of the economic system.

Another way to put Premise Eight: Any economic or social system that does not benefit the natural communities on which it is based is unsustainable, immoral, and stupid. Sustainability, morality, and intelligence (as well as justice) require the dismantling of any such economic or social system, or at the very least disallowing it from damaging your landbase.

PREMISE NINE: Although there will clearly someday be far fewer humans than there are at present, there are many ways this reduction in population may occur (or be achieved, depending on the passivity or activity with which we choose to approach this transformation). Some will be characterized by extreme violence and privation: nuclear Armageddon, for example, would reduce both population and consumption, yet do so horrifically; the same would be true for a continuation of overshoot, followed by a crash. Other ways could be characterized by less violence. Given the current levels of violence by this culture against both humans and the natural world, however, it's not possible to speak of reductions in population and consumption that do not involve violence and privation, not because the reductions themselves would necessarily involve violence, but because violence and privation have become the default of our culture. Yet some ways of reducing population and consumption, while still violent, would *consist* of decreasing the current levels of violence—required and caused by the (often forced) movement of resources from the poor to the rich—and would of course be marked by a reduction in current violence against the natural world. Personally and collectively we may be able to both reduce the amount and soften the character of violence that occurs during this ongoing and perhaps long-term shift. Or we may not. But this much is certain: if we do not approach it actively—if we do not talk about our predicament and what we are going to do about it—the violence will almost undoubtedly be far more severe, the privation more extreme.

PREMISE TEN: The culture as a whole and most of its members are insane. The culture is driven by a death urge, an urge to destroy life.

PREMISE ELEVEN: From the beginning, this culture—civilization—has been a culture of occupation.

PREMISE TWELVE: There are no rich people in the world, and there are no poor people. There are just people. The rich may have lots of pieces of green paper that many pretend are worth something—or their presumed riches may be even more abstract: numbers on hard drives at banks—and the poor may not. These "rich" claim they own land, and the "poor" are often denied the right to make that same claim. A primary purpose of the police is to enforce the delusions of those with lots of pieces of green paper. Those without the green papers generally buy into these delusions almost as quickly and completely as those with. These delusions carry with them extreme consequences in the real world.

PREMISE THIRTEEN: Those in power rule by force, and the sooner we break ourselves of illusions to the contrary, the sooner we can at least begin to make reasonable decisions about whether, when, and how we are going to resist.

PREMISE FOURTEEN: From birth on—and probably from conception, but I'm not sure how I'd make the case—we are individually and collectively enculturated to hate life, hate the natural world, hate the wild, hate wild animals, hate women, hate children, hate our bodies, hate and fear our emotions, hate ourselves. If we did not hate the world, we could not allow it to be destroyed before our eyes. If we did not hate ourselves, we could not allow our homes—and our bodies—to be poisoned.

PREMISE FIFTEEN: Love does not imply pacifism.

PREMISE SIXTEEN: The material world is primary. This does not mean that the spirit does not exist, nor that the material world is all there is. It means that spirit mixes with flesh. It means also that real world actions have real world consequences. It means we cannot rely on Jesus, Santa Claus, the Great Mother, or even the Easter Bunny to get us out of this mess. It means this mess really is a mess, and not just the movement of God's eyebrows. It means we have to face this mess ourselves. It means that for the time we are here on Earth—whether or not we end up somewhere else after we die, and whether we are condemned or privileged to live here—the Earth is the point. It is primary. It is our home. It is everything. It is silly to think or act or be as though this world is not real and primary. It is silly and pathetic to not live our lives as though our lives are real.

PREMISE SEVENTEEN: It is a mistake (or more likely, denial) to base our decisions on whether actions arising from them will or won't frighten fence-sitters, or the mass of Americans.

PREMISE EIGHTEEN: Our current sense of self is no more sustainable than our current use of energy or technology.

PREMISE NINETEEN: The culture's problem lies above all in the belief that controlling and abusing the natural world is justifiable.

PREMISE TWENTY: Within this culture, economics—not community well-being, not morals, not ethics, not justice, not life itself—drives social decisions.

Modification of Premise Twenty: Social decisions are determined primarily (and often exclusively) on the basis of whether these decisions will increase the monetary fortunes of the decision-makers and those they serve.

Re-modification of Premise Twenty: Social decisions are determined primarily (and often exclusively) on the basis of whether these decisions will increase the power of the decision-makers and those they serve.

Re-modification of Premise Twenty: Social decisions are founded primarily (and often exclusively) on the almost entirely unexamined belief that the decision-makers and those they serve are entitled to magnify their power and/or financial fortunes at the expense of those below.

Re-modification of Premise Twenty: If you dig to the heart of it—if there is any heart left—you will find that social decisions are determined primarily on the basis of how well these decisions serve the ends of controlling or destroying wild nature.

WE SHALL DESTROY ALL OF THEM

Much sheer effort goes into avoiding the truth: left to itself, it sweeps in like the tide.

Fay Weldon[1]

I'VE BEEN THINKING AGAIN ABOUT THOMAS JEFFERSON'S LINE: "In war, they will kill some of us; we shall destroy all of them." Part of the reason the civilized consistently defeat the indigenous (and the rest of the natural world) is that Jefferson's line isn't rhetoric, but is every day and in every way—from the most forceful to the most mundane—made manifest in the real world. This was obviously true when Jefferson said it, as Indians complained time and again that no matter how many white soldiers or settlers (land thieves) they drove off or killed, more came to replace them. It's true today as the indigenous inevitably lose land to the civilized—oh, the indigenous may stop an occasional oil well, but in time the dominant culture will take it all. It's true right now in Iraq, where the Iraqis may kill a few U.S. soldiers, but the U.S. military has shown itself willing to destroy as many Iraqis as are necessary to maintain control of the region. It's true on the home front: Part of the power of the police at protests is that each and every police officer (and each and every protester) knows that while one or two unruly protesters may throw a couple of rocks or bottles that bounce off cops' body armor, if need be the police could, and more importantly would, destroy every last protester. If (or rather I should say when, because it's going to start happening soon) some of us start liberating rivers by taking out dams, the fear is always that while we might be able to knock down a few, in time we'll all be dead or in prison. The same, we fear, will be true when we start sinking factory trawlers, pulling down cell phone towers, ripping up roads or parking lots, demolishing Wal-Marts, burning logging trucks, dismantling corporate headquarters, and so on.[2] We may get some of them, but we're pretty sure they're going to destroy every last one of us. Or at least they'll try. The fear this implacability raises is palpable, and paralyzing. If they're going to overwhelm us anyway with their machine-like grinding away at everything we hold dear, why don't we just accept the goods and services the system uses to reward those whose silence makes them collaborators in the ongoing destruction of the living planet? If wild salmon and wild humans are doomed species—and if I've bought into a civilized morality that conveniently declares withdrawal a virtue (I just read a quote from an American Buddhist in an interview ostensibly about activism but more accurately about justifying inaction in the face of immorality: "What I do for peace and justice is split wood"[3]) and fighting back a sin—shouldn't I at least make myself as (physically) comfortable as possible in the meantime?

Jefferson's statement is true in all parts of the culture. A corollary to the fourth premise is that whenever violence goes up the hierarchy (the "wrong" way) it's usually limited and immediate, while violence going down the hierarchy (the "right" way) is systematic, repeated (often incessantly[4]), and has at least the potential to be (and most often is) absolute and wide-ranging. Nearly always it is in one way or another relentless. This is as true on the personal and familial level as it is on the social and cultural. It is true on the interspecies level, where the occasional shark, for example, may take a bite—most often accidentally—out of someone swimming in its home,[5] while systematic violence against sharks—violence that is destroying them—never ceases.

It doesn't have to be this way. Thomas Jefferson and the rest of the civilized need not have a monopoly on an absolute determination to win. The stakes—life on the planet—are high enough I believe it's time we as well as they append "at all costs" to our determination to win. Hell, I think it's long past time we determined to win at all.

What this means is that I've been wondering what our resistance would look like and what it would accomplish—what the world would look like—if those of us who care about life on the planet leveled the playing field. What if we said Jefferson's statement back to those who are killing the planet, and what if we meant it?

What if we said to the police, "You will beat and shoot some non-resisting protesters (or how about the regular old everyday people cops kill: between four and six Americans die every day because they encounter police[6]), but we shall hold you accountable, and shall destroy all of you who do that"? What if we said to those who are killing rivers, "You will be able to stop some of us, but we shall destroy all dams"? What if we said to those in power, "You will be able to imprison or kill some of us, but we shall destroy all harmful economic activities"? What if we said, "In the war you are waging against the world, you will kill some of us, but mark our words, we shall destroy all of this civilization that is killing the planet"?

Even more important, what if we meant it?

Even more important than that, what if we put it into action?

❆ ❆ ❆

I'm not stupid. I know that part of the reason those in power can say that in war they shall destroy all of us is that they have millions of people with guns who are willing to do exactly that.[7] Determining that we will stop those who are destroying the planet is not sufficient for us to win, but it is a necessary first step. And so far most of us have not even made that step. Far from most of us being determined to win at all costs, I'm not even sure most of us want to win at all. Many pacifists over the years have chided me for speaking of winning. It's divisive and dualistic, they say: "If someone wins, then someone has to lose. If we're all creative enough we can find ways so all of us win." Tell that to the marlins, the tiger salamanders, the orangutans. It's easy to speak of everyone winning when you make yourself blind to the suffering of those you exploit and those you allow to be exploited. There are already winners and there are already losers, and expediently ignored in all this talk of everyone winning is that the world is already losing. Further ignored is that when the world loses, we all lose. And also expediently ignored is that, as Tecumseh's father Pucksinwah said, you cannot make peace with a culture, a people, that is trying to devour you.[8] War has long since been declared and is being waged against the world, and a refusal to acknowledge this war does not mean it's not happening. That American Buddhist can split all the wood he wants, and it's not going to save a single species.[9]

A primary reason so many of us do not want to win this war—or even acknowledge that it's going on—is that we materially benefit from this war's plunder. I don't know how many of us, even those of us who pretend to fight for the natural world, would be willing to give up our automobiles and cell phones, our hot showers and frozen yogurt, our electric lights and microwave ovens, our grocery and clothing stores,[10] and far more fundamentally the system that leads to these monumentally expensive technologies, and even more fundamentally than that our identity as civilized beings, even if all of these artifacts, this system, this identification, are killing us and more importantly killing the world. The right-wing bumper sticker says, "You can have my gun when you pry it from my cold, dead fingers," but it's not just guns: we're going to have to pry rigid claws off steering wheels, cans of hair spray, TV remote controls, and two-liter bottles of Jolt Cola. Each of these individually and all of these collectively are more important to many people than are lampreys, salmon, spotted owls, marbled murrelets, sturgeons, tigers, our own lives.

Of course we don't want to win. We'd lose our cable TV.

I want to win.[11] With the world being killed, with the forests gone or going, with the oceans being vacuumed and toxified, I want to win, and will do whatever it takes to win. I will match the insane determination and fiery hatred of the dominant culture with a grounded and relentless determination and a just as fiery and fierce love—and hate—of my own.

A friend recently gave me a mug with a photograph on it of Geronimo and some other Indians, all holding guns. The caption is not my second favorite that often accompanies this photo—"My heroes have always killed cowboys"— but instead my favorite: "Homeland Security: Fighting Terrorism since 1492." This friend said the mug made both her and another friend sad, but she couldn't figure out why. Her friend replied, "Because they didn't win."

I told her the mug didn't make me sad at all.

She asked why.

I said, as I had to Ward Churchill, "*Yet.* We didn't win *yet.*"

"Do you think we'll win?" She asked.

"I have no doubt," I said. "We are going to win."

☾ ☾ ☾

Those who are uncomfortable with winning and losing have a point, though, which is that in normal relationships winning and losing (except in games) can be very harmful. If I disagree with someone I care about, if I care more about winning than I do the subject at hand, we will have one sort of discussion (and a pointless, stupid one). If on the other hand I care more about dialog and the subject than I do about winning, we will have a different, altogether more pleasant conversation. In any sort of loving relationship, winning and losing shouldn't really matter.

The problem, however, is that we can't generalize this understanding to exploitative situations. If someone breaks into a woman's home and is trying to rape her, winning—defined as surviving, or not being raped, or however the woman wishes to define it—is far more important than maintaining the relationship.

The question really boils down to this: If the relationship is more important to you than the subject at hand, then it's inappropriate to speak of winning. If the subject is more important to you than the relationship, then it becomes eminently appropriate to speak of winning, and to try to win.

☾ ☾ ☾

When I was a child, growing up in Colorado,[12] sometimes windstorms frightened me. Not the sort of windstorms I experience now, living in a redwood forest, where the branches dance and clash, and the trunks groan and creak and mew like kittens or sing like whales.[13] That never scares me: the wind only delights and entrances me, even when, as happens, it causes trees to fall. Most of the trees, too, do not seem to mind, but seem at least to me to be experiencing the invigoration of a rigorous cleansing that takes away, literally in this case, the deadwood.

Those are not the storms I experienced as a child. Sometimes during winters in Boulder the jet stream drops to the earth and gets funneled through canyons to explode onto the plains at over a hundred miles an hour. Anemometers blow away. So do picnic tables, swing sets, mobile homes. My childhood home lost part of its roof, and a family friend heard a ripping, bursting blast, then rushed to her living room and was surprised to find herself looking up at the stars: her roof was gone. I remember one day in fourth grade sitting in school when suddenly a window gave way. I dove to the floor against the wall, though not before shards of glass pierced my back.

But these are not the images that scared me.

My childhood home had picture windows all along its west side, facing the mountains perhaps five miles away. In the kitchen, a sort of breakfast nook extended a few feet from the straight line of that western contour. When the winds would come, often for days at a time, I would sit, only briefly, because I could not bear it, and watch the entire west wall of that breakfast nook suck in and out a couple of inches with every gust. The plate glass windows along the rest of that side would bend and bow, and I would picture them shattering, first one alone, and then all at once. I would hurry downstairs to my room, get into bed, pull the covers high, and listen to the wind roar and the house moan.

Two nights ago I dreamt I was back in that house. I was upstairs. It was night. I could hear the wind. I saw a few aquariums that everyone had forgotten for months or years or longer, and I hurried to feed the fish in case they were still alive. Some were. That made me happy. Just as I closed the lid on the final tank I heard the wind rise to its fiercest pitch. I began to rush to that side of the house to try to hold the structure in place, but before I got there the house imploded, and I was thrown back against what had been the other wall.

I live now near the ocean, and sometimes I walk to the end of a long rocky peninsula, to watch waves crash onto tumbled boulders at water's edge, maybe thirty feet away and twenty feet below me. I do this when the waves are neither so low that they lap against this meeting place of sea and land, liquid and solid, nor so high that they could roll over me and carry me away. I make sure to stand behind row after row of rocks so if I get surprised I will not die. And then I watch, with eyes that grow wider as each wave comes in higher and higher against the horizon then smashes and scatters against the rocks below. Sometimes a slight mist makes it up to me. A couple of times, though, the ocean has

reminded me that it is not there for my entertainment, that it has power and purposes all its own. Once I misjudged the height of a wave till the very last
moment, then crouched behind the rock I used as a shield. I felt a heavy spray
wash over me, and thought I'd made it, until suddenly I felt the entire weight
of the ocean fall upon my back and throw me to the ground. Another time I saw
a wave even larger than this approaching from a fair ways off. I turned to run,
but in the maze of boulders made it only a few steps before the wave washed over
the end of the whole peninsula and threw me hard against a rock on the far side
of the little basin where I'd stood.

The same night I dreamt about my childhood home imploding I dreamt of
the ocean. I was in my childhood home, in my childhood bedroom, only now
the house was high on a hill overlooking the sea. I looked out the window at the
waves below. On the horizon I saw a wave far larger than any I had seen before.
It seemed to pick up speed as it got closer. I stood, transfixed, as it climbed the
hill toward the home of my childhood, and as it smashed into the house, shat-
tered and scattered windows, tore apart walls, ripped the house from its foundation, and carried it all away. I was left standing alone on the hill. I saw a car.
I got in, and drove to the ocean's edge.[14] I saw a wave far larger than the one
before and tried to drive away. The car got stuck in the sand. I got out and began
to run, but I knew it did not matter how far or how fast I ran, for the wave would
keep coming and would cover the entire world.

I used to think that Thomas Jefferson's statement—"In war, they will kill
some of us; we shall destroy all of them"—was simply appalling, and revealed
the abusive and genocidal mindset behind civilization's destruction of everyone
and everything it touches. Beyond that, my mind would shut down: I was so disturbed by his statement and the reality behind it that I wasn't able to think
clearly. I recognized its accuracy, but then before allowing myself to fully inter-
nalize the implications I put it in a box and went about my business. A couple
of years later his statement climbed out of the box, now grown into the realization that not only is the statement appalling, disturbing, accurate, and all that,
but that if we are to survive we need to be able to match it in its single-minded
attention to winning. But the implications of this realization once again scared
me too much, and so I put *this* understanding back into the box. There it
remained again for a couple of years, until it has come out again more recently,
now even more grown. Yes, Jefferson's statement is still appalling, disturbing,
accurate, and so on, and yes, if we are to survive we must combat it by an equal
or even greater determination, but now I understand that Jefferson's statement
has meanings I feel certain he could never have understood, meanings that

reveal the endpoint of civilization, for Jefferson's statement is—in a way he *never* intended—in at least one sense a simple statement of natural fact.

Before we talk about that we have to return for a moment to the Jews who participated in the Warsaw Ghetto uprising. When I wrote earlier about those Jews who resisted having a higher rate of survival than those who went along, and then at least implied that if we resist the ongoing holocaust around us we also may ultimately have a higher rate of survival than those who do not resist, I was in one sense cheating. There was a major difference between them and us. The Jews weren't attempting to take on the Nazis by themselves. By the time they rebelled, the Germans were losing ground in the East. Those Jews who were able to escape the Ghetto into the forests would have had to hang on for maybe fifteen months until the Soviets arrived to take the region from the already overextended and demoralized Germans. Never mind that the Soviets were as bad in their own way as the Nazis. Those in the resistance still had some hope of ultimate rescue, and the Soviets really were by this time crushing the Germans. We, on the other hand, don't have a huge army of Russians or anyone else to liberate us and the planet from the modern global empire. We've got to liberate ourselves and the planet on our own.

Maybe.

Maybe not.

Maybe the difference is not so great as it seems, and this is where my new-found understanding of Jefferson's statement comes in. Jefferson said his statement about the civilized destroying Indians, but if instead we invert his meaning so the statement describes the natural world destroying civilization, the statement becomes even more true than Jefferson ever intended: "In war, they will kill some of us; we shall destroy all of them." If you wage war on the natural world, you may be able to kill the passenger pigeons, the tigers, the salmon, the frogs, but the natural world shall surely destroy all of you. Every last one. Civilization may have the power to destroy much of the natural world and many tribes of wild nonhumans and humans, but the wild earth will ultimately destroy every last tank and gun and airplane, every last electrical wire, every last cell phone tower, every last rail line, every last factory trawler, every last logging truck, every last skyscraper, every last dam, every last civilized human being who opposes it.

Don't bet against it.

Thus the dreams. Both dreams share a central image of the natural world destroying my childhood home, which is not only the abusive family in which I was reared, but also the abusive culture that housed me. In the dream of wind, the fish still surviving in forgotten aquariums tells me that despite civilized efforts to manage and imprison wild creatures, and more broadly the wild, and despite all of us forgetting that these creatures and the wild even exist, and despite our consequent starving of wild creatures and the wild,[15] some of the wild sustains. Some of the wild will survive. In the dream of the ocean, me getting into a car and driving to the water's edge immediately after the structure in which I was reared was torn from its foundations tells me that many of us will not give up on the technologies of civilization even after the fundamental power of the natural world rips our cultural home to shreds. Many of us will ride our technology into even further danger, as I rode it right down to the beach, and then when it is far too late many of us will try to ride our technology to safety, even as huge waves loom over us,[16] huge waves that will engulf the entire world, and leave no place of safety for those who have declared war on the wild. In the end, the world will do whatever is necessary to destroy those who are trying to kill it.

❨ ❨ ❨

This understanding is not, of course, based only on dreams, but on simple logic, on the long history of civilizations foundering as they destroy the landbases on which they depend, and on what is clearly before us. You cannot destroy a world and live on it. It is only this culture's monumental arrogance, abusiveness, narcissism, and stupidity that causes so many people to believe that they can ignore the needs of the world, that they can manipulate it, that they can poison it, that they can blithely exploit and consume it, that they can take from it without giving back, and that the world will, like a good victim, continue to support those who are killing it, and that it will never fight back.

❨ ❨ ❨

It doesn't really matter in this case whether you are a mechanistic sort who does not believe that the world can and does and will fight back, or whether you are more of an animistic sort, who, as I do, experiences the world as full of volition. If you're the former, you can talk about ecosystems collapsing and losing their ability to support human (and if you're in an especially expansive mood, nonhuman) life. If you're the latter, you can talk about the planet withdrawing its support for this wretched culture, and fighting back, undermining and ultimately destroying that which is attempting to kill it.

WINNING

I attribute my success to this: I never gave or took an excuse.

Florence Nightingale[17]

WHY I WANT TO WIN, TAKE ONE. Just today I read the following article. This was in the British Newspaper *The Independent*. The headline reads: "Why Antarctica will soon be the *only* place to live—literally." The article: "Antarctica is likely to be the world's only habitable continent by the end of this century if global warming remains unchecked, the Government's chief scientist, Professor Sir David King, said last week.

"He said the Earth was entering the 'first hot period' for 60 million years, when there was no ice on the planet and 'the rest of the globe could not sustain human life'. The warning—one of the starkest delivered by a top scientist[18]— comes as ministers decide next week whether to weaken measures to cut the pollution that causes climate change, even though Tony Blair last week described the situation as 'very, very critical indeed'.

"The Prime Minister—who was launching a new alliance of governments, businesses and pressure groups to tackle global warming—added that he could not think of 'any bigger long-term question facing the world community'.

"Yet the Government is considering relaxing limits on emissions by industry under an EU scheme on Tuesday.

"Sir David said that levels of carbon dioxide in the atmosphere—the main 'green-house gas' causing climate change—were already 50 percent higher than at any time in the past 420,000 years. The last time they were at this level—379 parts per million—was 60 million years ago during a rapid period of global warming, he said. Levels soared to 1,000 parts per million, causing a massive reduction of life.

"'No ice was left on Earth. Antarctica was the best place for mammals to live, and the rest of the world would not sustain human life," he said. Sir David warned that if the world did not curb its burning of fossil fuels 'we will reach that level by 2100'."[19]

This article, describing the end of most life on this planet, did not make the front page of the *San Francisco Chronicle*, or any other paper I know. In fact it was not considered newsworthy enough to print at all. The front page of the *Chronicle*, did, however, carry an article about how "Google finally unveiled its long-awaited stock offering last week . . ."[20]

You and I both know that the threat of the end of much life on this planet will not cause those in power to sufficiently change what they are doing, and we both know that this threat will not cause most people in this culture to care sufficiently to act. We also know that most of those who act will not do so with a level of urgency commensurate with the situation.

But we also know that there are those who will. And these people are going to succeed. They are, with the planet's help, going to win.

Civilization needs to be brought down now.

(((

If you are not yet convinced that the mass of Americans will never be on our side, try this: Find a person on the street. Sit next to someone on the bus. Walk into your boss's office. Go to your neighbor's house. Talk to your parents. Tell them about the article I just mentioned. Make clear that Great Britain's chief scientist says that if we maintain the current trajectory most of this planet will likely become uninhabitable for humans (and surely for most other mammals as well, and surely for most other creatures as well). You needn't even mention toxification of the total environment, irradiation of the total environment, biodiversity crash, or anything else. Just mention this. That's part one. Part two: Ask the person, What are you going to do about this? What are you willing to do to make sure that the planet remains habitable? Part three: Repeat this process with person after person after person.

Now my question for you: Do you think that those people who are not willing to take drastic action—*whatever* action is necessary—in order to make sure the planet remains habitable are reachable by any means? If they will not fight back with all the world at stake—literally, physically, in all truth—when will they ever fight back?

(((

WHY I WANT TO WIN, TAKE TWO. It's a beautiful day, and solitary bees are flying low to the ground, buzzing around their homes, then crawling underground to deliver food to their unhatched babies. Small black spiders scurry everywhere, and I see an ant carrying an impossibly large piece of wood from who knows where to who knows where for who knows what reason. There's a slight breeze, and the tips of redwood branches sway softly. A small blue butterfly lands on my elbow. I walk to the pond. Tadpoles hang beneath the surface, and if I get too close they dive and wriggle their fat bodies into the mud.

Caddisfly larvae, looking for all the world like clumps of wet duff (probably because their armor *is* clumps of wet duff) trundle along reeds. Bright blue dragonflies dip their abdomens into the water, laying eggs, and tiny mayflies hover there, too. A couple of mayflies must have been caught earlier in spiderwebs, for now they're motionless, suspended.

I sit cross-legged on the ground a couple of feet from the edge, and begin to edit this morning's work. A quick movement catches me, and I see that a gray jumping spider has landed on my hand. Fearful of accidentally crushing it, I try to wipe it off on a piece of grass. It slips around my hand, always away from the grass and toward me. I let it stay. It turns to look at me, and I look back at it. I lift my hand so I can better see its gray face and many black eyes. It shifts, too, to keep my face always in view. I shift my hand. It shifts its body. I put my hand back on my knee, and begin to write with my other. The spider moves to the edge of my right hand that is closest to my left, clearly considers the distance, and finally jumps. It makes it. I stop writing. It peers again at my face, then walks to my wrist. I'm wearing a long-sleeved shirt, and the spider crawls in and out of the folds, stopping now and again to look up at me. It gets to my shoulder. It stops. It looks at me. I look at it, eyes straining to focus this close. I don't know how long it stays there. Maybe five minutes. Maybe ten. Then it makes its way back down to my wrist, to my hand, and jumps off into the grass.

Life is really, really good.

That's why I want to win.

《 《 《

If we're going to win, we're going to have to be smart. If we are stupid or careless, those in power will kill or imprison us.[21] Being smart sometimes involves just being smart, and sometimes it's even simpler, and involves thinking things through clearly. But sometimes before we can think things through clearly, it is helpful to have some tools or common language we can use to facilitate that process.

《 《 《

Before we talk about how to win, we have to talk about what we mean by winning. And before we talk about what we mean by winning, we have to talk about the difference between strategy and tactics.

Strategic goals are your largest-scale objectives. For the rebels in *Star Wars* this might have meant overthrowing the Empire, or at least liberating the rebel planets from its oppression. A strategic goal of the Nazis in World War II was to gain access to necessary resources, such as oil. A strategic goal of the United States right now is to gain access to necessary resources, such as oil.

Strategic goals can also be slightly smaller. While the largest strategic goal of the United States in the Civil War was to defeat the Confederacy—disallow the states that constituted the Confederacy from seceding from the United States—there were certainly many intermediate strategic goals: dismember it by gaining control of the Mississippi, cut it off from the rest of the world through a blockade of its ports, and so on.

Individuals can have strategic goals, too. One person's strategic goal may be to make a lot of money. Another's may be to get married. Still another's may be to bring down civilization.

One's strategies will be the methods or means by which one hopes to accomplish one's strategic goals. Or, as my dictionary puts it on the grandest scale, a strategy is "1(a)1: the science and art of employing the political, economic, psychological, and military forces of a nation or a group of nations to afford the maximum support to adopted policies in peace and war: (2) the science and art of military command exercised to meet the enemy in combat under advantageous conditions." If one has as a strategic goal of gaining and maintaining access to necessary resources, such as oil, one's strategies will be *how* one goes about achieving this goal. One example would be invading Iraq. Another would be maintaining "friendly" governments (which by virtue of their "friendliness" by definition become freedom-loving™ and democratic™) wherever there happens to be oil. Another would be the Nazis invading Russia to get at the oil fields in the Caucasus (and to push the Russians farther from the Ploesti oil fields in German-allied Rumania).

Intermediate between strategic and tactical goals are operational goals. For the rebels in *Star Wars*, an operational level goal was to take out the Death Star. For the Nazis in World War II one operational goal was to attempt to secure Rostov so they could push on toward Baku and Tblisi in the Caucasus. Another operational goal was to suppress partisans so rail lines remained open. An early operational goal for the Allies in the Normandy invasion was to establish enough of a beachhead to allow troops massed there some level of mobility. And so on.

Tactical goals are the smallest scale. If we're going to take Rostov, we may need to take this particular bridge to our left. And that factory to our right has snipers in it who must be killed or driven out. Tactics are *how* you accomplish these tactical goals. How are you going to secure this bridge? How are you going to take out the snipers in that factory? Who will provide covering fire as this unit moves forward? So far as the *Star Wars* example we've been discussing, the rebels have to get a ship close enough to the exposed tube to drop in a proton torpedo. How can Luke fly in without getting killed?[22]

The person who wants to get rich thinks strategically, operationally, and tactically as well. So does the person who wants to get married. We've all heard, for example, of men or women who tactically feign excitement over their potential partner's hobbies (the man goes dancing, the woman goes to football games, both feign an interest in sexual intimacy, and each feigns an interest in getting to know *who* the other person *is*[23]) in order to achieve their strategic goal of marriage. This reveals the importance of precision in defining one's goals, since while these tactics may lead to marriage, nobody said anything about a happy marriage.

I thought about strategic, operational, and tactical goals when I was in college. Strategic goal: Become a writer. To do this I would need the time to find out who I am and what I love, the time to think about what I want to write, the time to practice writing. This meant that part of my strategy had to involve gaining large blocks of uninterrupted time for me to think and to not think, to feel and then to feel some more.

The question became, how do I get that? Operational goal: Find a way to not get a job. (This became in a sense its own strategic goal: Whether or not I ever became a writer, I vowed to never work a job I didn't love: Why would anyone ever do anything so silly as sell one's life for mere money?) Because my first degree was in mineral engineering physics from the Colorado School of Mines, I knew I could get a high paying job right after college. One possible means to achieve my operational goal[24] would be to get a job for an oil company, work hard and live cheaply, then retire in a few years with a nest egg that would, the plan went, last me long enough for me to start making money writing. Another plan I considered, and you may not believe this, was to join the military.[25] The navy has something called the Nuclear Propulsion Officer Candidate (NUPOC) Program where they pay college juniors and seniors signing bonuses of something like 10K, and then pay another $2,500 per month while they go to school. In exchange the kids have to spend five years in the navy and another three years in the reserves after they graduate. In the navy they act as nuclear officers on submarines.[26] I thought I'd have a heck of a nest egg when I got out. I'd save my money through college, and then of course I'd save all my money when I was on the submarine: what could I buy there? I was very excited about this. I'd waste eight years of my life, and then I'd be ready to go. How bad could it be?

I found out. The Navy flew me to Charleston, South Carolina, where recruiters gave me and a bunch of other students tours of submarines. That's when I understood.

Do you ever have those moments where suddenly you make a quantum jump in understanding, where you see the world so differently that you cannot imagine how you could have perceived it any other way before? Do you have those times when this new understanding makes you feel as though up until that moment you must have been deluded or asleep or just plain stupid?

I had a couple of those on the submarine. The first came as we stood cramped in a narrow hallway, and one of the recruiters spoke excitedly about the nuclear missiles this sub could launch at targets thousands of miles away. Suddenly, and remember that I warned you my realization made me feel I must have been stupid, I understood that this program had nothing to do with Jacques Cousteau. I had somehow been under the impression that the United States government would pay me to go to school so that afterwards I could go down in a submarine to research the habitat needs of squid and octopi. I hadn't considered that I might actually be doing my part to blow up the world.

Soon after, I realized that while most of the sailors I talked to hated their jobs and would gladly have blown up the submarine if they could have gotten away with it and if it would have meant they could have gone home (and still received a paycheck), many of the officers actually got off on the power that they held, not only over the sailors but over the technologies at their fingertips. They clearly enjoyed the fact that they—though they didn't use this language— were part of a team that together could more or less end life on the planet. I felt stupid about this as well: Given how much I had read about the military, the banal nature of their arrogance and depravity surprised me far more than it should have.

A question not often enough discussed among those killing the world and too often discussed among those of us trying to stop them has to do with the relationship between strategy and tactics on one hand, and the morality of those strategies and tactics on the other. The first option of those in power is to explicitly delink the latter from the former. Indeed, as George Draffan and I discuss in *Welcome to the Machine*, the civilized generally consider themselves to be highly rational, and a great working definition of rationalization is that it is the deliberate elimination of information unnecessary to achieving an immediate task. This information to be eliminated can certainly include morality. As we say in that book: "If your goal is to maximize profits for a major corporation, all you need to do is ignore all considerations other than that. If your goal is to maximize gross national product (that is, the rate at which the world is converted into products), then all you need to do is ignore everything that might stand in the way of production." The second option of those in power, when de-linking morality and strategy (or tactics, what have you) falls too absurd, is, as mentioned earlier, to invoke claims to virtue. Given the insanity of most of the members of this culture, it is almost not possible for claims to virtue to be too absurd to be accepted by many. The point is that by *claiming* the moral high ground on the strategic level those in power then exempt themselves from morality on the tactical level: Under the sign of the Cross (or capitalism, freedom™, democracy™, or civilization) we just this once (time and again) go forth and conquer, and in doing so we may end up committing any number of what would be considered, if someone else did them, atrocities.

Those of us at least pretending to oppose the horror that is this culture play a complementary role in this same game. We de-link morality and strategy as surely as do those in power. They carry strategy and tactics without effectively concerning themselves about morality. We carry morality without concerning ourselves with effective strategies and tactics.[27] As always, this is a convenient game for all who play it. We all get what we want (or have been enculturated to want). Those in power get to extend their perceived control. We get to feel moral. All players get to reap the material rewards of playing this game. And of course wild humans and wild nonhumans are still driven extinct.

We members of the too-loyal opposition often spend far more time discussing morality than we do either strategy or tactics (or heaven forbid, actually accomplishing something). More to the point, when we do discuss strategy or tactics, nine times out of ten (or more realistically 999 times out of 1000) we don't *need* to discuss morality because we've so thoroughly internalized premise four of this book that the possibility of violating it under any circumstances is entirely incomprehensible. Our morality has become morality™, defined, designed, and pre-packaged for us by those who are killing the planet. And our strategy and tactics have become strategy™ and tactics™, also defined, designed, and pre-packaged for us by those who are killing us, defined, designed, and pre-planned to be ineffective.

It's all a very interesting, bizarre, unfortunate, and self- and other-destructive manifestation of the same old fragmentation that characterizes this culture. Those in power define and carry for us morality on the largest scale, ignoring morality on the smallest scale, and we carry and abide by (their) morality on the smallest scale, ignoring morality on the larger scale.[28] Because those in power come to represent all that is good and great, because they are advancing civilization, because they are developing natural resources, because they are bringing democracy™, freedom™, the free market™, Christianity™, and McDonald's, they may lie and kill to achieve their ends. We, on the other hand, must always act honorably (or rather, honorably™). We must never lie. We must never kill. They do the dirty work. We carry the morality. The world burns. In the meantime, those of us on the inside of this game continue to have our participation purchased through the carrot of the material benefits of exploitation and the stick of repression.

It should come as no surprise that the mass of us abandon responsibility for interpreting large (and small) scale morality to our betters. That's how the system *works*. It cannot work without this. This abandonment is central to *every*thing that is wrong with this culture, and it is central to the explorations in this book. From birth we are trained to abandon this responsibility by almost every authority figure, at almost every moment. Priests mediate between us and God, and translate God's words and intents and desires for us (no, not *for* us, but *instead of* us, *against* us) so that we will not be tempted to misunderstand, to come to any conclusions not in line with their own, to interpret our own morality ourselves, or get it directly from God, or get it directly from the land. Scientists, too, stand between us and God—now called knowledge™—and interpret God's (or nature's, or knowledge's) will for us. Instead of us. Against us. Judges mediate between us and God—now called justice™— and interpret God's (or the law's) will for us. Instead of us. Against us. Teachers mediate between us and education, determining for us what and how and when we shall learn. The list goes on.

In each of these cases we give up our own authority, our own responsibility to make choices.

Violence—the defensive birthright of every being, no matter how peaceful— falls prey within this culture to the same fragmentation, the same insane division of labor. Soldiers and police (and rapists and abusers) carry violence for the rest of us. *They* are violent. *We* are not. We are, instead, moral. But the distinction isn't really that distinct; it's all part, as I've been saying, of the same sick game. Civilization—like any other abusive relationship—is based on force, on violence, on theft, on murder, on exploitation, and whether or not the pacifist (or you or I) pulls a trigger matters not a whit for culpability.

Hate is the same. Soldiers hate. They are trained to hate. That's what boot camp is *for*.[29] We, on the other hand, do not hate.[30] Hatred is for killers. We do not kill.

Thus we do not hate. The soldiers carry our hatred for us, and they do our killing for us. So do the police. We carry morality for them. And the poor groan under the weight of the soldiers' boots—surrogates for our own—stomping on their faces.

We are cut off from our own violence, and we are cut off from our own nonviolence. We have neither. We are split off shells, partial people pretending to be whole but only completed by our split off and disavowed twins.

It's all about disconnection. This culture is based on disconnection. Man (strong) versus woman (weak), man (good) versus nature (flawed), thought (honest) versus emotion (misleading), spirit (pure) versus flesh (polluted), love (good) versus hate (bad), serenity (good) versus anger (bad), nonattachment (good) versus attachment (bad), nonviolence (righteous) versus violence (evil), and so on *ad nauseum.* So often I've heard pacifists and others say we need to get rid of all dualism, that by speaking of those who are killing the planet as my enemy I am perpetuating the same dualisms that got us here. But striving to eradicate dualism is perpetuating the same dualism! This time it's nondualism (good) versus dualism (bad). It's all nonsense. The problem isn't that there are pairs of opposites. Opposites simply exist. Nor is the problem that there are values assigned to these opposites. We can—and I certainly would—argue against the values chosen by this culture for each of these poles, but the truth is that the different poles do have different values. And that leads to the real problem, which is the word *versus.* Yes, men and women are different. But they are not in opposition; instead they work together. Yes, humans are different than nonhumans (as it would also be true that salmon are different than nonsalmon, and redwoods are different than nonredwoods). But they are not in opposition; instead they work together. Thought is different than emotion. But they are not in opposition; instead they work together. Spirit is different than flesh. But they are not in opposition; instead they work together. Love is different than hate, serenity is different than anger, nonattachment is different than attachment, nonviolence is different than violence. But they're not in opposition; each of these paired opposites works together. Dualism is different than nondualism. But they are not in opposition; instead they work together. Duh.

What happens if you reconnect? What happens if you make choices as to when you should think, and when you should feel? What happens if your thoughts and feelings merge and diverge and flow in and out of each other, with each one taking the fore when appropriate (and sometimes when inappropriate, since perfection does not exist in the real world, and emotions and thoughts each sometimes make mistakes: That's life) and with them working sometimes together and sometimes in opposition? What happens if you make choices as to when you should think and feel dualistically—in opposition to some other— and when you should work with this other? What happens if you sometimes make choices and sometimes you do not? What happens if when appropriate you are violent ("Hear! Hear!" say the wolverine and the shrew), and when not appropriate you are not?

The "answer" is not to try ever more desperately to eradicate hate from our own hearts, to carry more and more of the love that is split off from the rest of the culture—as if it's the case that if we can only carry enough love to make up for everyone else that things will be all right, or even that any love we feel might in any way counter someone else's hate—and split off ourselves the hate we do not allow ourselves to feel. The "solution" is to reintegrate, to feel what we feel, to determine our own moralities (large and small scale) and to act on them.

In any case, I decided that while my larger goal of finding a way to live without a wage job was certainly moral, the proximate strategy of joining the navy would be an immoral way to achieve it. Robbing banks or shoplifting from Wal-Mart would undoubtedly have been more moral. I did neither. As I wrote about in *A Language Older Than Words*, I became a beekeeper, partly because I love bees, and partly because as a beekeeper I would work hard part of the year and I would have time to just be for part of the year.

Now that we've defined what are strategies and tactics, but before we start trying to determine what ours are, we need to talk about what we want to accomplish. What do we want?

By now I'm pretty sure you have a good idea what I want. A world not being killed. A world being renewed. A world with more wild salmon each year than the year before. A world with more wild humans each year than the year before.

What do you want? The question is not rhetorical. Don't just pass it by and move to the next chapter. Stop. Put down the book. Go outside. Take a long walk. Look at the stars. Pet the bark of trees. Smell the soil. Listen to a river. Ask them what they want. Ask your heart what you want. Ask your head. Ask your heart again. Then figure out how you're going to get it.

IMPORTANCE

If all mankind were to disappear, the world would regenerate back to the rich state of equilibrium that existed ten thousand years ago. If insects were to vanish, the environment would collapse into chaos.

Edward O. Wilson[31]

THERE ARE THOSE WHO THINK IT'S NOT ENOUGH TO TAKE DOWN civilization, that for wild nature to survive not only this culture but all of humanity must be eradicated. Those who believe this seem at least to me to fall into two very different categories, the first of which I understand but don't agree with, and the second of which I find intolerable.

The first consists of those people—and there are quite a lot of them—who feel that humans have fucked up so badly that we've forfeited the right to live. Looking at factory farms, vivisection labs, factory trawlers, clearcuts, plowed soil, dams, rates of rape, child abuse, inaction in the face of global warming, the irradiation and toxification of the planet, the pesticide industry, my damn neighbor who cuts every tree on his property,[32] then sprays Roundup on every "weed" he sees, Monsanto (corporate creator of Roundup), zoos, off-road vehicles (and their callous and arrogant drivers),[33] CIA torture manuals, U.S. bioweapons laboratories, a U.S. government that invades countries left and right in the name of peace and democracy, and the whole sorry history of lies, invasions, conquest, land theft, exploitation, rapine, rape, torture, murder, genocide, and ecocide that characterizes so much of the last many thousand years, it can sometimes be pretty damn easy to forget about all the other cultures eradicated by civilization—cultures where people did not act as we do—and start thinking the world would be a lot better off without humans. The widespread understanding of how bad things have become might help explain[34] how Margaret Atwood's not unsympathetic portrayal in *Oryx and Crake* of someone who creates a virus to destroy all humans became a bestseller and a finalist for the Booker Prize. I'm frankly surprised that no one versed in genetic engineering or biowarfare has pulled an *Oryx and Crake* or *Twelve Monkeys* and tried to get rid of us all for the good of nonhumans everywhere. Probably quite a few lab rats, monkeys, beagles, and cats would, if necessary, volunteer to sacrifice themselves for this greater good, to save their brothers and sisters from being tortured and their cousins from being extirpated. I'm surprised also that no bioengineers have yet released pathogens targeting the civilized: if those in power can develop them for use against the "lesser breeds," you'd think it would be technologically feasible to search for the opposite markers, and take out those people and classes who are doing the damage.

As I said, I don't conflate civilization and our species, and I don't agree that the problem is humanity, although I often despair at the slight odds of most of those in this culture remembering how to be human.

The second group consists of those who insist that humans are inherently destructive. This group doesn't seem so much to believe that humans have committed unforgivable sins—which of course we have, and which of course we all continue to do by not stopping those who are actively committing them—as that humans are, in our very essence, destructive. This seems nothing more than the same old Biblical tradition of Original Sin dressed in modern scientific clothing. I almost never see this attitude among honest-to-goodness activists. I see it all the time among people who pretend to care about the natural world, but want to use this as an excuse for inaction: *There's no use fighting our biology, so there's no use fighting the destruction. We as a species are just too damn smart for our own good. We're killing the planet, but so do indigenous peoples. Humanity is just an experiment that failed. It will be sad but interesting to watch our passing, and the passing of the planet. And pass me the remote control, will you? I'm getting bored.* I also see this a lot among academics who, like their couch potato counterparts, pretend to care but who are actually apologists for the current way of being.

I have before me an essay published a couple of years ago in the *Atlantic Monthly* which purports to show that all humans destroy their environment. The article is pretty ludicrous, and makes such extraordinary claims as Indians "shaped the plains into vast buffalo farms" (an especially insane assertion when you consider that prior to the arrival of horses, the population density in the Great Plains states was approximately .001 humans per square mile.[35]) The article asks, "Is it possible that the Indians changed the Americas more than the invading Europeans did?" and answers in the affirmative. The article goes on to call "the Amazon forest itself a cultural artifact—that is, an artificial object," and says that "the lowland tropical forests of South America are among the finest works of art on the planet." According to the article, one can consider basically all of the nonflooded Amazon forest "of anthropogenic origin—directly or indirectly created by human beings," a "built environment." What hubris. What ignorance. Back in North America, the author, Charles C. Mann, states, "Far from destroying pristine wilderness, European settlers bloodily *created* it [by killing the Indians who had themselves evidently been messing up the landscape]. By 1800 the hemisphere was chockablock with new wilderness."

Remember, all writers are propagandists, and remember also that if your writing rationalizes this culture's insane worldview and destructive activities, no amount of illogic will serve to discredit your commentary: It is not necessary that the lies be particularly believable, but merely that they be erected as barriers to truth, which must at all costs be avoided. Here's yet another example from Mann. He states that passenger pigeons—which prior to being slaughtered wholesale by the civilized had been unimaginably abundant, with five times as many passenger pigeons as all other species of North American birds combined[36]—were not actually all that common prior to the arrival of Europeans (never mind the first European explorers who described their massive flocks— SHHHHH!). Here's his logic [sic]. Passenger pigeons were easy to hunt (he even approvingly cites someone calling them "incredibly dumb," a phrase I would reserve for those who exterminated them and those who rationalize their extermination), which means Indians—because they are humans, and are therefore destructive—must have hunted them heavily enough to reduce their numbers. How do we know the Indians reduced their numbers? Because there aren't many pigeon bones in pre-Columbian Indian trash heaps, meaning that the Indians must not have hunted them very much.

I kid you not. That is his logic [sic]. Pigeon populations were kept low by Indians who hunted them lots, and the way we know that the populations were low is that we can't find much evidence that Indians hunted them.

The questions become: Why does Mann lie like this? and Why do so many people accept these lies and illogic? Mann makes clear his reasons. He asks, "[If] the work of humankind was pervasive, where does that leave efforts to restore nature?" and then answers his own question: "Guided by the pristine myth, mainstream environmentalists want to preserve as much of the world's land as possible in a putatively intact state. But 'intact,' if the new research is correct, means 'run by human beings for human purposes.'"

Bingo. Mann's startling "new research" is the same old rationale for exploitation we found in Genesis. Convenient, isn't it, that the whole world was made to be run by human beings for human purposes?[37] Meet the new God, same as the old God.

He continues, "Environmentalists dislike this, because it seems to mean that anything goes. In a sense they are correct. Native Americans managed the continent as they saw fit. Modern nations must do the same."[38]

He is, of course, full of shit. But you already knew that. There is a world of difference between indigenous peoples forming long-term relationships with their landbases, and ExxonMobil drilling for gas. It is precisely the difference between the give and take of making love with a much-cherished partner, and rape. I would feel sorry for Mann and those like him were they not raping, torturing, and murdering the planet.

You may have heard of something called the Pleistocene Overkill Hypothesis, although most likely you heard it with the final word missing to lend it more credence, as though it were something more than someone's almost entirely unsupported idea. The Pleistocene Overkill Hypothesis, discredited as it is, remains the ace in the hole for those who wish to believe all humans are destructive. The hypothesis blew open when Paul S. Martin noticed that many large mammals such as woolly mammoths, dire wolves, saber-toothed cats, giant ground sloths, giant beavers, and so on in North America went extinct about 10,000 years ago. Martin then posited (incorrectly) that this coincided with the arrival of the first humans in North America. Further, he then posited (also incorrectly) that humans hunted these animals to extinction. He punched some numbers into a computer, ran a simulation, and just like that, the computer told him the Indians were ecocidal. If a computer says it, it must be true.

Once, when I was sharing the stage with an American Indian, someone in the audience brought up this exact point: "You were just as destructive as we are," he said.

The Indian responded, "Then why were there buffalo here when your people arrived?"

"Your people were in the process of killing them off, too."

"We must have been awful slow, because there were between thirty and sixty million of them left." He paused, then said, "You know, it's always the same with those who are destroying life. First you ignore the damage entirely. We can kill all the bison we want, you say, and it doesn't matter. We can kill all the passenger pigeons we want, and it doesn't matter. When it doesn't work to ignore the damage, then you deny it's happening. The herds are just as big as they were last year, you say. It's harder to hunt them, but the herds *must be* just as big. When it doesn't work to deny that the damage is happening, then you attack those who try to tell you about it. You attack their reputations: Oh, they're just Indians, they aren't scientific, they don't know anything about population dynamics. They're just environmentalists, they're too emotional, and you know they'll lie to protect some piece of ground. When it doesn't work to attack the messenger, you pretend the damage isn't really damage. Who needs bison anyway? And what does it matter if there are a few chemicals in every stream? You already have chemicals in your body, and they haven't killed you yet, have they? Global warming is good, isn't it? When it doesn't work to pretend the damage isn't damage, then you try to blame someone else. The Indians, not the whites, were the ones who overhunted the bison and killed them off. When it doesn't work to blame someone else, and you finally have to acknowledge that you are the one who did this damage, then you say that someone made you do it. You wouldn't have killed the bison except the Indians wouldn't give you their land, so you had to starve them off. If they wouldn't have fought to keep their land, you wouldn't have had to kill the bison. It's the Indians' fault. And when it doesn't work to blame your destructiveness on someone else, then the last resort is for you to say that everyone does it, so you cannot be held accountable. Indians destroyed their habitat, too, you say, so it must be natural for us to do it. It's all crazy. No matter what we say, you've got an answer for it, and no matter what we say, you keep on destroying our home."

I can't tell you how many times at how many talks and in how many discussion groups I have had people—nearly all, once again, nonactivists—tell me that humans are by nature destructive. When I counter that on civilization's arrival in North America the continent was extraordinarily fecund, their response is always the same: "Pleistocene Overkill." When I comment on the fact that whether or not this occurred, it was a hell of a long time ago, and that to rely on something that may or may not have happened 10,000 years ago to defend current behavior is really pathetic, their response is always the same: "Pleistocene Overkill." When I state that even if this did occur, which is doubtful, that perhaps those Indians who participated learned their lesson, changed their ways, and that's how they were able to live in relative equilibrium with their surroundings for the most recent one hundred centuries, their response is always the same: "Pleistocene Overkill." When I say that I think it's obscene to use the alleged extermination of a few dozen species to excuse the destruction of the entire planet, their response is always the same: "Pleistocene Overkill." When I comment on the racism inherent in saying that the indigenous would be just as destructive as the civilized if only they had the curiosity and intelligence to invent backhoes, chainsaws, napalm, and nuclear weapons, their response is always the same: "Pleistocene Overkill."

I would love to debunk the Pleistocene Overkill Hypothesis myself, but many far better scholars than I have already done it, perhaps none so engagingly as the President of the Society of Ethnobiology Eugene S. Hunn, who wrote, in a paper presented at the Ninth International Conference on Hunting and Gathering Societies (and I quote this at length because I am so damn sick of having to respond to the "Pleistocene Overkill" crowd):

"Pleistocene Overkill. This beast, like Dracula, will not die, despite a broad consensus of archaeologists knowledgeable with respect to the evidence, or lack of evidence, to support the hypothesis, that it is a just-so story with no empirical support. The apparent coincidence of the arrival of the first proficient hunters on the New World scene and the demise of some 35 genera of charismatic megafauna is not sufficient grounds to convict 'Clovis Man' of their atrocity. The temporal priority of Clovis is now widely dismissed, on the strength of finds in southern Chile that date to the late Pleistocene at 14,000+ BP. Furthermore, the association of Paleo-Indian kill sites with the extinct megafauna is scant. More telling, in my opinion, are the theoretical and empirical reasons to reject the clever computer simulation devised to show how it could have happened.

"Martin's 1972 simulation reported in *Science* is not the only such attempt at a virtual reenactment of the crime nor the most sophisticated, but it is clear from that simulation that it requires assumptions about human behavior that are most improbable. For example, Martin's model assumed that: 1) the Paleo-Indian population would double every 20 years; 2) '. . . a relatively innocent prey was suddenly exposed to a new and thoroughly superior predator,[39] a hunter who preferred killing and persisted in killing animals as long as they were available . . .'; 3) 'Not until the prey populations were extinct would the hunters be forced, by necessity, to learn more botany . . .'; and 4) '. . . on the front one person in four destroy one animal unit (450 kilograms) per week, or 26 percent of the biomass of an average section in 1 year in any one region. Extinction would occur within a decade . . .' Are these reasonable assumptions?

"Demographic studies of Kalahari San hunters demonstrate that the average birth-interval for this group of nomadic hunters is four years, which with expected child mortality results in stable or very gradually increasing populations over millennia. Population growth is limited not by the availability of food but by the difficulty of carrying infants.

"Hunters who prefer killing and persist in killing until the last animal is gone exist only in the tortured imaginations of misanthropic scholars. While it is questionable to what extent Native American hunters harvested their prey in accord with contemporary principles of maximum sustained yield, the ethnographic evidence affirms that subsistence hunting is an activity demanding not bloodlust but sophisticated knowledge of animal behavior and local landscape, subtle logic to decipher signs, great patience, and typically an attitude of humility and reverence toward the prey animal as an animate being and moral 'person.' Martin would have us believe that it could be adaptive for every adult male in a community to kill one 450 kg 'animal unit' per week, which comes to 16 kg per person per day, or perhaps half that dressed meat, which comes to some 30,000 calories per person per day, 15 times the daily requirement. If even 10 percent were consumed, and nothing else, the Paleo-Indians would have been too fat to have waddled to Tierra del Fuego in the time allowed. So we are to believe instead they wasted over 90 percent of this meat. This requires that we assume that it is extraordinarily easy to dispatch a 450 kg 'animal unit,' so much so that one disdains to preserve the meat.

"Such profligacy is perhaps characteristic of the 'economies of scale' associated with recent industrial production, but bears no resemblance to the practical realities of a hunter-gatherer way of life. Rather, we should expect the economies associated with the 'domestic mode of production' to operate in subsistence hunting-gathering communities. One works only as hard as is needed to feed one's family and contribute in addition to the reproduction of one's community. Furthermore, the ethnographic evidence is overwhelming that hunting-gathering economies are based on a division of labor between men and women, and that in all but the arctic and sub-arctic extremes, women contribute very substantially to the diet by collecting edible plant foods. 'Learning botany' is not a measure of last resort for the starving hunter, as Martin would have it, but in every case is an integral element of a hunting-gathering subsistence strategy.

"Finally, just how stupid were the American Pleistocene megafauna that they would fail to recognize a new superpredator and learn to avoid him before it was too late? Continental megafauna evolved in the presence of fierce predators, such as saber-toothed cats and short-faced bears, and thus are hardly to be compared with the naive predator-innocent animals of isolated islands such as the Galapagos.

"Thus, not only is there no credible archaeological evidence to support the Pleistocene Overkill scenario—at least as a major factor in the extinction of more than a few of the species lost, but also it flies in the face of all that anthropologists have learned about the actual practice of hunting-gathering by contemporary and historic hunting-gathering societies."[40]

And now to the point: Rarely do any of those who chant "Pleistocene Overkill" take their logic to its end, which is that if you truly believe that humans are inherently destructive (in other words, you're not just saying that to rationalize this culture's destructiveness as well as your own inaction in the face of evil), that is, if you truly believe that humans are "thoroughly superior predators" who always have and always will destroy their habitat (and the habitat of others), and if you believe that the civilized are more destructive because, well, the civilized are better at everything (because it's such an advanced state of human society [and maybe because we the civilized are so damn smart]), and if you care in the slightest about the natural world, you need to get rid of all humans before they destroy it all. It's your logic, not mine.

☾ ☾ ☾

It all reminds me of a story told to me by a friend in prison. He was sick of a prisoner friend of his always telling him how much he hated black people, and how he wanted to kill them. My friend knew where a shank was buried on the yard. He brought his friend to it, uncovered it, put it in his friend's hand, and said, "I'm sick of you talking. Either do something or shut the hell up."

The friend shut the hell up.

IDENTIFICATION

There is one thing stronger than all the armies in the world, and that is an idea whose time has come.

Victor Hugo

The earth has Fates all her own. The earth has purpose. And we can only partially know what that purpose is. So I want to reverse our understanding and usage of that word, just like we need to reverse the notion of who is doing environmental work, or any of the other work on this planet. We think we're doing it all. But the animals are doing the real work of holding it all together, and keeping us on our path. As are the plants. It's as if we think the stars and sun and moon and the earth itself aren't doing any work. It's as if we think that all of nature is unintelligent except for us. Well, the earth has intelligence and purpose and fates all her own. And those are really the primary fates.

This calls upon us to have tremendous responsibility: it *is* our responsibility to participate in those Fates. We're part of the earth. It is up to us to be in alignment with those purposes, not to go against them, nor to sit back and pretend. You have to give back. You have to participate in those fates, because they are your Fates as well.

Jane Caputi[41]

THE OTHER NIGHT I DREAMT I GOT OFF AN AIRPLANE AND WALKED through one terminal to another, carrying a metal suitcase. My first flight had been delayed, and now I was late for the next. I arrived at the gate. The person behind the counter asked for my ID. I pulled out my wallet and handed her my driver's license. For some reason she didn't like the numbers on it and so wouldn't accept it. No problem, I thought, and pulled out another driver's license. I evidently had about ten. Unfortunately she didn't like any of them. Me and my metal suitcase were going to have to find another way to get where we were going.

That dream has me thinking about identity. Who am I?

The whole question of identifying myself as a consumer, or as a voter, or as a writer, or as a revolutionary has me thinking again about definitions. Just as it seems clear to me that the only level of technology that is sustainable is the Stone Age, for reasons I hope I've made clear, it also seems clear to me that only a certain sense of self is sustainable, too.

The eighteenth premise of this book is that *this culture's current sense of self is no more sustainable than its current use of energy or technology.*

If you perceive yourself as a consumer, consume you will. If you perceive yourself as a "new and thoroughly superior predator" you will act like one.[42] If you perceive yourself as a member of a species that can act no other way than to destroy your landbase, that is what you will do. If you perceive yourself as the "apex" of evolution, you will try to climb to the top of something that has no top, and you will crush those you perceive as being beneath you. If you believe you are separate from your landbase, you may believe you can destroy your landbase and survive, and you may very well destroy it. If you perceive yourself as entitled to exploit those around you, you will do so.

Not one of those self-perceptions is sustainable. If you perceive yourself in any of those ways, you and those who perceive the world like you will not survive long into the future. I don't really care about that: if that's how you perceive the world, then good riddance to you. The problem is that before you go down you will cause a lot of unnecessary misery, and you will take down a lot of others with you.

☾ ☽ ☾

WHY CIVILIZATION IS KILLING THE WORLD, TAKE TWENTY-ONE.
ORVs. Off-road vehicles. Whenever I begin to feel even the slightest bit optimistic
about our future, all it takes to bring me down is these three little letters: ORV.

Here's a headline from a recent *San Francisco Chronicle* (page B-1, above the
fold, full picture): "Curbing off-road recreation: Asbestos, rare plants threaten
freewheeling bikers in the Clear Creek Management Area." The article begins:
"If there is a Garden of Eden for off-road enthusiasts, it might be the Clear
Creek Management Area, about 50,000 miles of dirt-bike bliss 55 miles south of
Hollister. . . . But as in Eden, it appears that new information could bring an
end to, or at least lead to limits on, the freewheeling good times at Clear Creek."
The "new information" is that off-road vehicle use destroys soils and kills
plants,[43] in this case the rare San Benito evening primrose, among others.[44]

Remember that the first rule of propaganda is that if you can slide your
premises by people, you've got them. Remember also that abusers must always
position themselves as the real victims. Here, it is not the drivers of off-road
vehicles who threaten (or rather kill) plants, but instead the rare plants who
"threaten freewheeling bikers."[45]

If riding ORVs is explicitly conflated with bliss (dictionary definition: "complete happiness") and being in the Garden of Eden (dictionary definition: "paradise"), if irresponsible and inherently destructive behavior is redefined as (or
considered to be) "freewheeling good times," there is no realistic chance this
culture will stop killing the planet.

I think about this a lot. If we must fight as hard as we do to stop destructive activities that are purely and merely recreational, and if we so often lose
those fights anyway, even with the continued existence of species themselves
at stake, there is no realistic chance that by using current strategies and tactics we will be able to stop destructive activities that serve utilitarian functions. This is not to say that utility or production is more important than
recreation or leisure, but to point out that ORVs are peripheral to the system,
and if we cannot halt destructive activities that are entirely expendable to the
system's functioning, we will not be able to stop activities on which the system relies. If we cannot by accepted means stop ORVs, we will never by
accepted means stop logging, mining, or oil extraction. If we are still fighting
over rodeos or the wearing of fur,[46] to use different examples, we will never stop
vivisection, factory farming, or factory fishing.[47] If we cannot by accepted
means stop that which is both ridiculous and nonessential to the system, we
will never by accepted means stop that which, while still ridiculous, is fundamental to the system's continuation.

❧ ❧ ❧

WHY CIVILIZATION IS KILLING THE WORLD, TAKE TWENTY-TWO.
ORVs. No, I'm not including them again just because they're so pointlessly
destructive.[48] Instead I'm including them because of the U.S. Supreme Court.
Yesterday the Supremes unanimously ruled that citizens do not have the right
to sue federal agencies that fail to enforce environmental laws. Yes, you read that
correctly. The case concerned ORVs. Here's the story. Citizens had filed suit
against the Bureau of Land Management (BLM)[49] in Utah for failing to keep ORV
users out of two million acres of potential wilderness (note that if ORV users
or other despoilers can punch roads into these wilderness areas, even illegally,
the areas can then be precluded from wilderness protection). The BLM is
required by law to protect these lands from ORV users (or, as someone from the
Chronicle might write, is required to expel them from paradise), but as Supreme
Court Justice [*sic*] Antonin Scalia put it, any land-use plan requiring that the BLM
protect a certain area is merely a "statement of priorities" and not a legally binding commitment. The essence of the Supreme Court ruling is that citizens *can*
sue to force federal agencies to take some legally required action, but *cannot* sue
to force them to act with haste or effectiveness. Thus if the BLM is required to
protect a piece of ground from ORV users, it seems that so long as the BLM
comes up with some plan—any plan, no matter how ludicrous—citizens have
no recourse through the judicial system. *We stand by our vital commitment to protecting these grounds,* the BLM could say, *and will spend exactly three dollars per
year on enforcement. We will be sure to stop all destructive ORV use on these lands
by 2050, or when the last drop of oil has been used, whichever comes last.* And
there's nothing you could do to stop them.

Or rather there is nothing you could do using the courts.

Here's the real story. You and I both know that the primary purpose of regulatory agencies is to provide buffers between citizen outrage and those who
destroy their landbases and their lives,[50] to provide "statements of priorities"
without providing legally binding commitments. Anyone who has ever
attempted to protect a piece of ground from the Forest Service or Bureau of
Land Management has run into this: *every* time someone finds a way to use the
agencies' own rules to stop the destruction, the agencies change the rules.

Scalia was being disingenuous when he talked about the BLM making a
"statement of priorities." Priorities are evident in action (or inaction). Any statement made by the BLM or by anyone else that is not backed up by action is
merely a smokescreen, something to create the distractions Lundy Bancroft said

are characteristic of abusers: "In one important way, an abusive man works like a magician. His tricks largely rely on getting you to look off in the wrong direction, distracting your attention so that you won't notice where the real action is. . . . His desire, though he may not admit it even to himself, is that you wrack your brain in this way so that you won't notice the patterns and logic of his behavior, the consciousness behind the craziness."[51]

The same is true here. The BLM states that protecting wild places is a priority but does not do so. The Supreme Court pretends we live in a place where citizens can act to protect their landbases, but tries to prevent them from doing so. The list goes on.

This Supreme Court ruling reminds me of the famous line by John F. Kennedy: "Those who make peaceful revolution impossible make violent revolution inevitable." I would modify that to apply to our landbases: "Those who make peaceful protection of our landbases impossible make violent defense of our landbases inevitable."

⟨ ⟨ ⟨

In related news, Humboldt county, one county south of here, has the second highest per capita rate of rape in California. Last year sixty-one rapes were reported to the Humboldt County Sheriff's Department, which arrested precisely two rapists. The department arrested at least eight times that many tree-sitters, who were attempting to stop Pacific Lumber from illegally deforesting the county.[52] In just one not-atypical example, twenty-nine sheriffs, three trainees, and four California Highway Patrol provided the on-ground muscle for the PL contractors who pulled down the tree-sitters. Just as with the BLM, a primary purpose of the Sheriff's Department is to provide the illusion of protection while enforcing, as always, Premise Four of this book. Priorities are evident in action.

I don't think there are many here among us who would object in this case to me modifying Kennedy's statement again, and applying it to rapists: Those who make peaceful protection of our bodies impossible make violent defense of our bodies inevitable.

⟨ ⟨ ⟨

Let's bring this back now to the toxification of our total landscape and of our bodies by chemical corporations and those who run them. Those who make peaceful protection of our bodies impossible make violent defense of our bodies inevitable.

❨ ❨ ❨

What is done to the earth is done to ourselves. It really is that simple. We cannot live without the earth. The earth can live without us. It is an open question at this point whether it can live with us. It certainly cannot live with us as we are now.

Because of civilization, almost 1,400 square miles of land per year are converted to desert, more than twice the rate from thirty years ago (and essentially infinitely more than the rate before civilization). In about twenty years, two-thirds of the arable land in Africa will be gone, as well as one-third in Asia and one-fifth in South America. Even the capitalist press acknowledges, "Technology can make the problem worse. In parts of Australia, irrigation systems are pumping up salty water and slowly poisoning farms. In Saudi Arabia, herdsmen can use water trucks instead of taking their animals from oasis to oasis—but by staying in one place, the herds are getting bigger and eating all the grass. In Spain, Portugal, Italy and Greece, coastal resorts are swallowing up water that once moistened the wilderness. Many farmers in those countries still flood their fields instead of using more miserly 'drip irrigation,' and the resulting shortages are slowly baking the life out of the land." The corporate press further acknowledges that prior to civilization even some of what are now the most inhospitable deserts were habitable, saying that "much of the Middle East, the Mediterranean and North Africa were once green. The Sahara itself was a savanna, and rock paintings show giraffes, elephants and cows once lived there."[53]

As industrial civilization kills the land, so too it kills the oceans. Each summer a dead zone covers 8,000 square miles in the Gulf of Mexico. Another blankets Chesapeake Bay. Another the Baltic Sea. Altogether, there are almost 150 dead zones, places where the water contains too little oxygen to sustain life. This number has doubled each decade since the 1960s. The cause? Industrial agriculture.[54]

And of course dead zones are not even the greatest threat to the oceans. Far greater is deep-sea trawling. Damage is severe enough to cause the Canadian Broadcasting Corporation to report, "Every year, the giant nets that trawler ships pull across the bottom of the sea devastate an area of the global seabed twice the size of the United States, scraping up everything from coral to sharks."[55]

Industrial civilization is killing the land. It is killing the water.

❨ ❨ ❨

It would be a mistake to think this culture clearcuts only forests. It clearcuts our psyches as well. It would be a mistake to think it dams only rivers. We ourselves are dammed (and damned) by it as well. It would be a mistake to think it creates dead zones only in the ocean. It creates dead zones in our hearts and minds. It would be a mistake to think it fragments only habitat. We, too, are fragmented, split off, shredded, rent, torn.

Just last night I saw a television commercial put out by BP, the corporation formerly known as British Petroleum. The corporation now claims that BP stands for Beyond Petroleum, and runs public relations campaigns extolling its renewable energy research. For example, BP has made a lot of noise about the fact that in 1999 it paid $45 million to buy Solarex, a corporation specializing in renewable energy. This may seem like a lot of money until we realize that BP paid $26.5 billion to buy Arco in order to expand its petroleum production base, and until we realize further that BP will spend $5 billion over five years to explore for oil just in Alaska, and until we realize even further that BP spent more in 2000 on a new "eco-friendly" logo than on renewable energy.[56] As Cait Murphy wrote in *Fortune*, "Here's a novel advertising strategy—pitch your least important product and ignore your most important one. . . . If the world's second-largest oil company is beyond petroleum, *Fortune* is beyond words." BP's regional president Bob Murphy acknowledges that BP is "decades away" from moving beyond petroleum, which means that the whole Beyond Petroleum name change is meaningless: by that time we'll *all* be beyond petroleum, since the accessible oil will all be gone. Further exemplifying the meaninglessness of the name change, a resolution calling for BP to do more to slow global warming was opposed by the board and defeated. BP's chair Peter Sutherland told shareholders that "there have been calls for BP to phase out the sale of fossil fuels. We cannot accept this, and there's no point pretending we can."[57]

In other words, BP's name change is a "statement of priorities" and not a legally binding commitment. Or more to the point, it's another one of those smokescreens.

This particular type of smokescreen has been most fully developed by a public relations consultant with the appropriately named Peter Sandman. He has been nicknamed the High Priest of Outrage because corporations hire him to dissipate public anger, to put people back to sleep. Sandman has explicitly stated his self-perceived role: "I get hired to help a company to 'explain to these confused people that the refinery isn't going to blow up, so they will leave us alone.'"[58]

He developed a five point program for corporations to disable public rage.

First, convince the public that they are participating in the destructive processes themselves, that the risks are not externally imposed. *You asked for it by wearing those clothes*, says the rapist. *You drive a car, too*, says the PR guru. Second, convince them that the benefits of the processes outweigh the harm. *You could never support yourself without me*, says the abuser. *How would you survive without fossil fuels?"* repeats the PR guru. Third, undercut the fear by making the risk feel familiar. Explain your response and people will relax (whether or not your response is meaningful or effective). *Don't you worry about it, I'll take care of everything. Things will change, you'll see*, says the abuser. *We are moving beyond petroleum and toward sustainability*, says the PR guru. Fourth, emphasize again that the public has control over the risk (whether or not they do). *You could leave anytime you want, but I know you won't*, says the abuser. *If we all just pull together, we'll find our way through*, says the PR guru. Fifth, acknowledge your mistakes, and say (even if untrue) that you are trying to do better. *I promise I will never hit you again*, the abuser repeats. *It is time to stop living in the past, and move together into the future*, drones the PR guru.[59]

Speaking to a group of mining executives, Sandman, who also consults for BP, stated, "There is a growing sense that you screw up a lot, and as a net result it becomes harder to get permission to mine." His solution is not actually to change how the industry works, of course, but instead to find an appropriate "persona" for the industry. "Reformed sinner," he says, "works quite well if you can sell it. . . . 'Reformed sinner,' by the way, is what John Brown of BP has successfully done for his organization. It is arguably what Shell has done with respect to Brent Spar. Those are two huge oil companies that have done a very good job of saying to themselves, 'Everyone thinks we are bad guys. . . . We can't just start out announcing we are good guys, so what we have to announce is we have finally realised we were bad guys and we are going to be better.' . . . It makes it much easier for critics and the public to buy into the image of the industry as good guys after you have spent awhile in purgatory."[60]

In the ad I saw last night, an off-camera interviewer asks a woman, "What would you rather have: a car or a cleaner environment?"

The woman pauses, seemingly thoughtfully, before at last saying, "I can't imagine me without my car. Of course I'd rather have a clean environment, but I think that that compromise is very hard to make where we are."[61]

The ad ends with a voiceover saying what BP is doing to make the world a better place.

Look what just happened here. What are the premises of this advertisement? The first is that the advertisement is presented in the form of an interview, and it can be easy to forget that it's a paid advertisement. A BP website devoted to the ads stresses that the ads were culled from hundreds of interviews with "random strangers." I'm sure they did interview hundreds: that gave them a larger sample from which to draw the responses that most closely fit their needs. Consider, had the directors of the advertisement happened to ask me this question, I doubt they would have wasted film on my answer, and certainly they would not have paid to put it on television. Instead the "interviews" the director decided to use were chosen precisely because they brilliantly and succinctly put in place Sandman's five points: we participate willingly, the benefits outweigh the harm, the risk is somehow familiar, we have control over the risk, and BP is working to solve the problems.

But there is more going on here. First, the ad pretends that the "environment" is something "out there" that is separate from ourselves. Consider if the "interviewer" had asked, "Given that our own well-being is inextricably linked to the health of our landbase, would you rather have a healthy landbase or an entire culture based on the 'comforts and elegancies' that come from destroying this landbase?" And then consider if he would have followed up by asking, "If you choose the latter, what does it say about you as a person?" Or what if instead the "interviewer" had commented that just in the United States about 30,000 people die each year from respiratory illnesses caused by auto-related airborne toxins, and that 65 percent of all carbon monoxide emitted into the environment is from road vehicles, then asked, "What would you rather have, car culture or your life?"

Second, what are the implications of the "interviewer" using the adjective *cleaner* to describe "the environment" this woman would allegedly gain were she to stop driving. This presumes that "the environment" is already clean, and that the current situation is the default. How would the ad run if we change the question to, "What would you rather have: a planet that is not being made filthy and in fact destroyed by automobiles and other effects of civilization, or your car?"

A deeper, more invisible unstated premise of this ad is that a non-clean planet is her fault (and by extension ours, insofar as the director of the ad is able to get us to identify with this woman). "What would you rather have: a car or a cleaner environment?" It's her choice. It's her car. If only she would sell her old Honda Civic, the implication goes, everything would be okay. But she can't do that. As she says, "I can't imagine me without my car." She, and once again by extension each of us, is supposed to identify more with the artifacts of civilization, with machines, than with a landbase. This is what we are trained to do. We are also trained to lack imagination. If our imaginations had not already been clearcut we could not—would not—live the way we do. And further, we are also trained to be narcissistic enough to believe that if we personally cannot imagine something that it must not be possible.

This identification with the artifacts of civilization is precisely what each of us *must* break. If she cannot imagine herself without her car, I wish her luck in imagining herself without her planet.

And another premise. She states, "Of course I'd rather have a clean environment, but I think that that compromise is very hard to make where we are."

This "compromise" would only be difficult for those who have already had their sanity effectively destroyed. The world does not need us or our cars. We need the world.

❨ ❨ ❨

These corporations just never stop. I saw another ad by BP. The "interviewer" asks a man, "So what would you say to an oil company?"

The man responds, "I'd say: Your products are a necessary evil, but we all use it, we all partake in it, we all enjoy it. Let's figure out ways together to make it a little more environmentally safe."[62]

You can parse this one. Go ahead, tear it apart. It's fun. And once you're done tearing apart their discourse, we can proceed to tear apart something else.

❨ ❨ ❨

BP is not alone. There are other corporations just as bad. Many of them.

For example, one corporation, ExxonMobil, has by itself released 5 percent of this culture's carbon emissions. According to a recent study, "From 1882 to 2002, emissions of carbon dioxide from Exxon and its predecessor companies, through its operations and the burning of its products, totaled an estimated 20.3 billion metric tons. . . . That represents 4.7 percent to 5.3 percent of global carbon dioxide emissions during that time."

It is not surprising, then, that Exxon has, to quote the same report, "been active in undermining climate science and policy making for years, in particular in lobbying against the Kyoto Protocol, the main international agreement to tackle climate change."

It is probably also not surprising that last year after shareholders voted down by a margin of four to one a resolution calling for the corporation to switch to renewable energy sources, CEO and Chair Lee Raymond stated, "We don't invest to make social statements at the expense of shareholder return."[63]

ABUSERS

Violence against women and violence against the Earth, legitimated and promoted by both patriarchal religion and science, are interconnected assaults rooted in the eroticization of domination. The gynocidal culture's image of woman as object and victim is paralleled by contemporary representations that continually show the Earth as a toy, machine, or violated object, as well as by the religious and scientific ideology that legitimates the possession, contamination, and destruction of Mother Earth.

Jane Caputi[64]

WE HAVE BEEN TOO KIND TO THOSE WHO ARE KILLING THE PLANET. We have been inexcusably, unforgivably, insanely kind.

I understand now. For years I have been asking whether abusers believe their lies, and I'm finally comfortable with an answer.

This understanding came in great measure because I finally stopped focusing on the lies and their purveyors and I began to focus on the abusers' actions. I realized, following Lundy Bancroft, that to try to answer the question of whether the abusers believe their lies is to remain under the abusers' spell, to "look off in the wrong direction," to allow myself to be distracted so I "won't notice where the real action is." To remain focused on that question is exactly what abusers want.

Bancroft helped me realize some very important things. He writes specifically about abusers, emphasizing perpetrators of domestic violence, but what he says applies as well to this whole culture of abuse, and to perpetrators of the larger scale abuse I've been writing about.

His central thesis seems to be that the primary problem is not that abusers particularly "lose control" or that they are particularly prone to "flying into a rage," but instead that they feel entitled to exploit, will do anything in order to exploit, and will exploit precisely as much as they can get away with.

Bancroft excels at exploding misconceptions. When a woman stated that her abusive partner Michael loses control and breaks things in a rage, only to feel remorse afterwards, Bancroft asked whether the things that were broken were Michael's or hers. She answered, "I'm amazed that I've never thought of this, but he only breaks my stuff. I can't think of one thing he's smashed that belonged to him." Bancroft asked who cleans up. She does. He responded, "Michael's behavior isn't nearly as berserk as it looks. And if he really felt so remorseful, he'd help clean up."[65]

I remember a time my father was berating and beating my teenaged sister, and her boyfriend showed up an hour early for their date. My father immediately ceased calling her a slut, dropped his hands to his sides, smiled, and walked to greet her boyfriend as if nothing had happened. His rage was not out of control, but something he was able to turn on and off like a light switch.

Or picture this. My father hits my mother. He has hit her many times before.

But this time she slips into another room, calls the police. She comes back out. My father hits her again and again. He is interrupted by the doorbell. He points one finger at her, runs his other hand through his hair, walks to the door, opens it. There are two policemen. My father is cool, calm, as though nothing has happened. My mother is frantic, frightened, having just been beaten. The cops sympathize with my father for living with someone so emotional—they also sympathize because their allegiance already runs to the abuser (see, for example, the arrest rates for rapists in Humboldt County)—and they leave. The door closes. My father resumes beating my mother. His rage, once again, could be turned on and off.

My mother can perhaps be forgiven for her naïveté in relying on authorities to assist her. She was, after all, nineteen years old, with two children and pregnant with a third. But at this point, especially on the larger scale, the rest of us should not be so naïve.

Abusers are not out of control. They are very much in control. I never understood that till I read Bancroft's book.

Similarly, I speak of this culture's destructive urge, and how those in power destroy those things they cannot control. I have written of clearcuts, of devastated oceans, of murdered poor and extirpated species. But corporations and those who run them do not flail willy-nilly at everything around them. Like Michael, they do not destroy what belongs to them. And of course they do not clean up their messes, no matter how much remorse they may feign, and no matter how much they may claim to have moved beyond petroleum, or into new forestry, or whatever other words they may wish to throw around.

Bancroft asks the abusers he works with what are the limits of their violence. He might say, "You called her a fucking whore, you grabbed the phone out of her hand and whipped it across the room, and then you gave her a shove and she fell down. There she was at your feet, where it would have been easy to kick her in the head. Now, you have just finished telling me that you didn't kick her. What stopped you?" His point is not so much the question as the answer. He says the abusers "*can always give . . . a reason.*"[66] Some of the reasons: "I wouldn't want to cause her a serious injury." "I realized one of the children was watching." "I was afraid someone would call the police." "I could kill her if I did that." "The fight was getting loud, and I was afraid neighbors would hear." The most frequent response is, "Jesus, I wouldn't do *that.* I would never do something like that to her." Only twice in fifteen years has Bancroft heard the answer, "I don't know."[67]

His point is that when abusers are committing their atrocities, they remain acutely aware of the following questions, "Am I doing something that other people could find out about, so it could make me look bad?[68] Am I doing something that could get me in legal trouble? Could I hurt myself? Am I doing anything that I myself consider too cruel, gross, or violent?"[69]

These questions are asked word-for-word in corporate boardrooms. I spoke at length a few years ago with a former corporate lawyer who recovered her conscience, quit, and began working against the corporations. "The people who run these corporations," she said, "know exactly what they're doing. They know they're killing people. They know they're destroying rivers. They know they're lying. And they know they're making a lot of money in the process."

Bancroft continues, "A critical insight seeped into me from working with my first few dozen clients. *An abuser almost never does anything that he himself considers morally unacceptable.* He may hide what he does because he thinks *other* people would disagree with it, but he feels justified inside. I can't remember a client who ever said to me: 'There's no way I can defend what I did. It was just totally wrong.' He invariably has a reason that he considers good enough. In short, *an abuser's core problem is that he has a distorted sense of right and wrong.*"[70]

This is true on the larger social scale. Clearly, a culture killing the planet has a distorted sense of right and wrong. Clearly a police department that arrests tree-sitters yet neither deforesters nor rapists has a distorted sense of right and wrong.

Bancroft asks his clients whether they ever call their mothers a bitch. When they say they don't, he asks why they feel justified to call their partners that. His answer is that "the abuser's problem lies above all in his belief that controlling or abusing his female partner is justifiable."[71]

Once again, the connections to the larger cultural level should be obvious. In some ways this is a restatement of premise four, but it's different enough and important enough to become the nineteenth premise of this book: *The culture's problem lies above all in the belief that controlling and abusing the natural world is justifiable.*

It all comes down to perceived entitlement. As Bancroft states, "*Entitlement* is the abuser's belief that he has a special status and that it provides him with exclusive rights and privileges that do not apply to his partner. The attitudes that drive abuse can largely be summarized by this one word."[72]

This same attitude applies on the larger social scale. Of course humans are a special species, to whom a wise and omnipotent God has granted the exclusive rights and privileges of dominion over this planet that is here for us to use. And of course even if you subscribe to the religion of Science instead of

Christianity, humans' special intelligence and abilities grant us exclusive rights and privileges to work our will on the world that is here for us to use. And of course among humans, the civilized are especially special, because we are such a high stage of social and cultural development, with especially exclusive rights and privileges to use the world as we see fit. And of course among civilized humans, those who run the show are even more special, and so on.

The flattering belief that one is entitled to exploit those around him is a major reason abusers so rarely stop their abuse. Although this is, according to Bancroft, "rarely mentioned in discussions of abuse," it "is actually one of the most important dynamics: the *benefits* that an abuser gets that make his behavior *desirable* to him. In what ways is abusiveness rewarding? How does this destructive pattern get reinforced?"[73]

He also states, "When you are left feeling hurt or confused after a confrontation with your controlling partner, ask yourself: What was he trying to get out of what he just did? What is the ultimate benefit to him? Thinking through these questions can help you clear your head and identify his tactics."[74]

My father tells my sister to do the dishes. She complains that she has never seen him do them. He stares at her. She does them. He points out a place she missed on a plate. He hits her. Never again will she suggest he do dishes, unless she is willing to accept the consequences.

My father wants sex. My mother tells him no. He stares at her. He pouts. Later that day he hits her because of something unrelated. But this happens again later that week, and again the next week, and the week after, until finally she makes the connection. Never again will she tell him no, unless she is willing to accept the consequences.

As Bancroft writes, "Over time, the man grows attached to his ballooning collection of comforts and privileges."[75]

This takes us right back to William Harper's 1837 defense of slavery: "The coercion of Slavery alone is adequate to form man to habits of labour. Without it, there can be no accumulation of property, no providence for the future, no taste for comforts or elegancies, which are the characteristics and essentials of civilization."[76]

On the larger scale, too, each time we are left confused or hurt by the lies or other tactics of those in power—as ExxonMobil changes the climate, as Boise Cascade deforests, as Monsanto poisons the world, as BP lies about its practices, as politicians lie about everything—we need to ask Bancroft's questions: What are those in power trying to get out of what they just did? What is the ultimate benefit to them?

❨ ❨ ❨

One of the bad things about abusers as compared to other sorts of addicts is that at least substance abusers sometimes "hit bottom," where their lives become painful enough to break through their denial. No such luck with those who abuse others.

Bancroft states that partner abuse "is not especially self-destructive, although it is profoundly destructive to *others*. A man can abuse women for twenty or thirty years and still have a stable job or a professional career, keep his finances in good order, and remain popular with his friends and relatives. His self-esteem, his ability to sleep at night, his self-confidence, his physical health, all tend to hold just as steady as they would for a nonabusive man. One of the great sources of pain in the life of an abused woman is her sense of isolation and frustration because no one else seems to notice that anything is awry in her partner. *Her* life and her freedom may slide down the tubes because of what he is doing to her mind, but *his* life usually doesn't."[77]

❨ ❨ ❨

Many Indians have asked these questions about the civilized. I have asked these same questions about CEOs, corporate journalists, politicians. How do these people sleep at night?

Soundly, in comfortable beds, in 5,000 square foot homes, behind gates, with private security systems, thank you very much.

❨ ❨ ❨

It is others who lose sleep over their activities.

❨ ❨ ❨

Within an abusive family dynamic, everything—and I mean everything—is aimed toward protecting the abuser from the physical and emotional consequences of his actions. All members are enculturated to identify more closely with the family structure and its abusive dynamics than with their own well-being and the well-being of their loved ones and other victims. Because the dynamic is set up to foster the well-being of the perpetrator, every action, then, by every member of the family—and more to the point every member's every thought and non-thought and feeling and non-feeling and way of being and not-being—has as its goal the protection of the abuser's well-being. This "well-being" is a particular sort, devoid of relationship and accompanying emotions, heavy on the kind of external rewards abusers reap because of their abuse (and of course precisely the kind of external rewards emphasized by a grotesquely materialistic culture), and most especially focused on allowing the perpetrator to avoid confronting his own painful emotions, including the pain he inflicts, the pain he received as a child (and adult) that caused him to separate from his own emotions (to identify not with himself but with an abuser and an abusive dynamic), and the pain of living in an abusive dynamic where rewards gained through abuse never quite compensate for the emptiness of living a "life" devoid of real relationship.

In my book *A Language Older Than Words* I detailed, among other things, the importance of amnesia or selective memory to the survival of abused children. If you are powerless to prevent yourself from being harmed or to defend yourself in any way, it serves no purpose to consciously remember the atrocities. In fact it can be lifesaving to read and then identify more closely with the perpetrator's emotions and state of being than one's own. After all, the child's emotions don't matter, but the child needs to be capable at all times of reading and if possible placating the powerful adult's emotions. But I did not mention the function this induced amnesia serves for the perpetrator: it allows him to confront neither the emotional consequences nor the emotional motivations for his abusive behavior.

Everyone at every moment acts to protect the abuser. Think about it in your own life. How many times has someone abused you and you did whatever was necessary to make sure the other person did not feel bad? What did you do to take care of the other person? Here is a story a woman just told me. She was sitting in a bar with her sisters, drinking Coca Cola. A man struck up a superficial conversation with her. Soon she walked into the bathroom. When she emerged from her stall, he was waiting for her. She asked what he was doing. He forced her against the wall, pushed his hips hard into her. She somehow slipped from his grasp, and returned to the main room. He followed. He remained within ten feet of her. She stayed for another hour. Now here's the point: Not only did she not make a scene, but she did not even leave. Even as she was slipping away from his attempted rape and all through the next hour she was thinking, *I don't want to hurt his feelings.*

I cannot tell you how many times I have similarly betrayed myself to protect an abuser.

Years ago, in the midst of one of those abusive relationships I mentioned earlier, a friend was counseling me through the latest incident of abuse. At one point I said, "I don't think she meant to hurt me. Here's what I think she was thinking—"

My friend cut me off: "If I was interested in what she was thinking, I would talk to her. But I'm not, so I won't. I'm interested in what you were thinking, and feeling."

I didn't have an answer. I had no idea. I was too busy taking care of the other person's feelings.

To care about another, to have compassion for another, is beautiful and life-affirming. To care about and have compassion for another who is abusing you is a toxic mimic of real compassion, and is one of the obscenities spawned by a culture of abuse.

The same thing happens all the time on the larger scale. I also cannot tell you how many times I have been told that I must have compassion for CEOs, who are human too, and who once were children. We must never hurt their feelings, nor especially their person. We must always be polite to those who are killing us. If we insist on using any hint of violence, we are told, if we absolutely must kill them back, we must kill them only with kindness. This is supposed to somehow be effective at something. But the only one it helps is the perpetrator.

Bancroft states that one of the most common forms of support for abusers is the person "who says to the abused woman: 'You should show him some compassion even if he has done bad things. Don't forget that he's a human being, too.'" Bancroft continues, "I have almost never worked with an abused woman who overlooked her partner's humanity. The problem is the reverse: *He* forgets *her* humanity. Acknowledging his abusiveness and speaking forcefully and honestly about how he has hurt her is indispensable to her recovery. It is the *abuser's* perspective that she is being mean to him by speaking bluntly about the damage he has done. To suggest to her that his need for compassion should come before her right to live free from abuse is consistent with the abuser's outlook. I have repeatedly seen the tendency among friends and acquaintances of an abused woman to feel that it is their responsibility to make sure that she realizes *what a good person he really is inside*—in other words, to stay focused on his needs rather than her own, which is a mistake."[78]

We have all been trained to identify more closely with the abusive personal and social dynamics we call civilization than with our own life and the lives of those around us, including the landbase. People will do anything—go to any absurd length—to hide the abuse from themselves and everyone around them. Everything about this culture—and I mean everything—from its absurd "entertainment" to its equally absurd "philosophy" to its politics to its science to its interspecies relations to its intrahuman relations is all about protecting the abusive dynamics.

R. D. Laing named three rules that govern abusive family dynamics, that allow the family to not acknowledge the abuse:

Rule A: Don't.

Rule A.1: Rule A does not exist.

Rule A.2: Never discuss the existence or nonexistence of rules A, A.1, or A.2.[79]

These rules hold true for the culture. We see them every day in every way, from the most intimate to the most global. This culture collectively and most of its members individually will give up the world before they'll give up this abusive structure.

(((

A few years ago I asked the great thinker and writer Thomas Berry what transformation would be required for us to have a sustainable sense of self (and by extension a sustainable culture).

He responded, "We have to get beyond the artificial division we've created between the human community and the rest of planet. There is only one community, and it lives and dies as a unit. Any harm done to the natural world diminishes the human world, because the human world depends on the natural world not only for its physical supplies but for its psychic development and fulfillment. This is most important, because people talk about the need to destroy the natural world in order to advance the human world. Well, anything that diminishes the wonder and fulfillment we receive from the natural world spoils the human enterprise. We may get a pile of possessions, but it won't mean much if we can't go to the mountains or the seacoast, or enjoy the songs of birds or the sights and scents of flowers. What does it do to our children when they cannot enjoy such things?"

He continued, "In back of this, and really what I'm concerned with, is the question of how we experience the universe. My proposal—and this is why a cosmological worldview is so important—is that a cosmological order is what might be called the great liturgy. The human project is validated by ritual insertion into the cosmological order. Our job is to participate in the great hymn of praise that *is* existence."

❨ ❨ ❨

This culture won't change on its own. The demands it makes on the natural world and on the humans it exploits won't diminish until the culture is destroyed. As Bancroft writes, "An abusive man expects catering, and the more positive attention he receives, the more he demands. He never reaches a point where he is satisfied, where he has been given enough. Rather he gets used to the luxurious treatment he is receiving and soon escalates his demands."[80]

The same is true on the larger scale, as no comforts or elegancies, no feeling of power over another, no accumulation of property can make up for a failure to participate in the great liturgy. It's an attempt to use increasing amounts of emptiness to plug a great void (or, as R. D. Laing wrote, "How do you plug a void plugging a void?"[81]). It's an attempt to cure loneliness through power. But loneliness can only be cured through relationship,[82] and relationship is precisely what exploitation and abuse destroy.

There can be no compromise with the insatiable. They'll ask, then negotiate, then demand, their threat of violence informing all interactions, and in the end they'll take. But that will not be the end, because they'll not be satisfied. They'll begin again, by asking, then negotiating, then demanding, then taking. And then they'll ask, negotiate, demand, take, until there's nothing left. And yet they'll keep on pushing.

Because Bancroft's book is in some ways self-help, he puts all this slightly differently: "*Objectification is a critical reason why an abuser tends to get worse over time.* As his conscience adapts to one level of cruelty—or violence—he builds to the next. By depersonalizing his partner, the abuser protects himself from the natural human emotions of guilt and empathy, so that he can sleep at night with a clear conscience. He distances himself so far from her humanity that her feelings no longer count, or simply cease to exist. These walls tend to grow over time, so that after a few years in a relationship my clients can reach a point where they feel no more guilt over degrading or threatening their partners than you or I would feel after angrily kicking a stone in the driveway."[83]

Or perhaps he means that abusers would feel no more guilt over threatening their partners than civilized humans would feel blasting stones from a quarry, or damming a river, or deforesting a hillside.[84] Stones, rivers, trees, forests, their feelings, far beyond not counting, have within this culture long since ceased to exist.

❨ ❨ ❨

Thomas Berry said to me, "We have lost touch with the natural order of things. For example, which day of the workweek it is may be more important to many of us than the great transition moments in the seasonal cycles. And which hour of the day it is—will I get to work on time? Will I avoid rush hour traffic? Will I get to watch my favorite television program?—may be more important to us than the transitional moments in the diurnal cycles. We have forgotten the great spiritual import of these moments of transition. The dawn is mystical, a very special moment for the human to experience the wonder and depth of fulfillment in the sacred. The same is true of nightfall. And it's true when we pass from consciousness to sleep, where our subconscious comes forward. That this is a special moment of intimacy is particularly apparent to children. They often know that the moment of falling asleep is the magic or mystical moment when there is a presence. Parents talk to their children in a very special way at this time. It's very tender, sensitive, quiet. It's the great transitional moment in our day-night cycle.

"There are magical moments in the yearly cycle, too. There is the winter solstice, the moment when the transformation takes place between a declining and ascending sun. It's a moment of death in nature, a moment when everything is reborn. We have lost touch with this intimate experience.

"In the springtime, humans are meant to wonder and to ceremonially observe succession, leading to the fulfillment of summer, and the beginning of the movement again toward death. At the harvest there is another time of gratitude and celebration. I think the Iroquois thanksgiving ceremony is one of the greatest festivals in the religious traditions of humankind. Different elements are remembered and thanked: the water, the rain, the wind, the fruitfulness of the earth, the trees. The Iroquois articulate fifteen or more specialized powers that humans need to commune with and be grateful for.

"All of this is cosmological. Such experience evokes a sense of wonder at the majesty of things. We participate in the world of the sacred, the world of mystery, the world of fulfillment. To recognize our fulfillment in these moments is to know what it is to be human.

"We can say the same for places as for moments. To be fully human is to fully experience the spectacular formations of the planet: particular mountains, particular rivers, certain rock structures.

"We no longer do this. We don't experience the natural world surrounding us. We deny ourselves our deepest delight by not participating in the dawn, the dusk, the solstice, the springtime."

❆ ❆ ❆

Unfortunately, abusers don't particularly care about what they're losing. Bancroft writes about this, too: "It is true that partner abusers lose intimacy because of their abuse, since true closeness and abuse are mutually exclusive. However, they rarely experience this as much of a loss. Either they find their intimacy through close emotional connections with friends or relatives, as many of my clients do, or they are people for whom intimacy is neither a goal nor a value (as is also true of many nonabusers). You can't miss something that you aren't interested in having."[85]

This transposes easily to the larger scale with only a few substitutions. "It is true that the civilized lose intimacy with their landbase because of their exploitation of it, since true closeness and exploitation are mutually exclusive. However, they rarely experience this as much of a loss. Either they find their intimacy through close emotional connections [*sic*] with other humans, or they are people for whom intimacy with the land is neither a goal nor a value (as is true of nearly all of the civilized). You can't miss something that you aren't interested in having."

I've heard many environmentalists state that if only they could get CEOs and politicians out of their boardrooms and legislative halls (or out of their penthouses and vacation homes) long enough to breathe clean forest air and to feel duff beneath their feet, long enough to stop thinking about stock prices and start thinking about spotted owls, that the CEOs would undergo magical transformations and suddenly no longer want to destroy the homes of their newfound forest friends.

It ain't gonna happen. This false hope ignores many things. It ignores the fact that when Europeans first encountered a wildly fecund North America, they were not entranced by it, they did not fall in love with it, they feared and hated it, and they began to dismantle it, a dismantling that continues its acceleration to this day. It ignores the fact that many loggers spend much of their adult lives in forests, claiming to love these forests they're destroying. It ignores the fact that CEOs and politicians, like other abusers, are financially and socially well-rewarded for maintaining their disconnected state. It ignores the fact that if some individual does have an epiphany, he will simply be replaced and the destruction will continue apace. And most of all it ignores the fact that, as mentioned before, the culture's problem lies in the belief that controlling and abusing the natural world is justifiable.

❧ ❧ ❧

Where does this leave us?

Well, if you agree with my thesis—which I think I've more than amply supported—that the motivations, dynamics, and damage of abuse play out not only in the bedrooms of little girls and boys, not only in the black eyes and bruised and torn vaginas of women, not only in the fragmented and fearful psyches of the traumatized, but also in blasted streams and dammed rivers, poisoned oceans and extirpated species, and in enslaved, domesticated, or destroyed humans (and nonhumans, and landscapes), then it means that asking, cajoling, or even sending lovingkindness™ to abusers is at best a waste of time. Bancroft again: "You cannot get an abuser to change by begging or pleading. The only abusers who change are the ones who become willing to accept the consequences of their actions."[86] And yet again: "You cannot, I am sorry to say, get an abuser to work on himself by pleading, soothing, gently leading, getting friends to persuade him, or using any other nonconfrontational method. I have watched hundreds of women attempt such an approach without success. The way you can help him change is to demand that he do so, and settle for nothing less."[87]

Let's apply this on the larger scale: We cannot get large-scale abusers to stop exploiting others by pleading, soothing, gently leading, getting people to persuade them, or using any other nonconfrontational method. It won't work.

But you knew that already.

Bancroft continues, "It is also impossible to persuade an abusive man to change by convincing him that *he* would benefit, because he perceives the benefits of controlling his partner as vastly outweighing the losses. This is part of why so many men initially take steps to change their abusive behavior but then return to their old ways. There is another reason why appealing to his self-interest doesn't work. The abusive man's belief that his own needs should come ahead of his partner's is at the core of the problem. Therefore when anyone, including therapists, tells an abusive man that he should change because that's what's best for *him*, they are inadvertently feeding his selfish focus on himself: *You cannot simultaneously contribute to a problem and solve it.*"[88]

Let's once again explicitly make the connection to the larger scale. It is impossible to persuade the civilized to change by convincing them that they would benefit and simultaneously allowing them to remain within the framework and reward system of civilization, because the civilized perceive the benefits of controlling those around them (including humans and nonhumans;

including the land, air, water; including genetic structures; including molecular structures) as vastly outweighing the losses. This is part of why so many of the civilized initially take steps—or at least mouth rhetoric and pretend to take steps—to change their abusive behavior but then return to their exploitative ways. There is another reason why appealing to the self-interest of the civilized doesn't work (apart from the fact that the entire economic system, indeed all of civilization, is based on this limited and unsustainable sense of self which leads people to believe it's in one's self-interest to exploit others, indeed, which causes it to be, within this limited sense of self, *actually in* one's self-interest to exploit others): the belief of the civilized that their own needs should come ahead of the landbase's is at the core of the problem. Therefore when people, including activists, tell a civilized person—for example, a CEO or politician—that he should change because that's what's best for *him*, they are inadvertently feeding his selfish focus on himself: *You cannot simultaneously contribute to a problem and solve it.*

Let's go one more time. Bancroft: "An abuser doesn't change because he feels guilty or gets sober or finds God. He doesn't change after seeing the fear in his children's eyes or feeling them drift away from him. It doesn't suddenly dawn on him that his partner deserves better treatment. Because of his self-focus, combined with the many rewards he gets from controlling you, an abuser changes only when he has to,[89] so the most important element in creating a context for change in an abuser is placing him in a situation where he has no other choice. Otherwise, it is highly unlikely that he will ever change his behavior."[90]

Pay careful attention. No other choice.

No, really. Pay careful attention. No other choice.

No, now *really* pay attention. No other choice.

None.

Let's transpose this to the larger scale. Those who are killing the planet won't change because they feel guilty or drop their addiction to consumerism or find God, or Nature. They don't change after seeing the fear in factory farmed or vivisected animal's eyes (or in the eyes of the poor) or feeling wild creatures drift away from them. It doesn't suddenly dawn on them that the landbase deserves better treatment. Because of their self-focus, combined with the many rewards they get from controlling those around them, these abusers change only when they have to,[91] so the most important element in creating a context for change in those who are killing the planet is to place them in situations where they have no other choice. Otherwise, it is highly unlikely that they will ever change their behavior.

No other choice.

None.

❨ ❨ ❨

The answer that allowed me to move past the question of whether abusers believe their lies is this: it doesn't matter. It doesn't matter at all. What matters is stopping them.

❨ ❨ ❨

Last night I dreamt I was on a ship with thousands of other people. A few men gathered us into a huge ballroom. We knew, even though they never said a word, that they were going to kill us. We huddled against walls or crouched on the floor, waiting to die. The men had guns, but I wondered why we didn't rush them, didn't fight back. There was no way they could kill us all unless we chose not to fight, in which case we would all surely die. Yet there we stayed.

I got up. The men with guns didn't notice. The captives who huddled or crouched hissed at me to get back on the floor. I was endangering them, they said, by standing up. They'll notice us, they said, and get upset. *Upset?* I thought. *They herded us here with guns. They take us three at a time to another room. We hear gunshots. They come to get three more. And you're worried that I'll endanger us?*

I wanted to fight but couldn't do it alone. I knew none of those on the floor would join me. They were all going to die. I made my way slowly to a door, then left the room, went down a hall, and emerged on the deck. A woman swam through the ocean toward the boat. She climbed the side. She was beautiful. She was nude. We didn't speak. We knew what we each had to do. We looked at each other for a moment before she leapt back into the ocean. I went room to room on the ship, searching for people who had not already entered the ballroom.

I half awoke, then lay there in the moonlight, slowly shifting focus from the dream and all it meant to a muffled sound above my head. The sound and its meaning became less fuzzy, then more clear, till I knew the sound was wings fluttering against glass. I sat up, turned around, reached up, and cupped a moth in my hands. I used my thumb to open a window and let it out. I went back to sleep, back to the dream.

The woman swam again to the ship, climbed aboard. She smiled. She had brought help. We were ready. We knew what we had to do. We knew what we wanted to do.

I woke up.

A THOUSAND YEARS

I have noticed in my life that all men have a liking for some special animal, tree, plant, or spot of earth. If men would pay
more attention to these preferences and seek what is best to do
in order to make themselves worthy of that toward which they
are attracted, they might have dreams which would purify
their lives. Let a man decide upon his favorite animal and
make a study of it, learning its innocent ways. Let him learn to
understand its sounds and motions. The animals want to
communicate with man, but Wakantanka [the Great Spirit]
does not intend they shall do so directly—man must do the
greater part in securing an understanding.

Brave Buffalo[92]

IN *STRANGELY LIKE WAR*, GEORGE AND I CITED RAY RAFAEL, WHO has written extensively on the concept of wilderness: "Native Americans interacted with their environment on many levels. Fortunately, they did so in a sustainable way. They hunted, gathered, and they fished using methods that would be sustainable over centuries and even millennia."[93]

So here's a question I've been asking lately: How do I want the land where I live to be in a thousand years? The answers to that question depend of course on answers to: How does the land want to be in a thousand years? And those answers depend on answers to: How was the land prior to the arrival of civilization?

We can safely say the land itself knows better than we what it wants and what is best for it. The questions then become: How well can we perceive what it wants, and how can we help it get there?

Before all you biocentrists freak out at me putting my own desires into this discussion, let's first consider: every being affects its surroundings. If I purchase land and set it aside, that has an effect on that land (many effects, actually, on the land and many other things, including my psyche and the economy). If I purchase land and clearcut it, that will have different effects (on the land, my psyche, the economy, and other things). A deer eating foliage affects the land (and many other things). A mountain lion eating deer affects the land (and many other things). Someone blowing up a dam affects the land (and many other things). Someone refusing to blow up a dam affects the land (and many other things). Note that I'm not saying that all effects are equal, I'm merely saying that we're all part of a web of relationships.

Let's also consider that we all have preferences. Paving over paradise and putting in a parking lot reveals one set of preferences. Ripping up asphalt reveals another. Doing nothing and letting paradise be paved reveals another.[94] And doing nothing and letting plants rip the pavement back out expresses yet another.

Not only are all writers propagandists, by dint of what they do or don't include, but we—all of us—actualize preferences through every act we take or don't take. Further, just as those writers who claim to be "objective" are the least honest and most foolish, any of us who claim to not impose some preference by our every action or inaction are wrong and just plain silly.

But the real reason you biocentrists don't need to freak out is that when you take a long-term perspective, the dissonance between anthropocentric and biocentric viewpoints disappears, or at least becomes much less. I am excluding the perspective of those who eagerly look forward to a future ever more dominated (and ruined) by technology. Those who advocate a technologically controlled future are not only *not* taking a long-term perspective (peak oil, anyone? How about overshoot and crash?) but they're simply insane. They are not in touch with physical reality (that's what "high technology" *does*—it separates us from physical reality). They aren't even truly anthropocentric, but rather technocentric (or maybe power-centric, or control-centric).

A truly anthropocentric perspective, especially in the long term, is biocentric. It must be, since the anthro relies on the bio. No bio, no anthro. Any anthro who isn't bio must be really stupid. Or made stupid by a stupid culture.

I don't believe many farmers (consciously) want to denude the topsoil and poison the land where they live. Yet that's what they do. Their enacted preference is for production over the health of the land. If they were to shift their preferences, if they were to act as though their descendants would still be on the land in a thousand years, they would act dramatically differently than they do now. Indeed, they would not—could not—practice industrial farming. There are questions as to whether they could farm at all. The same is true for other "professions."

Living on the land in a way that doesn't harm the land is not, I believe, generally at wide variance with what the land itself wants. Further, the only way to live on the land in the long run is, obviously, to live on the land in a way that the land *does* want. Over time, if you want something different than what the land wants, you may harm some of it, but it will destroy all of you.

Although it's obvious that we are living in a way that the land does not want, discerning what it does want is not always simple, at least for those of us who have been made mad and stupid by civilization. Yesterday afternoon I stood on a bluff overlooking the ocean with my friend Karen Rath. She's a longtime environmentalist, another person who knows that this whole rotten system of civilization will soon collapse. Indeed, like so many others, she longs for it.

We were talking about some species of native plants who live in the dunes north of where we stood. The plants are endangered, being choked out by an exotic invasive, European beachgrass. She asked if I knew how the grass got there.

I shook my head.

"It was planted maybe a hundred years ago by some of the farmers."

"Why?"

"They wanted to stop the dunes from wandering."

"What?"

"They're alive, you know."

"Dunes? Yes, I know."

"They move all over. They didn't want them to move. They planted European beachgrass. It took over."

The dunes to the north look yellow, even in the distance. "All this?" I asked.

"There are still some native beach grasses, but they're getting crowded out."

"This creates a problem," I said.

"No shit."

"No," I responded. "Another problem." I told her about what I was writing, about how if you plan on living someplace forever, then your decisions will generally be in line with what the land wants.

She nodded.

"But these farmers," I said, "mess up my whole theory. Couldn't you say that even if they were planning on their families living here a thousand years that they might still have planted the beach grass? Or what about the guy who released starlings in Central Park? You've heard about that, right? The guy wanted to bring over to the U.S. every creature mentioned by Shakespeare. So he brought a few starlings. And some woman missed dandelions, so she brought over a couple for her backyard. Might not all of those people have still introduced these exotics? Maybe she wants dandelions in a thousand years, he wants starlings, and the farmers want frozen dunes."

"First," Karen said, "if the farmers were interested in living here for a thousand years, they wouldn't use so many pesticides, or any at all, really. The farmers use everything from metam sodium to methyl bromide to all sorts of other nasty carcinogenic chemicals. Second, dunes are not the only things the farmers wants frozen in place. You know the Westbrooks, right?"

"Yes, I do." The Westbrooks are one of the wealthiest families in Del Norte County. They own or harm thousands of acres here. As is true for essentially all fortunes within this culture, this family fortune was founded on, and continues to rely on, the exploitation and despoliation of the land and the impoverishment or elimination of native human and nonhuman cultures.

"They've already frozen the Smith River," she said.

"What?" I exclaimed again.

She said, "You know just by Yontocket, where the Smith River approaches the ocean, then heads straight north to parallel the beach for a few miles before finally turning back west right near Westbrook's resort? In the big flood of 1964, the Smith cut through those dunes and made a new mouth near Yontocket. Westbrook went down with a bulldozer and closed it off. Do you know why?"

"Because he hates wild nature?"

"Good guess. But the more immediate answer is that he has a fishing derby every year up by his resort, and you can't have a fishing derby without a river right there."

"He moved the river for a fishing derby?"

"The third thing is that what he and the other farmers are doing is wrong. You need to add that to your discussion of what you do with the land where you live. They were trying to immobilize sand dunes, just like he did the river. It's all about control and enslavement. That's wrong. What you do has to be right."

I nodded.

She said, "Something else is missing. You talk about the importance of thinking and feeling forward a thousand years to help you make decisions now, and that's a great thing, but you haven't mentioned going back a thousand years, too."

"I thought I did. I mentioned that we need to think about what the land was like before it was trashed by civilization."

"That's true, but there's something more even than that. The Tolowa, whose land this is, would have known to not introduce that beachgrass because they've lived here long enough to learn what's appropriate. It's just like any relationship: it takes time to get to know the other. In the case of the land it can take many generations. And I don't mean many generations of exploiting it. I mean many generations of paying attention. How can you know the patterns of the river's movements unless you and your people have lived here long enough to watch it move? How can you know anything unless the stories and songs you heard as a child, passed down from generation to generation, taught you the subtleties of the land's long and short cycles?"

I immediately thought about the conflicts I have over removing Himalayan blackberries from the land where I live. I know they're exotics, and I know they're invasive, and I know they crowd out native plants. But I also know that when I've started hacking at them some Lincoln's sparrows have yelled at me (this concern is mitigated by the sad knowledge that at this point there are more Himalayan blackberries than there are songbirds to live in them).

Of course I also know that a big part of my conflictedness about this is the phobia toward responsibility that is the hallmark of this culture. The planet is being killed, yet no one is responsible. Loggers wouldn't cut down trees, they say, if their jobs didn't depend on it. Cops wouldn't protect illegal logging, they say, if that wasn't the word from on high. CEOs wouldn't deforest, they say, except that shareholders demand it. And shareholders don't make any decisions at all. I want the blackberries gone; I just don't want to take responsibility for killing them. I want the dams gone; I just don't want to take responsibility for demolishing them. I want civilization brought down; I just want peak oil or some other means to do it for me.

Larger issues of irresponsibility notwithstanding, I've noticed in my four years on this land that where the forest has come back, it has begun to shade out the blackberries, meaning that if I wait long enough, so long as the blackberries don't halt the return of the forest by killing little trees, that whether or not I kill the blackberries won't matter. If the land wants forest, forest it shall have.

Further, I've long known that blackberries—as is true for most invasive plants—most readily move in to areas that have already been disturbed. When I walk in old growth forests, I don't see Himalayan blackberries. I am not denying that Himalayan blackberries cause damage, because they do, but it's almost always secondary. With every swing of my machete and every closing of my clippers, I can almost hear the blackberries cry out, "Scapegoats. We're scapegoats, and you're a hypocrite. If you really want to remove destructive exotics, we should be low on your list. What about bulldozers? Backhoes? Cars? Pavement? Number one would be *homo domesticus* (called by some *homo stupidus*)—civilized humans.[95] Take your machete elsewhere, and go after real sources of destruction."

You don't *even* want to hear what invasive pampas grass says about this shifting or responsibility. It would burn your ears.[96]

If I had lived here long enough, I would have more information to better act. Perhaps Himalayan blackberries prepare the soil for the next stage of succession, and I should butt out. Maybe in fifty, one hundred, or five hundred years it won't matter: the forest will have returned in either case. And maybe these blackberries will provide much needed food for bears, birds, bumblebees, humans, and others (and homes for songbirds) through the crash. Or maybe not. Maybe the forest will succeed the Himalayan blackberries, but in the meantime the blackberries will crowd out rare native plants who under normal circumstances thrive in temporary forest openings. I don't know. And that's the point.

One reason I've been asking the question about what I want the land to look like and to be in a thousand years is that having some sense of that will help me make decisions now.

I know, to start with, that it's a lot easier to not put in exotics than it is to take them out later. I wish people had thought of this before they brought in Himalayan blackberries. A couple of years ago my mom made the offhand suggestion that I put koi in the pond outside my home. I don't want goldfish here in a thousand years, and they're easier to not put in than to take out later, so I didn't put them in. I wish the entire culture followed this one simple rule.

It's a lot easier to not introduce toxic radiation into a landscape than it is to clean it up, and since I don't want this land to be irradiated in five hundred years, I'll not introduce radioactive materials. It's the same for carcinogens.

On the other hand, I do want coho salmon. It seems clear to me that if I want for there to be coho salmon a thousand years from now in the stream behind my home (as there was a thousand years ago), I need to do what I can to make sure the stream provides habitat. The same is true for Lincoln's sparrows, silver-haired bats, northern red-legged frogs, and Del Norte salamanders. The same is true for redwoods and Port Orford cedars. The same is true for deer, elk, bears (brown and grizzly), wolves, and everyone else whose home this was and shall be. What do each of these want in their home? How can I help make this land as inviting for them as it was before the arrival of civilization?

A lot of this may seem obvious, but even more obvious is that this sort of basic common sense, and these sorts of questions, are almost entirely absent within this culture, and carry almost zero weight in policy decisions at the corporate, governmental, or cultural level, and at the level of personal decisions made by almost every civilized person (see, for example, the neighbors I mentioned above who move to redwood country and cut down paradise to put in a lawn).

I'm sure you see the flaw in my thinking. If I want coho salmon in this stream, I cannot hole up in my home and work only on restoring this particular piece of ground. It could be the best salmon habitat in the world, and if the oceans are dead, there'll be no salmon. If the rivers are dammed, there'll be no salmon. If rivers are clogged with silt from logging, there'll be no salmon. Similarly, if I want birds to fly through these forests a thousand years from now, I've got to stop deforestation not only here but elsewhere, and to stop pesticide use not only here but elsewhere. I need to take out cell phone towers. And it turns out that skyscrapers kill up to a billion birds per year in the United States alone. That needs to stop, too.

At this late stage, it's not nearly enough to "think globally, act locally." With the world being destroyed by a global monoculture, my new motto has become, as alluded to early in this book: Dismantle globally, renew locally.

❮ ❮ ❮

This doesn't mean, of course, that we shouldn't dismantle locally as well. We absolutely need to. Everywhere is local to somewhere, and if all dams, for example, need to come down, then that means local ones do, too. Perhaps the motto should be: "Renew locally, dismantle both locally and globally." Unfortunately, while more accurate this isn't quite so catchy.

What are you waiting for?

DAMS, PART I

The world will be saved, if it can be, only by the unsubmissive.

Andre Gide[98]

WHY CIVILIZATION IS KILLING THE WORLD, TAKE TWENTY-THREE.
Dams.

Not just because they imprison rivers. Not just because they kill fish. Not just because they drown forests. Not just because they leach mercury from the soil and cause it to enter the food stream. Not just because they inundate the homes of humans and nonhumans alike (the World Commission on Dams estimated in 2000 that 40-80 million people worldwide have been displaced by dams[99]). Not just because they lead to mass wastage of water (see, for example, Las Vegas, golf courses, cotton and alfalfa fields in Arizona, and so on). Not just because they're ugly. Not just because they're ubiquitous (quick, name three undammed rivers). Not just because they're often intentional instruments of genocide and ecocide. Not just because they're often promoted as environmentally "clean." All of these could certainly be considered good enough examples of how and why civilization is killing the world. None of these are what I'm talking about right now.

Instead I'm talking about the business of dam removal. Emphasis on business. Utter lack of emphasis on dam removal. Here is part of an article that appeared this spring on the front page of the *San Francisco Chronicle*, which describes a beautiful canyon in northeastern California as "the best hope for two endangered populations of Sacramento River salmon—the winter run and spring run." The article continues, "Five years ago, a consensus was reached to resuscitate the salmon runs: remove five of the eight small PG&E hydropower dams on Battle Creek and outfit the remaining three with fish ladders. It was a revolutionary concept in the 150-year history of water development in California; it would mark the first time that dams would come down rather than go up. [That's actually not entirely accurate: there have been many other dam removals, including twenty-two just along the Klamath between 1920 and 1956 at a total cost to the state of $3,000.[100]] But today the projected price tag for a Battle Creek restoration has skyrocketed, from $26 million to about $75 million, and not a single dam has been removed."[101]

So much has been spent to provide so little help to fish that even a spokesperson for a landowners group made up mainly of cattle ranchers (not generally known as militant environmentalists) said, "Everyone up here is absolutely appalled at the cost over-runs, especially considering how little has been accomplished."[102]

The spokesperson is wrong. Much has been accomplished. In fact the process is accomplishing precisely what it's supposed to. The point was never to save salmon. Part of the point is to pretend to save salmon but the real point is, as always within this culture, to make money. And that, it is doing.

Seventy-five million dollars to remove from five to eight dams. That's between 9 and 15 million per dam. These dams are not big. The article states they're twenty to thirty feet high, but the accompanying photos suggest they're smaller. I'm guessing fifteen feet high by three feet thick by a hundred feet across (although of course span is far less important for demolition than height or thickness, since you only have to breach the dam in one or two places, with the water doing the rest).

I don't know if it will matter to readers that my first degree was in mineral engineering physics when I say that I could take down these dams for far less than 9 to 15 million each. I took (and hate to admit, enjoyed) classes in statics, fluid mechanics, strengths of materials, and so on. I have a working knowledge of engineering, physics, chemistry. I could do it no problem.

Wanna hear my plan?

Choose a date in mid-October, when water is lowest. For weeks beforehand keep sluice gates wide to lower the water even further. Announce the dam's removal date long in advance, and tell salmon lovers from all over the West (and especially the tribes whose lives have been intertwined with these fish forever) that you're going to tear down the dam. Ask them to bring sledgehammers. Ask them if they've heard of Amish community barn raisings, and tell them we're going to have a community dam demolition. Those without sledges can bring wheelbarrows, shovels, picks. Those without tools can bring sandwiches and big coolers of juice. I can guarantee hundreds, if not thousands, of people would show up to work shoulder to shoulder, bashing away at this barrier that separates fish—and humans—from their home. Chips would fly as fast as jokes, chunks would fall to the ground below the dam to be picked up by sweaty men and women smiling as they work together to make something beautiful, to liberate someone they love, to help the river once again to be wild. It's hard work, but as we all know, working hard with friends is more fun than any party ever could be. And the work is productive. For the first time in many of their lives, these people are doing work that does not harm but helps the land. Gouges in the dam grow deeper, wider. As people atop the dam stop for breaks, to eat the delicious homemade food brought by others (everything from vegan potato salad to fried free-range chicken to smoked salmon to watermelon to the best watercress sandwiches you've ever dreamed of), others jump up eagerly to take their place. Few have ever before experienced this sort of communal coming together, only its toxic mimic at football games, parties, and political rallies. Some sing while they work. Some are silent. Some just grunt with every swing of the heavy sledge.

Finally we reach the water level.

To be honest, I'm not sure what we'd do next. I haven't done this before. The question is: How do you knock away concrete below water level without getting washed away yourself? I'm sure there are answers. I just don't know them.[103] If we have a big backhoe or a wrecking ball, we're still in good shape. We just have to stand aside and knock the damn thing down. If we don't have access to those infernal machines, then I'm not sure if we should stop and let the river do the rest of the work, or if we should continue to weaken the dam, lower it little by little, until the river rises up to finish its unshackling. But I do know that three or four of us engineers could figure it out pretty quickly. Or maybe not even engineers but just human beings. Or maybe it would be different for every dam, depending on the circumstances (you didn't think we'd do this only once, did you?). Or maybe some readers will be able to supply—and more importantly, actualize—some answers. It is, after all, a communal project, where we each bring our skills.

I also know that it's not really a technical problem. Although often presented as such, the primary obstacles to dam removal are almost never technical, any more than the primary obstacles to deconstructing the rest of civilization are primarily technical, any more than the primary obstacles to stopping abusers are primarily technical, any more than the primary obstacles to losing weight or quitting smoking are primarily technical. The primary obstacles are perceptual, emotional, moral, spiritual, inertial.

We would figure it out, and we would remove the dam. Together as a community.

And it wouldn't cost the state a fucking dime. I'm sure tribes and salmon organizations would cover the costs of gas for people to get there and for food to keep them full.

Which of course is why it won't happen this way, at least not with state approval. This would accomplish something for the river, for the fish, for the people and communities involved, but it would accomplish nothing for the engineering firms that take in millions to produce neatly bound feasibility studies.

And that, of course, is the point.

❨ ❨ ❨

I want to nip something in the bud. If you're a part of the dam removal study industry (emphasis on study, utter lack of emphasis on dam removal) and you want to write to tell me you're offended by my portrayal of how easy it would be to dismantle a dam ("The sediment, my boy, the sediment!"), don't give in to the temptation. I may have already heard from one of your colleagues (either publicly, expressing dismay, or privately, expressing solidarity). There is a certain type of dam removal expert more concerned with legalities than living rivers, with science than salmon, with process than justice. These experts plague me like pacifists, cautioning me to be cautious, systematically telling me that the system works if only we'll let it, that if we just have patience those in power will see the light and remove the dams of their own accord. If you are one of those experts and have made it this far in the book, don't worry, I'll address your concerns—or more likely just piss you off more—in the next few pages.

❨ ❨ ❨

I want to talk about the Elwha River on the Olympic Peninsula in Washington. Prior to its damming, all five North American species of Pacific salmon ran the river, as well as sea-run cutthroat trout, steelhead, and char. Some of the salmon weighed more than a hundred pounds, the largest salmon ever seen by humans.[104] The lives of the Clallam Indians (as well as many tribes of nonhumans) were centered around the 400,000 salmon who came up the river each year.

Now, about 3,000 fish come up the river annually. The reason? Dams.

The first dam on the Elwha was built by Thomas Aldwell, a Canadian backed by investors from Chicago. Aldwell summed up his relationship to the land in language that well manifests this culture's collective desires: "There is something about belonging to a place. You want to control more and more of it, directly or indirectly . . . land was something one could work with, change, develop."[105]

The dam was illegal. In its very first session, many years earlier, the Washington State Legislature had passed laws prohibiting anyone from blocking fish passage up any river or stream. As David R. Montgomery dryly notes in his extraordinary *King of Fish*, "Though the intent of such laws seems clear, they were generally ignored or circumvented in short order."[106]

The city of Seattle, for example, dammed the Cedar River in 1901, and "the dam stood in unchallenged violation of state law for over a century."[107]

The stated purpose of the Elwha dam was to produce hydroelectricity. Never mind that there were no markets, because what was at stake here was not mere electric power, but heaven on earth. As one article promoting hydroelectricity put it: "Should any considerable portion of that enormous power ultimately be developed and utilized, who will attempt to foretell the innumerable benefits which will accrue therefrom to mankind? It would completely revolutionize economical industrial conditions. The cost of living would be greatly reduced. Not only the necessaries but the luxuries of life would be easily within the reach of the poor as well as of the rich. With the many electrical appliances already invented for the use, convenience, and benefit of mankind, and with the inventions an inventive age will produce for the betterment of humanity, Bellamy's ideal commonwealth may not be as far in the future as the pessimist might imagine."[108]

I suspect however, the real reason for the dam's construction was that stated by Aldwell. If what you want is to control more and more land, what you'll do is attempt to control more and more land.

Dam construction began in 1910. By 1911, a Clallam county game warden wrote to the State Fisheries Commissioner, "I have personally searched the Elwha River & Tributarys [sic], above the dam, & have been unable to find a single salmon. I have visited the Dam several times lately, was out there yesterday and there appears to be thousands of salmon at the foot of the Dam, where they are jumping continually trying to get up the flume. I have watched them very close, and I'm satisfied now, that they cannot get above the dam."[109]

I'm not sure why the game warden needed to watch them so close before he could be satisfied. No matter how strong or determined the salmon were, they weren't going to clear the dam: it's more than a hundred feet tall.

Fisheries personnel were assured by on-site engineers that a fishway would be built. It should come as no surprise to any of us that the engineers lied.

The response by the state was of course not to demand the illegal structure be torn down—or even that it not be fixed after it failed in heavy rains in 1912.[110] Remember, the property of those higher on the hierarchy is always worth more than the lives of those below. Their solution was to demand that a fish elevator be built that would trap fish at the base of the dam then carry them to the top and release them in the reservoir.

This absurd solution was ignored. A new governor came in, and with him a new fisheries commissioner, Leslie Darwin. Darwin had all the right rhetoric, saying, for example, "It seems to me to be a crime against mankind—against those who are here and the generations yet to follow—to let the great salmon runs of the State of Washington be destroyed at the selfish behest of a few individuals who, in order to enrich themselves, would impoverish the state and destroy a food supply of the people. Unfortunately, every pressure is exerted in behalf of those selfishly interested. These selfish interests have gone to almost unbelievable extent in certain instances in order to silence any opposition in their course, and have slandered and vilified those who opposed their plans and methods. These persons do not want the people of the state to know the truth of the matter, believing that if they do they will act to protect and conserve. It is my belief that had the people understood the situation, they would have acted long ere this, and would have prevented the practical destruction of some of our greatest salmon runs."[111]

So he took down the dam, right? Well, no. He did what he decried, and went "to an almost unbelievable extent" to exert pressure "in behalf of those selfishly interested."

As historian Jeff Crane notes, "Whereas Darwin had elsewhere willingly used dynamite to remove small earthen dams in an effort to enforce the law and restore salmon runs, he was more flexible with such a heavily capitalized project as the Elwha Dam; he struck a deal with a company that had been in violation of the law for five years, years during which the salmon runs were dealt serious harm."[112]

He took advantage of a seeming loophole in the law. It was generally illegal to obstruct rivers, but one *could*, it seems, block rivers to capture fish to kill and take their eggs for use in fish hatcheries. Here's how Darwin's scheme worked, once again according to Crane: "Darwin proposed a clever, pragmatic, and illegal plan. He suggested that by selecting a hatchery site at the base of the dam and making the dam the obstruction for the purpose of collecting eggs for the hatchery, it would be possible to obviate strict enforcement of the fish passageway law . . ."[113]

In other words, the dam was no longer to be considered a dam, but instead an obstruction to stop fish from moving upstream so they could be captured by the operators of the hatchery, who just happened to be the operators of the dam, which just happened to produce hydroelectricity for sale. It's still illegal, but that didn't seem to bother the bureaucrats. It still destroyed the salmon and other fish, but that didn't seem to bother them, either.

Darwin was so pleased with his idea that he later convinced the state legislature to change the law to allow hatcheries in lieu of fishways. Never mind, once again, that the hatcheries not only didn't help wild salmon but harmed them.

The whole system is based on lies. So long as someone tells us comforting lies, we will continue to allow them to control more and more of the land and air and water and our genetic materials and everything on the planet.

The lie having completed its purpose, the dam having been built, all pretense of operating a hatchery was dropped in 1922.

Oh, and all that electricity that was supposed to fuel utopia? The dam produced just enough to run a sawmill. The sawmill was used, of course, to deforest the region.

❨ ❨ ❨

It's now 2004. For more than seventy years two illegal dams have stood on the Elwha. There is the Elwha Dam and the more than 200-foot-tall Glines Canyon Dam (built 1927). For more than seventy years these dams—illegal dams—have killed salmon, shad, steelhead, cutthroat, and other fish.

After decades of outrage and pressure from the Lower Elwha Klallum Tribe (which traces its creation to the Elwha River) and others, in 1992 Congress passed and the President signed a bill authorizing removal of the dams. The dams—illegal dams—were to be purchased from the Virginia-based transnational paper conglomerate James River Corporation (212 pulp and paper facilities in eleven countries, including the U.S., Canada, Mexico, Scotland, France, Italy, Finland, and Turkey[114]). Yet the illegal dams continue to stand because no money was allotted to purchase them.

A major sticking point was that the dams—illegal dams—still provided electricity that was used by the sawmill that was still used to deforest the region. The sawmill is owned by the Japanese-based transnational paper conglomerate Daishowa,[115] infamous for clearcutting the homeland of the Lubicon Cree in Canada.

So, salmon would continue to suffer so that two distant transnational corporations could continue to profit from these structures that had been illegal for more than seventy years. And if this problem were to be solved, American taxpayers would have to pay to purchase these illegal and destructive structures from these transnational corporations.

This is how the system works. This is one reason the planet is being killed.

The dams were finally purchased in 2000.

Demolition was supposed to begin in 2004, but was more recently pushed back to 2007. Presuming deconstruction does begin then, here's how it's supposed to work. The larger Glines Canyon Dam should be pretty simple, as engineers cut successive Vs in the concrete, each time slowly lowering the water, until the river is free. Such a straightforward approach won't, unfortunately, work on the Elwha Dam, because of the way it was patched after failing. Instead engineers will divert the river around the dam, drain the lake (Lake Aldwell), and demolish the dam.

Once the dams are gone it will only take months, according to Brian Winter, former fisheries biologist with the Tribe and now working on dam demolition for the National Park Service, before remnant salmon come home, and begin once again to explore the full length of the Elwha.[116]

❨ ❨ ❨

Let's say you want to take out a big dam. Let's say you have the full power of the state behind you, which means you don't have to worry about pesky cops coming to drag you and your trusty sledgehammer away for knocking down this illegal structure. Of course when the structure was put up, the cops were nowhere in the area. In fact the Law Enforcement Officers were probably off arresting protesters who were trying to block access and stop the dam from being illegally built in the first place. We should change cops' title to Selective Law Enforcement Officers.[117]

How are you going to bring it down?

There are five major ways the state takes out dams, with the most common by far being the last one.

The first consists of digging around the dam to divert the river, then using heavy equipment to dismantle the dam. An example of this would be the twenty-three foot high and nine-hundred foot long Edwards Dam on the Kennebec River in Maine, which was taken down this way in just a few days in 1999.

The second method, usually used on huge earthen dams, is to breach the dam using heavy machinery, and then let the river flow around the rest of the structure, which you allow to remain standing, presumably as a monument to this culture's arrogance and stupidity. I'm still not sure how you breach a dam with water behind it. I'd like to learn, since this is a relatively inexpensive method.

The third method is an easy one. If you've got a barrage-type dam with radial gates, you can just open the gates and pretend the dam isn't there. Two examples of this would be the Nagara Estuary Dam in Japan and the Pak Mun Dam in Thailand.

The fourth method is the one we've all been waiting for: the big blast. Explosives are sometimes used to take out concrete dams. A few examples of this would include dams on the Clearwater (1963), Clyde (1996), Loire (1998), and

Kissimmee (2000) rivers. Of course even with explosives it still helps if you've got a friend with some heavy equipment.[118]

The fifth method of dam removal used by those with the full power of the state behind them is the one we've come to expect from those who are able to get the full power of the state behind them, which is to do nothing at all. Although they either don't know or won't admit it, this is a form of dam removal, too, because eventually every dam will fail. The only question will be what's left of the river when that finally happens.

❨ ❨ ❨

If you're like me, you're probably wondering how much explosives it takes to knock out a big dam. The answer may make you as happy as it made me. It doesn't take much at all.

Read that again. It doesn't take much at all.

Imagine the possibilities!

On February 23rd of this year, a hundred foot breach was blasted into the Embry Dam on the Rappahannock River near Fredericksburg, Virginia. The dam was a bit over twenty feet high and nearly eight hundred feet long.

Divers placed explosives, and just a little after noon the signal was given to set them off. Only 10 percent ignited, making a burst of smoke and water but leaving the dam standing. Ninety minutes later they tried again. This time it worked. The river rushed through the broken dam.

All it took was six hundred pounds of explosives.

That's it.

And that was a pretty big dam.

What are you waiting for?

❨ ❨ ❨

After the Embry Dam blew, I got an email from someone pointing out to me that the dam was removed by people working within the system. Seventy miles of river were opened to migratory fish such as American and hickory shad, blueback herring, alewife, striped bass, and yellow perch. "See," he wrote, "the system works!" Never mind that the dam hadn't produced electricity since the 1960s, so it took forty more years of killing the river to get the dam removed. He ended his note, "Please stop talking about people removing dams on their own. The system works. Trust it."

I couldn't believe he actually said that.

I responded, "I am of course happy that the dam is down. I have no problem with working inside the system when it works. That's preferable, *when it works.* The Embry was the second largest dam removed in the United States since the Edwards Dam in Maine in 1999. There are about 75,000 dams over six feet tall just in the United States (about 65,000 over twenty feet tall). There are 365 days in a year. 3652 days in ten years.[119] 36,525 days in a hundred years. 73,050 days in two hundred years. If we only take out one dam per day, it will take us about two hundred years to take out all the big dams in the United States. Then we can start on Mexico. Nobody knows how many small dams there are in the United States, which lets you know how out of control *that* situation is. The best estimate is about two million.[120] At one per day it would take us almost 5,500 years. Then we can start on Canada. And then Russia. Then China. And so on. If there are two million dams in the United States, I don't think my math is up to figuring out how many dams there are around the world. But I do know that worldwide, dams have been erected at a rate of about one per hour.[121] The dams need to come down. The dams will come down, and we can do our part. We've got a lot of work to do. I'll make you a deal. You work within the system and I will support you in that work. I'll help all I can. All I ask in return is that if others choose to work outside the system that you give them the same courtesy and support."

I didn't hear back from him.

❨ ❨ ❨

The good news is that even working within the system dam removals are on the rise. There have been about 120 fairly big dams removed in the United States in the past forty years.[122] The bad news is that at this rate of three per year, it would take about 25,000 years just to get the big dams out of this country, even if no new ones were built. I'm not sure salmon can hang on that long. I'm not sure we can either. The good news is that these numbers greatly underestimate the real total of dams removed. If you include small dams, one guess—and that's the best we can do is guess—suggests about 600 to 700 dams have been removed these past few years,[123] but that number is probably low. Just in Wisconsin, there have been about 800 or 900 dams abandoned, now waiting for someone to take them out. According to Emily Stanley, a biogeochemist at the University of Madison-Wisconsin, "Dams used to be like stairsteps along rivers, but they have been dismantled or blown out by floods, or the logging industry blew them up when they were done with them. There have been a lot more removals than we realize."[124] The bad news is that if we're going to include small dams, then the math gets hairy again as we move to exponential figures, with two million dams needing to be removed: that's 2×10^6 to you nerds out there, or 2E6 to you übernerds.

Now we get to the worse news we knew was coming all along. When people "within the system" talk about removing dams, almost never is it for environmental reasons. Sure, we radicals get to talk all we want about taking out the Elwha (which with its sister the Glines Canyon Dam are the *only* large dams scheduled to be removed for purely environmental reasons[125]), or the dams on the Lower Snake (which ain't gonna happen through legal means), or the Edwards, or the Glen Canyon, or Hoover, because we want to see rivers run free. But that's not why dams get removed.

Why do they get removed? Money. It's always money. Within this culture, money talks, the environment and everything else walks. Or gets entombed in concrete. Or dies.

Money is of course the primary reason dams are built. The fine, if a tad academic, folks at the World Congress on Dams make clear that all other considerations drop to the wayside: "Pervasive and systematic failure to assess the range of potential negative impacts and implement adequate mitigation, resettlement and development programmes for the displaced, and the failure to account for the consequences of large dams for downstream livelihoods have led to the impoverishment and suffering of millions, giving rise to growing opposition to dams by affected communities worldwide. Since the environmental and social costs of large dams have been poorly accounted for in economic terms, the true profitability of these schemes remains elusive."[126] In other words, when those at the top of the hierarchy tell you that a dam will be profitable we can guess who (always) profits and who (always) loses.[127] It's the same old story: the bank accounts of those higher on the hierarchy are worth more than the lives of those below. It is, as always, profits über alles. Here's how the World Congress on Dams people translate this into language surprisingly spry for still being acceptable within the Ivory Prisons of Academia: "Perhaps of most significance is the fact that social groups bearing the social and environmental costs and risks of large dams, especially the poor, vulnerable and future generations, are often not the same groups that receive the water and electricity services, nor the social and economic benefits from these. Applying a 'balance sheet' approach to assess the costs and benefits of large dams, where large inequities exist in the distribution of these costs and benefits, is seen as unacceptable [by some of us, I would add, though certainly not by those in power] given existing [*sic*] commitments [*sic*] to human rights [*sic*] and sustainable [*sic*] development [*sic*]."[128]

The next premise of this book is a softer restatement of the eleventh premise, that this culture is a culture of occupation, and that the government is a government of occupation. I'd call it a corollary of number eleven, but I think that even those who shy away from the "o" word will agree with the twentieth premise, which is: *Within this culture, economics—not community well-being, not morals, not ethics, not justice, not life itself—drives social decisions.*

Having said that I now want to modify it, make it, as happened to the motto above, less catchy but more accurate. First, the economics that drives social decisions is not any sort of real free market economics, where equal partners make equal decisions about transactions that affect them equally. It isn't *any* sort of human economics—of which a true free market economics would be one type—where humans make decisions based on human concerns. Rather it is an economics where, as above, the profits of those higher on the hierarchy trump all other considerations and where men (and in this more "feminist" age, women) with guns enforce these transactions over the often-dead bodies of their victims. So let's modify it to read: *Social decisions are determined primarily (and often exclusively) on the basis of whether these decisions will increase the monetary fortunes of the decision-makers and those they serve.*

Within this culture money is a stand-in for power, so we'll modify the premise again: *Social decisions are determined primarily (and often exclusively) on the basis of whether these decisions will increase the power—or its stand-in, money—of the decision-makers and those they serve.*

But we need to modify it more. As is true for choices made by perpetrators of domestic violence, underlying these social decisions is *always* an attitude of entitlement. Under this attitude, those higher on the hierarchy are *always* entitled to dam rivers: the only real basis on which decisions are made is whether those in power deem the action in question worth their while. Obviously. So we need to modify this premise more: *Social decisions are founded primarily (and often exclusively) on the almost entirely unexamined belief that the decision-makers and those they serve are entitled to magnify their power and/or financial fortunes at the expense of those below.*

Good, huh?

Now, I hate to do this to you, but I think we've got to modify it again.

These social decisions so often seem compulsive. A dam every hour? Even if dams were a good idea, don't you think that's a bit much? Two million dams in the United States alone? Couldn't dams, especially in such outrageous numbers, be considered concrete manifestations of a severe psychotic obsession? Wouldn't you consider it obsessive to kill 90 percent of the large fish in the oceans, 90 percent of the native forests, 90 percent of the native peoples? How about *all* of the passenger pigeons? We snicker or gasp or mainly feel superior when we read about some hermit living in a house full of newspapers running back to 1947, or saving toenail clippings in labeled and sorted jars of formaldehyde, yet how does each of those compare to these larger obsessions, these larger needs to control one's surroundings?

You know, the insanity says, *there's just something about land (women, water, power) that makes you want to control more and more of it, directly or indirectly. Land (women, water, power) is something you can work with, change, develop.*

Recall premise ten of this book, which is that this culture as a whole and most of its members are insane. The culture is driven by a death urge, an urge to destroy life. A few years ago I was talking with Ward Churchill about how stupid it was of the Nazis to keep meticulous records of their atrocities, even when these records led at least a few of them to the hangman. He responded, "What do you think GNP is?"

I'd long known that production is the conversion of the living to the dead, but I'd never made the connection that GNP is really nothing more than the sum of these atrocities, with ledger sheets being the enumeration of the awful details. Wall Street was formed as a market for slaves, and now functions as a market in slavery, and a market for futures built on planetary murder.

He continued, "Not only do they have to kill the wild and the people of the wild, but they have to record it all, and they have to do this without acknowledging to themselves that they are making lists of the murdered. So they hide that behind all sorts of fancy financial jargon, so many millions of board feet of timber, so many tons of fish, so many dollars in their bank accounts. But those are records of the dismemberment of the planet. And the accumulation of dollars is, as we both know, just an excuse for the primary destruction."

So let's modify it one more time. On a conscious level, social decisions are determined primarily (and often exclusively) on the basis of whether these decisions will increase the power—or its stand-in, money—of the decision-makers and those they serve. But these decision-making processes are in fact charades. This of course is known to any member of the public who has ever attempted to participate. It is known to the decision-makers themselves. The charades run deeper, however, than even the "decision-makers" comprehend, because these motivations of increasing power and/or money are not the primary motivations. The primary motivations—lying beneath, unacknowledged, often unperceived—are to control and to destroy. *If you dig to the heart of it—if there were any heart left— you would find that social decisions are determined primarily on the basis of how well these decisions serve the ends of controlling or destroying wild nature.*

Given even the conscious motivations driving social decision-making processes—and even moreso the unconscious motivations—it should come as no surprise that the financial concerns of those at the top of the hierarchy determine whether or not—the latter being more likely—dams will be removed through legal channels.

As the authors of one website on dam removal say, "[T]he impetus for dam removal comes not from environmental damage, but from the simple fact that dams are getting long in the tooth: 25 percent of America's 2 million dams are older than 50 years. Many of these codger dams have problems: They may have cracks. Water may have undermined the foundation. They may be so full of sediment that they cannot store water. They may have been built for a purpose that no longer exists. Or they may endanger swimmers or canoeists, who can get trapped and drown in 'scour holes' that appear downstream of dams. The dam-removal process often begins when a state inspector looks at a dam and insists on repairs. These often turn out to cost far more than removal, so repair can only be justified if the dam provides significant economic benefits. If a dam was built, for example, to power a grain mill that is long gone, or is supplying only a small amount of hydroelectricity, who would want to pay a million bucks to keep it going, when it could be ripped out for $50,000 or $100,000?"[129]

Or as another website devoted to dam removal put it: "Dams are removed for a variety of reasons, but most boil down to economic considerations. Dams have become unsafe (looming structural failures) or ineffective (loss of reservoir capacity through sediment buildup), or the original rationale for their existence has simply disappeared (irrigation for land no longer farmed), and in all these cases the owners have decided that the cost to repair/maintain the dam is no longer worthwhile."[130]

Money talks. The environment walks.

《　《　《

Those at the top of the hierarchy make social decisions based—at least consciously—on whether these decisions will increase their power and money.

There are those, however, who make decisions based on different criteria.

PRETEND YOU ARE A RIVER

Years ago, I was in northern California, and saw a huge bird on a fence. I heard the bird say, "We all used to speak the same language, but you've forgotten that we really are all one being. That's why you can't understand our language anymore."

Jane Caputi[131]

A human being, too, is many things. Whatever makes up the air, the earth, the herbs, the stones is also part of our bodies. We must learn to be different, to feel and taste the manifold things that are us.

Lame Deer[132]

PRETEND YOU ARE A RIVER. PRETEND YOU ARE THE MIST WHO FALLS so fine—so gentle—that nothing separates water and air. You are the rain who falls in sheets, explodes onto the ground to leave pocks and puddles. You are the ground who receives this water, soaking it up, taking it in, carrying it deep inside. You are the cracks and fissures where the waters accumulate, flow, fall to join more water, and more, in pools and rivers who move slowly through cavities, crevices, pores. You are the sounds and silence of water seeping or staying still. You are the meeting of wet and dry, the union of liquid and solid, where solids dissolve and liquids solidify. You are the pressure who pushes water through seams. You are the rushing water who bubbles from the earth.

You are a tiny pool between rocks. You overflow, find your way to join others who like you are moving, moving. You are the air at the surface of the water, the joining of substantial and insubstantial, the union of under and over, weight and not-weight. You are the riffle, the rapid, the tiny waterfall who turns water to air and air to water. You are the mist who settles on the soil. You are the plants who drink the mist, and you are the sun who warms and feeds them.

You are the fish who feed on insects who feed on plants who feed on soils who feed on fish. You are the fish who become soils who become plants who become insects who become fish who flow down the river.

You are the river who joins other rivers to become a new river who is all of the rivers and something else.

You are the river. You do not stop at the banks, where liquid turns to solid. You reach into the sky and into the soil. Water moves through rocks, comes up to form pools far from the fast flow where the rivers move together, seeps down to join still waters deep below the surface, waters who sleep and wake and sleep and mingle with the stones who are the river, too.

You are the river, who is married to the mountains you have known since they were young, who have given themselves to you as you have given yourself to them. You are the canyons you nestle into, each year deeper than the year before. You are the forests who give you their fallen trees, and the meadows you flood and feed and who feed you back their fruits and fine insects who fly to your surface to be taken in by the fish who with their own bodies again feed the meadows.

You are the river who feeds the ocean, who feels the tides pushing and pulling against your mouth, the waves mixing fresh and salt. You are that intermingling. That is who you are. That is who you have always been.

❨ ❨ ❨

You are the river. You have lived with volcanoes and glaciers. You have been dammed by lava and ice. You have carried log jams so large and so old they grow their own forests, with you running beneath. You have lived through droughts and floods.

You are the river. You miss the salmon. You miss the sturgeon. You miss the ocean. You miss the meadows. You miss the forests. You miss the beavers and otters and grizzly bears. You miss the human beings.

You are the river. You want them back. You want to feel the tickling of the sturgeon, the thrusting of the salmon. You want to carry food and soil to the ocean. You want to cover the meadows as you used to, and you want to give yourself to them and you want them to give themselves to you, as you have done forever, and as they have too.

❨ ❨ ❨

Now, pretend you are a forest. You are the bark of trees, and the hairy moss who hangs from them. You are the duff who becomes soil who becomes trees who become seeds who become squirrels who become owls who become slugs who become shrews who become soil.

You are the trees who cannot live without the fungi who cannot live without the voles who cannot live without the trees. You are the fire who cannot live without the trees who cannot live without the woodpeckers who cannot live without the beetles who cannot live without the fire.

You are the wind who speaks through the trees and the trees who speak through the wind. You are the birds who sing, and the birds who do not.

You are the salamanders. The ferns. The millipedes. The bumblebees who sleep on flowers, waiting for the morning to warm you up so you can eat and fly on home.

You, too, have lived through drought and flood, hot and cold. And you, too, miss the salmon. You miss the owls, the grizzly bears. You miss the rivers. You miss the human beings. You want them all back. You need them back, or you will die.

❨ ❨ ❨

When I talk about taking out dams, I'm not "just" talking about liberating rivers, and I'm not "just" talking about saving salmon. I'm talking about forests and meadows and aquifers and everyone else whose home this was long before the arrival of civilization. I'm talking about those whose home this *is*. You cannot separate rivers from forests from meadows, and it's foolish to think you can. If you kill rivers, you kill forests and meadows and everyone else. The same holds true for all parts of these relationships, in all directions.

I've long known that salmon feed forests, but I did not know how dependent forests are on these fish until I read a luminous essay called *The Gift of Salmon* by Kathleen Dean Moore and her son Jonathan. The essay begins with part of a letter Jonathan wrote to his mother from Alaska, where salmon have not yet been destroyed by civilization: "The creek is so full of sockeye, it's a challenge just to walk upstream. I stumble and skid on dead salmon washed up on the gravel bars. It's like stepping on human legs. When I accidentally trip over a carcass, it moans, releasing trapped gas. In shallow water, fish slam into my boots. Spawned-out salmon, moldy and dying, drift down the current and nudge against my ankles. Glaucous-winged gulls swarm and scream upstream, a sign the grizzlies are fishing. The creek stinks of death."[133]

The next summer, Kathleen went to visit the spot, now clean of salmon, and asked, "Where did the piles of dead salmon he witnessed go? What difference does their living and dying make to the health of the entire ecosystem?"[134]

As you know, salmon provide a tremendous influx of nutrients into the forest. They put on about 95 percent of their weight in the ocean, and carry this weight into the forest and die. Prior to the arrival of civilization—and dams—the amount of nutrients that flowed into forests this way was nearly unimaginable. Salmon, steelhead, shad, herring, striped bass, lamprey, eels, and many other fish ran the rivers to bring their bodies home. Researchers estimate that about five hundred million pounds of salmon (not including steelhead, lampreys, and so on) swam up the rivers of the Pacific Northwest (with some streams averaging more than three salmon per square yard over the whole stream). That's hundreds of thousands of pounds of nitrogen and phosphorous each year.[135]

When the salmon come in, it's time for a feast. Bears eat salmon. Eagles eat salmon. Gulls eat what the bears and eagles leave behind. Maggots eat what the gulls leave behind. Spiders eat the maggots-turned-flies. Caddisflies eat dead salmon. Baby salmon eat living caddisflies. In the Pacific Northwest, sixty-six

different vertebrates eat salmon. That includes salmon themselves: up to 78 percent of the stomach contents of young coho and steelhead consist of salmon
carcasses and eggs. Between 33 and 90 percent of the nitrogen in grizzly bears
comes from salmon, or at least it did when there were salmon for them to eat.
This was true as far inland as Idaho. As go the salmon, so go the bears. Phosphorous from pink salmon makes its way into mountain goats. Trees next to
streams filled with salmon grow three times faster than those next to otherwise identical streams. *Three times.* David Montgomery, in *King of Fish*, writes,
"For Sitka spruce along streams in southeast Alaska this shortens the time
needed to grow a tree big enough to create a pool, should it fall into the
stream, from over three hundred years to less than a century. Salmon fertilize
not only their streams but the huge trees that create salmon habitat when they
fall into the water."[136]

As go salmon, so go lakes. Kathleen Dean Moore notes that "the cycles of
salmon are mirrored by the growth of plankton, the foundation of the food
chain that nourishes life in a lake. The more salmon, the more zooplankton,
and the more algae flourish in the lake. . . . [Studies] show the precipitous
drop in plankton levels and lake productivity that mark the start of large-scale
fishing in the late 1800s. Over the last 100 years, fishing has diverted up to
two-thirds of the annual upstream movement of salmon-derived nutrients
from the local ecosystem to human beings."[137] Add in dams, industrial forestry,
and the other ways the civilized torment and destroy salmon, and rivers in
the Northwest starve: they only receive about 6 percent of the nutrients they
did a century ago.[138]

The forests need salmon. We need salmon. And salmon need us. As Bill Frank
Jr., Chair of the Northwest Indian Fisheries Commission stated, "If the salmon
could speak, he would ask us to help him survive. This is something we must
tackle together."[139]

I think they are speaking, if only we would listen. Here is what Jonathan
Moore wrote to his mother: "I have seen sockeye salmon swimming upstream
to spawn even with their eyes pecked out. Even as they are dying, as their flesh
is falling away from their spines, I have seen salmon fighting to protect their
nests. I have seen them push up creeks so small that they rammed themselves
across the gravel. I have seen them swim upstream with huge chunks bitten out
of their bodies by bears. Salmon are incredibly driven to spawn. They will not
give up."[140]

They are speaking. We must listen.

DAMS, PART II

I realize that if I wait until I am no longer afraid to act, write, speak, be, I'll be sending messages on a ouija board, cryptic complaints from the other side.

Audre Lorde[141]

THERE ARE THOSE WHO TELL US NOT TO LISTEN. THERE ARE THOSE WHO tell us we must be reasonable. There are those who caution patience. There are those who against all evidence tell us the system can be made to work. These people are wrong.

They are journalists and scientists and activists and engineers and technicians. They are the doomed men—the already dead though still breathing men—huddling against the walls of the ballroom in the ship, terrified lest anyone break their unacknowledged death watch.

Foresters preside over the murder of forests. Hydrologists preside over the murder of lakes and rivers. Of course they do not call it this. They call it *management.*

The doomed ones huddled in the ballroom will try to stop you through any means necessary. They have been listening too long to the echo-chambers of their own intra-human institutions, and like Jack of R. D. Laing's Jack and Jill they must stop anyone from listening to the natural world, lest they be reminded of what they have forgotten—that they and the institutions they serve and with which they identify are murdering the forests and rivers and plains and oceans and skies and aquifers and mountains and those who live in these places, those who *are* these places. They've forgotten also—and will stop anyone from reminding them—that they too were once capable of hearing the salmon and the spotted owl speak. They will kill you to maintain their enforced deafness, because otherwise they will lose their identity as journalists and scientists and activists and engineers and technicians; they will lose their identity as civilized; they will, from their perspective, die.

☾ ☾ ☾

Usually, though, the experts don't need to kill us. Instead, they just tell us to trust them, and so we surrender to them. We trust our health to the hospital industry, our safety to the police industry, our children to the education industry, our salvation to the church industry. We trust journalists to tell us what's going on locally and in the world, and we trust scientists to tell us how the world "works."

So far as taking out dams, we're told by experts in the employ of corporations or the government (of occupation) that we should leave dam removal in the hands of experts in the employ of corporations or the government (of occupation).

Here's one example of how it works. I give a talk, during which I describe civilization's murder of rivers. I detail how salmon and sturgeon have survived for millions of years, but they are not surviving civilization and its dams. I speak of the need to remove these dams, and I speak of the need to do this now.

A man who looks to be in his mid-fifties stands, says, "I'm a hydrologist. I trust that when you talk about people taking out dams you mean that metaphorically, that you're saying they need to remove the dams in their own hearts, the things that stop them from doing what they need to do."

I respond, "Sure, it works as a metaphor, but removing metaphorical dams doesn't do a damn thing to save salmon."

Six months later I give another talk in the same town. He comes again, makes the same plea. I respond the same way. This time his wife stands, too. She leans forward, grasps the back of the seat in front of her, and says, voice strong with emotion, "I'm also a hydrologist. I'm here to *strongly* urge you to not be irresponsible and take this into your own hands. I cannot tell you how much harm you're causing just by talking about this."

"Harm to whom?" I ask.

"To the rivers. There may be people who act on your words, and if they take out a dam they'll kill the river below. Dams fill with sediment, and if you suddenly remove the dam, water will surge down in a muddy flood, scouring the river."

How can I respond to that? I'm not an expert. Maybe she's right. I *have* seen rivers and streams devastated by sediment. I've seen pools filled in that before were deep and bright with the flash of fish rising to strike at flies. Admittedly, the sediment I've seen came not from dams being removed but from clearcuts causing hillsides to slump into streams, but it's a powerful image, and I see her point.

We're in a difficult spot. The rivers are in an even more difficult spot. The rivers are being killed, and if we do not remove the dams they will die. But if we—human beings, not experts—do remove the dams we will kill the rivers.

I don't know what to do. Maybe I should trust the experts. After all, they know more than I do.

((((((

I'm doing a radio call-in show in Olympia, Washington, and I mention my dam dilemma.

Someone calls in, says, "I've got exactly two words for you: Toutle River."

Silence on the line. Finally I say, "Thank you very much, but I have no idea what you're talking about."

"Three words then: Mt. St. Helens." It's clear he's enjoying this.

"Help me out," I say.

"When the volcano Mt. St. Helens blew back in 1980, the Toutle River just below it was not only scoured by sediment, it was boiled. A hundred foot wall of water, ash, and debris came down at a hundred to a hundred and fifty miles per hour, annihilating everything in its path. Two hundred square miles of forest were flattened. All animals were presumed dead. That's something like ten million fish, a million birds, fifteen hundred elk, two hundred bears, and so on. All visible mosses, ferns, and other plants disappeared. About fifteen miles of the river were gone. Not just scoured. Not just boiled. Gone. It looked like a moonscape. Some scientists suggested it would never recover, certainly not in our lifetimes. Others speculated that not even insects would come back."

"And?"

"The Toutle River is in great shape, except where the Forest Service used the volcano as an excuse to let the timber industry go crazy, and where the Corps of Engineers used it as an excuse to build more of their sorry structures. No, the scientists were uniformly wrong. Insects reinhabited quickly, as did plants and birds. Most of the amphibians are back. The fish are back. The mammals are back.

"Mt. St. Helens caused far more damage than any dam removal ever could. If the river is allowed to recover, it does fine."

❰ ❰ ❰

Someone else calls. He says, "I've got two more words for you."

"Is this an Olympia thing?" I ask.

He ignores me. "Missoula Flood."

"Go on."

"During the last ice age glaciers dammed the Columbia River and one of its major tributaries, the Clark Fork, creating a lake more than four times as big as Lake Erie. Eventually the water got deep enough, about 2,000 feet, to float the glacier, and that immediately busted the dam. This 2,000-foot wall of water rushed across what is now Idaho and Washington at about 100 miles per hour. The whole lake drained in a couple of days. I think the volume of water was on

the order of five or ten cubic miles per hour, more than all the other freshwater flows in the world combined. The flood was strong enough to lift and carry
100-ton rocks all the way to the ocean. Huge backwaters formed everywhere, as
the main channel couldn't hold all that water. A wall of water probably 400 feet
tall pushed 100 miles south of the main channel, to what is now Eugene, Oregon. Another wall pushed up the Snake River for about the same distance."

"Your point is . . ."

"The river and the salmon and the sturgeon survived that flood. The bust-
ing of Grand Coulee Dam would be tiny compared to that."

Silence.

He said, "One more thing. It's actually incorrect to talk about *the* Missoula
Flood. My understanding is that there were between forty and ninety of them.
The river survived them all. It's not surviving now."

❧ ❧ ❧

I contacted the hydrologist and asked him, based on his decades of experience
working within the system, to give me his best shot. "If you can convince me,"
I said, "that do-it-yourself dam removal is more harmful to rivers than waiting
for the government and corporations to take out dams (when the "owners have
decided that the cost to repair/maintain the dam is no longer worthwhile")—
or, put another way, if you can show me how acting with the approval of these
organizations is better for rivers than acting without it—I will stop calling for
people to remove dams on their own.

"I really want what is best for rivers. I don't trust organizations with members who say they wish salmon would go extinct so people can get on with living, but if the other options are worse, I will regretfully do that. I am open to being
convinced."

The hydrologist is a nice man. I like him very much. We've spoken a few
times, and notwithstanding our difference of opinion on dam removal we get
along. I believe he, too, really does want what is best for rivers, and I'm sure he
has accomplished much that is good. He wrote me a kind note telling me the
name of the book he said would convince me: "I think you will enjoy reading
Dam Removal: Science and Decision-making, 2002, by The H. John Heinz III
Center for Science, Economics and the Environment.[142] It has a bit of everything—well, practically everything—that goes into the decision-making on
dam removal. . . . I'm fairly happy about the result, except that it doesn't deal with
complex political issues that swirl around, for example, the Snake River, Colorado

River or Columbia River dams. But my point to you was that to blow up dams was a misleading image of 'jfdi' (just f...ing do it) and can create as many problems as building it in the first place!"

He continued, "Before I end my professional involvement in rivers and watersheds (which has been extensive globally), I really want to remove a dam—even a small one! I was largely responsible for winning a contract last year to plan the removal of a nine foot dam, but the client went and tied it into a proposal to increase the height of a water supply dam downstream, as mitigation. Quite spoiled my day/week/month. Ah well, I'll still enjoy it when (and if) they find the money . . ."

I'm sure you can see why I found his note both troubling and puzzling. His explicit goal was to try to convince me to work within the system, yet he was stating outright that in decades of globally extensive involvement with rivers and watersheds this obviously dedicated professional had not been able to remove a single dam, not even a small one. And the one dam removal he'd started to participate in had been stalled by—you guessed it—politics and money. Further, his *if* implies the distinct possibility the dam may not get removed at all. The failure to remove the dam will, I'm sure, do more to the river than spoil its day/week/month.

Nonetheless, I was happy to order the book, and I read it quickly. I wanted to understand. I soon saw that the book made his case no better than his note. In fact, far worse. I wrote him back, starting my note, "I was surprised to see that Kenneth Lay is one of the trustees of the Heinz Foundation," the organization that put out this hydrologist's best shot at convincing me to work within the system.

As you probably know, Kenneth Lay was the head of Enron, the fraudulent energy corporation responsible for the largest bankruptcy ever, costing investors around $30 billion. Lay and Enron were also responsible for the California energy crisis that cost the public billions more. You may recall that energy traders were caught on tape discussing how they had manipulated California's energy market. For example, one Enron employee was recorded saying, "He just fucks California. He steals money from California to the tune of about a million."

Another responds, "Will you rephrase that?"

"OK, he, um, he arbitrages the California market to the tune of a million bucks or two a day."

Another Enron employee was caught complaining about possible government fines: "They're fucking taking all the money back from you guys? All the money you guys stole from those poor grandmothers in California?"

"Yeah, Grandma Millie, man."

"Yeah, now she wants her fucking money back for all the power you've charged right up, jammed right up her ass for fucking $250 a megawatt hour."

Yet another Enron employee said on tape, "It'd be great. I'd love to see Ken Lay Secretary of Energy."

That very nearly happened. Lay made Bush's shortlist for that position. Enron contributed more than $3.5 million to Republicans between 1989 and 2001. Lay is a good enough friend and strong enough supporter of Bush that during the 2000 Presidential campaign Lay allowed Bush to use Enron jets. They're good enough friends that Bush nicknamed Lay "Kenny-Boy."

Bush is not the only member of his administration to have a relationship with Lay and Enron. As of 2002, fifteen high-ranking officials owned Enron stock. These included Defense Secretary Donald Rumsfeld, Karl Rove, deputy Environmental Protection Agency administrator Linda Fisher, Treasury Undersecretary Peter Fisher, and U.S. Trade Representative Robert Zoellick. Before taking over as Army Secretary, Thomas White was a vice-chair for Enron and owned $50 to $100 million in Enron stock.

It's not too much to say that Ken Lay and Enron set the Bush Administration's energy policies. Lay and Enron recommended policies, and Bush and company listened. Lay and Enron recommended people to implement these policies, and Bush and company listened.[143]

It was a damning enough indictment of the system the hydrologist wanted me to believe in that in his decades working on rivers and watersheds he hadn't removed a single dam. It was even worse that the document that was supposed to convince me not to act on my own was put out by an organization that had the head of *any* energy corporation on its board. No energy corporation—no for-profit corporation, but especially no energy corporation—will ever do what is best for rivers or fish (except incidentally, when the "owners have decided that the cost to repair/maintain the dam is no longer worthwhile"). But it's worse yet that the energy corporation in question is arguably the most spectacularly fraudulent corporation in recent history—quite an accomplishment—and one with close ties to George W. Bush, one of the most destructive enemies the natural world has today.

Amazingly, though, the document gets even worse. The Heinz Foundation had some help with this particular book: its co-producers were FEMA and EPRI.

Let's take these separately.

FEMA is the Federal Emergency Management Administration. It is perhaps most famously known for providing taxpayer-supported flood insurance for people who build in floodplains. This should make clear FEMA's relationship to dams: dams are often used to control floods. At the very least, FEMA is no friend to wild and unpredictable rivers. What a river might call spring cleaning, and what a meadow might call a welcome and necessary influx of nutrients, FEMA would call an emergency to be managed: a flood.

But FEMA is problematic for other reasons, too. The official FEMA website states: "DISASTER. It strikes anytime, anywhere. It takes many forms—a hurricane, an earthquake, a tornado, a flood, a fire or a hazardous spill, an act of nature or an act of terrorism. It builds over days or weeks, or hits suddenly, without warning. Every year, millions of Americans face disaster, and its terrifying consequences."[144] Although FEMA emphasizes its response to natural disasters, hints of FEMA's other purposes slip in. We may gain a clue as to which of these listed disasters FEMA focuses on when we learn that FEMA is part of the Homeland Security Agency.

FEMA's real focus, any right-wing paranoid conspiracy nut will tell you—and I have to emphasize that just because you're paranoid doesn't mean they aren't out to get you—is on laying the groundwork to put in place martial law and set up a nondemocratic shadow government. Given the rates of incarceration in this country as well as the absolutely cavalier divorce of the government from people's and communities' best interests, using FEMA for these purposes seems like overkill to me. That said, it can be pretty easy to dismiss the claims of the tinfoil hat folks when they state that FEMA spends something on the order of 6 percent of its budget on emergencies, and the rest on "the construction of secret underground facilities to assure continuity of government in case of a major emergency, foreign or domestic."[145] We can likewise scoff at the claim that an "Executive Order signed by then President Bush in 1989 authorized the Federal Emergency Management Agency to build 43 primary camps (having a capacity of 35,000 to 45,000 prisoners each) and also authorized hundreds of secondary facilities. It is interesting to note that several of these facilities can accommodate 100,000 prisoners. These facilities have been completed and many are already manned but as yet contain no prisoners."[146] That's all pretty funny, but we might stop laughing when we read the nutcases' pre-Ashcroft/Guantanamo claim that, "The plan also authorized the establishment of concentration camps for detaining the accused, but no trial."[147] And how hard will we laugh when we learn, "Three times since 1984, FEMA stood on the threshold of taking control of the nation"?[148] But thank our lucky stars we can stop paying attention and start laughing again when we read the claim that there "have been documented over 60 secret underground virtual cities, built by the government, Federal Reserve Bank owners, and high ranking members of the Committee of 300."[149]

Committee of 300 indeed.

Unfortunately, it would be a lot easier to dismiss all of this as paranoid fantasy if what we know of FEMA weren't so scary. For example, let's talk about former National Guard General Louis O. Giuffrida, who in 1981 was appointed by Ronald Reagan to head the organization. The two had already worked together. In 1971, when Reagan was governor of California, he and Giuffrida designed Operation Cable Splicer, which consisted of martial law proposals legitimizing the use of the military and police to detain political dissidents. Reagan may have chosen Giuffrida for this job because Giuffrida was experienced at planning to detain those who might get in the way of those in power: at the Army War College the year before, Giuffrida had advocated in writing that in the event of a national uprising at least 21 million "American Negroes" be arrested and transferred to relocation camps.[150] We're no longer in the realm of delusion, but history, which I suppose could be defined as the place where the delusions of the powerful combine with the force to make them happen. In any case, Giuffrida brought this same verve to FEMA, and joined like-minded people such as General Frank Salcedo, chief of FEMA's Civil Security Division, who in 1983 articulated his vision for FEMA as a "new frontier in the protection of individual and governmental leaders from assassination, and of civil and military installations from sabotage and/or attack, as well as prevention of dissident groups from gaining access to U.S. opinion, or a global audience in times of crisis."[151] A little later FEMA developed plans to seize power in these "times of crisis." FEMA soon led thirty-four other federal agencies (including the FBI, CIA, and the U.S. Treasury) in a massive exercise including, among many other things, plans to place 100,000 U.S. citizens into concentration camps. In a power struggle between FEMA and the FBI, FEMA was forced to turn over dossiers on more than 10,000 of these dissidents.[152] That same year FEMA drafted legislation that would be held in reserve so that in "times of crisis" Congress could have language ready to, according to Jack Anderson, the journalist who broke the story, "suspend the Constitution and the Bill of Rights, effectively eliminate private property, abolish free enterprise, and generally clamp Americans in a totalitarian vise."[153]

Floods and hurricanes, indeed.

I'll tell you something else that does not inspire me to confidence about FEMA's intent. Although Louis Giuffrida was in charge of FEMA for four years, from 1981 to 1985 (being forced to resign when it was learned he had used

$170,000 of taxpayer money to outfit his groovy bachelor pad in Maryland[154]), a search of the FEMA website for the word Giuffrida returns the error message: "No indexed terms." He's not there. Just like that Giuffrida disappears down the Memory Hole.

What is FEMA? Is it friendly folks who help us in times of disaster, or is it nasty plotters planning the police state, or is it somewhere in between, or is it somewhere else entirely?

I suspect the truth is close to what a friend responded when I asked her about it. She said her husband had worked often with FEMA because his job involved emergency response: "His impression is that they're largely incompetent. But the main thing he said is that what the agency accomplishes doesn't have so much to do with it having any sort of will (as was the case with the FBI under Hoover) but rather with the fact that FEMA is in a position to accomplish so many different things that any administration can mobilize it to do whatever it wants, whether that is to help those harmed by a hurricane or to imprison people who disagree with the rulers and generally facilitate a police state."

We don't really need to invoke the Committee of 300 to make FEMA a less than credible producer of a book on dam removal. Even in its capacity as insurer of floodplain dwellers FEMA spells bad news for the liberation of rivers.

If FEMA is less than credible on issues concerning dams, EPRI, the other creator of the book, has no credibility whatsoever. EPRI is the Electric Power Research Institute, self-described as "a non-profit energy research consortium for the benefit of utility members . . ." Yes, the hydrologist evidently believed that an organization created explicitly for the benefit of the electrical industry could be relied upon to be truthful concerning the relationship between dams (many of which provide hydroelectricity) and rivers. We may as well ask Jack the Ripper about gender relations. Does anyone want to guess what sort of recommendations EPRI makes concerning dams and the health of rivers? Let's let EPRI speak for itself as to what it does, and what motivates it: it "provides the knowledge, tools, and expertise you need to build competitive advantage, address environmental challenges, open up new business opportunities, and meet the needs of your energy customers. . . . Whether you are looking for ways to cut operation and maintenance costs, increase revenues, find cost-effective environmental solutions, or develop new markets and opportunities for the future, EPRI delivers solutions that work for you." In this description, I see plenty of explicit concern for money, for lowering costs and increasing revenues, for gaining "competitive advantage," for meeting the "needs" of energy customers. But I see no concern for the well-being of rivers and the fish whose lives depend on them. Dying rivers and extirpated fish are at best "challenges" to which EPRI must find "cost-effective environmental solutions."[155]

Another Heinz foundation trustee is Fred Krupp, head of the Environmental Defense Fund (which despite its name has received funding from such organizations as the far right Lynde and Harry Bradley Foundation). Here is what PR Watch, a project that "investigates and exposes how the public relations industry and other professional propagandists manipulate public information, perceptions and opinion on behalf of governments and special interests,"[156] has to say about Krupp: "One of [PR guru Peter] Sandman's protégés, Fred Krupp of the Environmental Defense Fund has carved out a niche for his organization as a 'pragmatic' dealmaker willing to sit down with corporations and negotiate environmental 'solutions.'"[157]

Instead of taking out a dam, I'm supposed to sit down with corporate heads and hammer out a deal where they pay for yet another study while the salmon go extinct. I guess then we could get on with living.

No, thank you.

Things keep getting worse. One of the book's authors is Thomas C. Downs of the law and public relations firm Patton Boggs (with clients including Angola's national oil company, Texaco, ExxonMobil, Shell, W. R. Grace, Peru, Qatar, and many others). This is Patton Boggs self-description: "Through nearly four decades of practice, we have established a reputation for cutting-edge advocacy by working closely with Congress and regulatory agencies in Washington, litigating in courts across the country, and crafting business transactions around the world. Patton Boggs began as an international law firm concentrating in global business and trade. Founded in 1962 by James R. Patton, Jr., and joined soon after by George Blow and then Thomas Hale Boggs, Jr., we have maintained our strong concentration in international and trade law with over 200 international clients from over 70 countries. Patton Boggs, for example, has participated in the formation of every major multilateral trade agreement considered by Congress." I can't speak for you, but I don't particularly want a law firm which lobbied for GATT, NAFTA, FTAA, etc.—and is proud of it—determining whether salmon survive. The website also states: "If the law appears to be the problem, Patton Boggs is well positioned to help effect a change. For example, in a dispute with the Department of Energy and a major aeronautics manufacturer over the threatened loss of valuable trade secrets and confidential data, the Department initially claimed it had no jurisdiction to consider the matter. We secured an amendment to an appropriations bill that not only conferred jurisdiction on the Department of Energy, but also directed it to solve our client's problem. Not surprisingly, our client obtained all the relief it originally sought."[158]

The question is, as always, what do you want? What are your goals? What is important to you? If your goals are to increase revenues, that is what you will do. If your goals are to cut costs, that is what you will do.

The natural world is not valued within this culture except as it can be considered resources convertible to cash.

There was something else about this book that saddened me—besides the content that followed the book's industry sponsorship. It was the book's dedication. The book was partly the result of a conference held on September 11-12, 2001. The dedication states, "None of us will forget where we were on September 11, 2001, nor will we forget the thousands of lives lost as a result of such senseless and brutal acts. We dedicate this report to the victims and their families and to the courageous firefighters, police, and rescue teams from New York, Washington, D.C., and Pennsylvania."

This was ostensibly a book about dam removal. It was not a book about New York City, airplane safety, or hijackers. Dead people in New York City, Washington, D.C., or Pennsylvania have nothing to do with this book. Dead rivers do. But dead rivers were not mentioned in the dedication. Because they are not important to these people. As I read this book, I kept wondering how it would have been different had the authors cared and dared to dedicate it to the salmon, steelhead, sturgeon, and lamprey killed by dams. The authors remember where they were on September 11, 2001, but do they even remember the date that the Grand Coulee Dam closed off the Columbia? How about the date the Iron Gate closed off the Klamath? They say they will never forget "the thousands of lives lost as a result of such senseless and brutal acts," yet they forget the many millions of humans and nonhumans destroyed by the senseless acts of civilization, and particularly by the senseless and compulsive acts of dam building. Even if we confine this to humans, how different would this book have been had they dedicated it not to the few thousand killed on 9/11 but to the 40 to 80 million people displaced by dams worldwide? *That* would have been a dam removal book worth reading.

But of course the authors forget all of this. The primary purpose of the book was never to truly explore whether removing dams is good for rivers, any more than the primary purpose of capitalist media is to convey information useful to individual and communal health, any more than the primary purpose of discourse within an abusive family is to facilitate healthy familial relationships. Henry Adams had it right when he wrote, "The press is the hired agent of a monied system, set up for no other reason than to tell lies where the interests are concerned." Indeed, telling lies where the interests are concerned is the primary function of all discourse within an abusive structure. This applies to books as well, including those put out by FEMA, EPRI, and the Heinz Foundation. The primary purpose of *Dam Removal* was to convince people that something is being done about the murder of the planet. If the interests and their experts were doing *nothing*, then we would know we have to stop the murder ourselves. But if they are doing something—anything—then both we and they can relax, because the experts are taking care of the problem. "See," they can say and we can hear, "we put out a book on dam removal. We're working on it. Have patience. Trust us."

I no longer have patience. I no longer have trust. I no longer have time. Nor do salmon, sturgeon, or the others.

It's a rigged game. It is now, and within this culture it always has been. So long as this culture stands it always will be. The primary basis for dam removal decision-making by the powers that be is cost-benefit analysis, and the analyses are always—*always*—stacked in favor of the powers that be. If you are one of them you count. If you're not, you don't.

☾ ☾ ☾

The game just got even more rigged. Today's *San Francisco Chronicle* carried an article headlined: "Bush would give dam owners special access: Proposed Interior Dept. rule could mean millions for industry." The article begins: "The Bush administration has proposed giving dam owners the exclusive right to appeal Interior Department rulings about how dams should be licensed and operated on U.S. rivers through a little-noticed regulatory tweak that could be worth hundreds of millions of dollars to the hydropower industry.

"The proposal would prevent states, Indian tribes and environmental groups from making their own appeals, while granting dam owners the opportunity to take their complaints—and suggested solutions—directly to senior political appointees in the Interior Department."

Later, it states, "The proposed rule comes at a pivotal time in the history of the hydropower industry. Most privately owned dams were built—and granted 30- to 50-year federal licenses—in an era before federal environmental laws required protection for fish and other riverine life. In the next 15 years, licenses for more than half of the country's privately owned dams will come up for renewal.

"The hydropower industry has complained that to comply with the law and renew their licenses with the Federal Energy Regulatory Commission, dam owners are being forced to pay large settlements to mitigate the environmental harm that dams cause fish and communities that depend on fish."

The purpose of the proposed regulation should be clear: "'It allows industry to go in and speak their piece without having to deal with the concerns of all the other stakeholders along a river,' said an Interior Department official who has worked for many years on the dam relicensing process and who asked not to be identified by name, also for fear of retaliation.[159]

"The hydropower licensing law was written in 1920, and the industry had few problems with it for nearly six decades—until tribes and environmental groups figured out how to use the law in a way that cost the industry a lot of money."[160]

Or in a way that would help fish. Or in a way that would help local communities. Or in a way that would help rivers. Or in a way that would help landbases.

Any time any of us figure out how to use their rules to stop the destruction of all we hold dear, those in power change the rules. Why would those in power allow activities that undercut their own power? The purpose of the rules was never to actually protect us or those we love, but rather to provide the illusion of protection. So long as we continue to mistake the illusion of protection for actual protection, all that we love will continue to be destroyed.

❨ ❨ ❨

Would you believe me if I told you that the game just got even more rigged? Did you believe that possible? Today there was an article in *The New York Times* entitled, "U.S. Rules Out Dam Removal to Aid Salmon."

The article begins, "The Bush administration on Tuesday ruled out the possibility of removing federal dams on the Columbia and Snake Rivers to protect 11 endangered species of salmon and steelhead, even as a last resort.

"In an opinion issued by the fisheries division of the National Oceanographic and Atmospheric Administration, the government declared that the eight large dams on the lower stretch of the two rivers are an immutable part of the salmon's environment."[161]

You read that correctly. According to the federal government, according to this government of occupation, dams are an immutable part of the Columbia. The artifacts of this culture are more important than the landbase.

How could we have ever been so foolish as to expect anything else?

Part of the federal report stated, "It is clear that each of the dams already exists, and their existence is beyond the present discretion" of federal agencies to reverse.

The authors of this opinion are correct. Each of these dams already exists. I suppose we should be glad they at least noticed. Yet if the federal government states explicitly that the existence of these dams "is beyond the present discretion" of the federal government to remove, perhaps just this once we ought to take them at their word, and when time after time they have done the wrong thing perhaps this once—and then every time—we should stop relying on them to do the right thing, and perhaps we should do it ourselves.

If we care about the salmon, it becomes increasingly clear what we need to do.

❨ ❨ ❨

Here is why we need to not wait for the government to remove dams. California is considering increasing the height of the already environmentally destructive Shasta Dam, over the objections of the Wintu people. Increasing the height of the dam will further inundate places that are sacred to them, such as the place where their young women go for their first menses. The Wintu will almost undoubtedly be steamrolled, in large part because Senator Diane Feinstein is pushing hard for the dam expansion. Why? She could not be more explicit than this: "I believe it is a God-given right as Californians to be able to water gardens and lawns."

This culture is insane. It must be stopped.

DAMS, PART III

For us, it is a dam of tears. We don't have water to drink, nor rice to eat. And we can't eat tear drops.

> *Paw Lert, a villager displaced by the Bhumipol dam*
> *who helped launch the Thai anti-dam movement, and who*
> *because of that was murdered by an unknown gunman.*[162]

I ALREADY KNEW THAT. I IMAGINE YOU DID TOO. I DIDN'T NEED TO LEARN yet again that the system serves the economically and politically powerful and will only rarely—and then incidentally—do what is right for the natural world, including rivers (including humans). I just wanted to know whether taking out dams helped or harmed rivers.

I need to be clear on the logic. From the perspective of a river the main difference between legal and illegal dam removal is that the former has the possibility of being done in stages (I guess another difference is that the former happens so damn infrequently, but so, unfortunately, does the latter). I think we can all presume the police would not stand by while we cut successive grooves in the Glen Canyon Dam (although as mentioned previously the selective law enforcement officers have certainly stood by, winking, while illegal dams have been constructed). Here is the real question I'm trying to get at: From the perspective of a river, is a catastrophic dam failure better than no failure at all?

More clarity on logic. The system is destructive. It is possible that the system is destructive and it is coincidentally more harmful to rivers for dams to fail than stand. It is also possible that the system is destructive and it is coincidentally more harmful to rivers for dams to stand than fail. The latter variable—the relative destructiveness of dam removal—is independent of the former constant—the culture's destructiveness.

That's why reading *Dam Removal* was useless to me.

I scurried to the library, and also spent many hours scouring the internet, and I found almost no research on the effects of catastrophic dam failures on fish and wildlife habitat below. I later learned from fisheries biologists that my inability to find studies came not because I'm a lousy researcher, but because the studies don't exist. "Every dam has a catastrophic failure review," one said, "but that's about humans. What happens to fish takes a back seat to how many bridges go out, and how many humans die." He's right. There are a *lot* of studies (many of them by our friends at FEMA) on the downstream effects of catastrophic dam failure on bridges and other pieces of infrastructure (of course, because the property of those higher on the hierarchy is always worth more than the lives of those below) revealing, no big surprise, that when a dam fails, towns below get flooded. But I found nothing detailing the effects on rivers. This is one of the most inexcusable, absurd, obscene lacks I have ever discovered. There are two million dams in this country, 75,000 of them over six feet tall. Every one of these dams will someday fail, yet before constructing these two million dams, nobody bothered to find out what would happen to the rivers when the inevitable happened.

What makes this even more inexcusable, absurd, obscene—*evil*—is that we can say the same thing about deforestation, the murder of the oceans, the manufacture of CFCs, the fabrication of plastics, the burning of oil, in fact all of civilization. Nobody bothered to find out what effects these would have on the natural world. The reason is clear: those who make the decisions don't care.

If the entire culture is predicated on an unexamined self-assumed right to exploit everyone and everything around you, why should you bother to think about the effects of your actions on others? Does a rapist care about the devastation he leaves in his wake? How about a child abuser? How about a CEO? How about Ken Lay? How about those who served him, who helped steal from "Grandma Millie"? This culture systematically inculcates us not to care about the damage the system causes—indeed, not to notice the damage at all. So why should we expect dam builders and dam defenders to care about fish? We shouldn't. It's not only a horrible mistake for us to believe they do, but it's a trap we fall into all too willingly, because it allows us to try to convince them, which means it allows us to play by the rules they set up, which means it allows us to pretend we are doing something while really we are doing what most of us do most of the time: protecting the abusers and maintaining an abusive social dynamic.

They do not care. If they did they would do something about removing the dams.

I care.

I called some fisheries biologists I'd heard care deeply about salmon. Mostly they worked for tribes whose lives have always depended on the fish, although some worked for the Park Service or Fish and Wildlife. I asked them all the same question: "If someone were to blow up a dam, what effect would that have on the river below?"

They all had the same response. They refused to answer. In retrospect I don't blame them. They probably thought I was either a fed provocateur or a terrorist (non-governmental variety), and a stupid one at that. Had I a Middle Eastern accent somebody probably would have called the feds merely because I asked the question, and I would have won an all-expenses paid one-way trip to beautiful Guantanamo. As it was, I was fortunate enough to merely receive a lack of answers.

I had sensed, however, that the people I talked to really did care about the fish, and really did want to answer, if only I would phrase my question better, so that a) it let them know I really wasn't as stupid and careless as this question, coming out of the blue, must have made me seem; b) it didn't scare them; and c) it gave them deniability so that if the Iron Gate suddenly did collapse, allowing that stretch of the Klamath to again run free—and if I were caught and didn't hold my mud—they wouldn't receive a series of visits from Uncle John Ashcroft and his black-sunglasses boys demanding to know why they'd encouraged some lunatic to blow up a dam, then promising them a dog-run right next to mine at Guantanamo-by-the-sea.

Here's the question as I eventually asked it: "I'm wondering if you can be very explicit about the damage caused to rivers by catastrophic dam failure, whether that failure is anthropogenic or natural. What are both short-term and long-term effects? How will the river be one day afterward, one year, one decade, fifty years, one hundred years? Are there gold-standard studies that have been done on this? To be clear: I want to know what precisely is the damage done by catastrophic dam failure.

"That leads to a thought experiment. One of the great things about being a writer is that I get to pursue all sorts of thought experiments to their endpoints (I guess we all get to do this, but it's what I get to do all the time). For part of one book I delved as deeply as I could into the cultural and economic causes of slavery. For part of another book I asked how the processes of schooling destroy our creativity, and what would a schooling that nurtured creativity look and feel like. In yet another I explored the history and future of surveillance.

"One of the things I'm doing in my current book is playing out short- to mid-term future scenarios and trying to explore what would be the right actions to take in those circumstances.

"So, here's the thought experiment: Pretend it's 2015 and the oil economy has collapsed. This brought down with it the electrical infrastructure. I'm putting forward this possible scenario because a) world oil production has probably peaked (or will peak very soon, as will natural gas production), so it's not unfeasible; and more to the point, b) I want to talk about community decision-making processes. In this scenario, the Corps of Engineers is no longer relevant. Nor is the U.S. government. Decisions affecting local rivers are made by those who live on these rivers (what a concept!). Under this scenario, dams are no longer useful for electricity or irrigation (for obvious reasons). Now, your community is going to decide whether to take out dams along this river. Because it is a communal decision, all humans along the banks of the river will be warned, so there will be no loss of human life. The people in your community ask you to speak for the fish and the other creatures of the river, for the river itself. The question for you is: Would it be in the interest of the salmon, lampreys, trout, and other river-based creatures (and the river itself) to remove the dam, even if it is done catastrophically, or would it be better to leave the dam standing, and to eventually let it fail on its own? Why in either case? Other stakeholders will discuss other perspectives on this, but community members really want to hear your interpretation, your understanding, of the river's perspective.

"The next question is: Would your answer be different for big dams rather than little ones?

"The next question: Would your answer be different if there were threatened nonanadromous populations in the river?" [Anadromous fishes are those who spend all or part of their adult life in saltwater and return to freshwater streams and rivers to spawn.]

"If the choice is to remove the dam, when would be best?

"The reason I'm doing this thought experiment is that just like in any good experiment (I knew my degree in physics would be good for something) I'm trying to reduce the variables and examine one question at a time. The one question here has to do with the relationship between dams and rivers. Given what seems very clear about the transitory nature of the oil economy, these questions of what communities want to do about the rivers that are their lifeblood will soon no longer be theoretical, and I would like to have some of these questions begin to percolate into public discourse, to be discussed as deeply, intelligently, and passionately as we can, so that as things become increasingly chaotic, people and communities have some analyses that may help them to make decisions in the best interests of their landbases."

It's an odd and overwhelming indictment of the self-censorship that characterizes this culture's discourse that I had to create this several-paragraph hypothetical frame simply to ask the first and in many ways only question that everyone who has ever been associated in any way with any dam anywhere in the world should ask at every moment, which is, "When this dam comes down, how will that affect the river?"

But creating this huge and in some ways ludicrous frame did get me answers. Did it ever!

The advocates for fish with whom I spoke were bursting to talk about the rivers they love, and how dams were killing the rivers. The words flowed in a rush, as though my mere rephrasing of the question to make it safe had allowed a dam to burst inside of them.

They in turn gave me that same gift, as the understanding I gained from these conversations burst cognitive and emotional dams inside me, destroyed old ways of seeing the world and made new ones. The conversations transformed me, made me stronger, more determined.

Maybe they will do the same for you.

❨ ❨ ❨

Rivers live for millions of years. If we were able to see rivers as they are, we would know that they are in no way static, that instead they writhe like snakes. They abandon old channels, cut new ones, refind old ones again.

These rivers who live for millions of years—these rivers who dance, sway, move to rhythms we might be able to hear in our dreams or if we listen in the dreams of the soil and the rocks and the salmon and the snails and caddisflies—are not only the water between their banks. A river is its entire valley, and the entire valley is the river.

Insects live in aquifers, and fish swim through gravel, ten, twenty, thirty feet below the "bottom" of the river. Coho swim in tiny ponds far from the river, and when you come back you are certain they have been eaten by raccoons, but you come back again the next day and there they are. Where were they? Swimming among the cobbles.

Rock, water, salmon, bear, eagle, insect, aquifer. These all live together. They are all part of the river. And they are all in constant flux.

Sometimes the flux is violent. One fisheries biologist told me, "Many people are upset when a river shifts channels during floods and leaves one dry and tears through another. They notice that fish are stranded (but become available for animals). They do not however typically think about the mice and salamanders and insects who are drowned or crushed. Plants are ripped out and washed to the ocean. I think about all of the flora and fauna, feel for all of them, but I've evolved into acceptance of the devastation (change) that occurs in rivers during floods (the river I work with and love is a really dynamic river). Many people want to stop rivers from migrating but migration forms new productive habitats!

"Based on this experience, I've developed a philosophy that in some instances it's better to remove a problem, accept the immediate impact to creatures, do what's possible and necessary to mitigate that impact, and open up the habitat to the organisms (human and nonhuman people) who belong there. I especially think of anadromous fish, as their nutrient input to upland environments is critical to supporting huge numbers of other animals and plants. I always apologize to everyone who will be impacted. In other cases, passive restoration (of landslides or mass wasting hillslopes and channels) is more appropriate— letting trees grow back for hundreds of years without interference."

Dams happen naturally. Landslides, lava flows, glaciers cover rivers. That happens. The problem is not that there is a dam on this or that river. The problem is that there are two million dams on almost every river and almost every stream.

And dams break naturally. It happens all the time. Witness the Missoula Floods, all forty of them. "Seventy thousand years ago," another person said to me, "a volcanic dam filled the entire Shasta Valley. But the water eventually wore through. And because of that the river is very productive. Six thousand years ago, Mount Mazama blew up and buried Upper Klamath Lake and the Williamson River in ten feet of volcanic ash. Because of *that* the river is very productive. When you think about it in geologic terms, that's how things happen. Streams and rivers get dammed, and then the water breaks through. That's what rivers *do*. Habitat is destroyed, and then habitat is created. Floods aren't really all that damaging to rivers, anyway. And that's what any catastrophic dam failure would be: a big flood. If you want to know what will happen when a dam blows, your best bet is to look at natural cataclysms that are on par with what you're talking about. Mount St. Helens closely mimics, although far exceeds, what would happen with a dam flood."

Another said, "If you take out a dam, yes, that's a major catastrophe, but you've got fish in the ocean who will come back in one, two, three, four years. And you've got fish who will reinhabit from other streams. If you don't take out a dam, the salmon will not make it. If you do, the salmon will, so long as there is still a living ocean."

And yet another, "Long term effects of dam removal: none. If the dams are going to come down eventually anyway, the river will eventually recover. That's what happens. Then the salmon reinhabit. That's what they did on the Columbia after the Missoula Flood. That's what they'll do here. About 2 percent of salmon don't go back to their original river, but find new places to spawn."

Yet another, "People don't understand that if you provide animals with high enough quality habitat, they can live there. If you destroy their habitat, they will die. And forests and rivers are dying. Below dams there is a tremendous starvation for sediment,[163] and above dams a tremendous starvation of nutrients from the oceans. Salmon, sturgeon, rivers, and so on have survived millions of years of volcanoes, glaciers, and so on, but they have barely survived one hundred years of this culture. They will not survive another hundred. Probably not another fifty. Maybe not another twenty. The dams need to go."

I asked them to speak for the fish and the other creatures of the river, to speak for the rivers themselves. I asked, "Would it be in the interest of the salmon, lampreys, trout, and other river-based creatures (and the river itself) to remove dams, even if done catastrophically, or would it be better to leave dams standing, to let them eventually fail on their own?"

They said, "Remove them," and "Yes, take them down," and "Yes, they need to come down now," and "Yes," and "Yes," and "Yes." They said, "If there's no way to take out the dam gradually, then yes, just get it done and over with. The short-term damage to the fishery would be worth the long-term gains, absolutely." They said, "Catastrophic dam removal can destroy short-term habitat and create long-term habitat." They said, "From the perspective of a river the only case I can think of where you might not remove a dam would be if there is a small population of a rare species found nowhere else, then a dam failure could cause its final extinction. That might be the case with the Missouri River and the pallid sturgeon. But what is killing that sturgeon? Dams. So I think even then it's not so much a question of not removing dams as it is just being more careful about when and how you do it." And the people said, "Yes," and "Yes," and "Yes." They said, "What we need to do is so very clear. People who oppose dam removal are shortsighted. They absolutely cannot see the long view." And they said, "Those who oppose dam removal are small-minded people who haven't thought about what and who these creatures are, and what these obstructions mean." They said, "Those who oppose dam removal have no faith in the natural world. They have no faith in its resilience and will to live. They think that more management is necessary because humans—always humans—know what is best for rivers, who somehow won't survive without our meddling. That's nonsense. We need to set rivers free and then trust that rivers know how to take care of themselves."

❨ ❨ ❨

Here is what a Yurok Indian man said in a recent editorial about removing dams on the Klamath River.

"When we speak out on issues concerning life on the Klamath River, we speak with conviction for all people and creatures living in or near the lands the Creator gave to us to cherish and protect forever. This is our sacred mission and the purpose given to us. This purpose is enshrined in our Tribal Constitution. We are speaking now.

"We call upon PacifiCorp and ScottishPower's shareholders to take a bold, historic step forward in the preservation of a great species on a great river, the Klamath River, and remove the dams. We believe that they are poised to do so and we call upon all of our friends on the North Coast of California to support them. Very few times in one's life is there an opportunity to realize something truly great. I believe such a time is now. Together, the Yurok people and all the people who love and cherish this earth can help renew the strength and vitality of the salmon of this river.

"The existence of dams, these weapons of mass destruction, harms the life cycles of our salmon brothers. That's right, I say 'salmon brothers.' It is our belief that before there were any people, we were all kindred spirits. Spirits became birds, mammals, reptiles or fish. No creatures are more or less important than the spirits who became people. Thus, we believe all creatures are related as brothers and come from the same Creator. It is hard for me to lift a fish out of the water that has been trapped in my net and not hear him call out to me for help. And with so few salmon in the river these days, it is always with great respect that he will be food for my family and my people. I thank him and the Creator for the sacrifice of his life so that we can eat.

"Lately, the heavy burden I feel as I lift up my nets is not the weight of the fish, but of the heavy sadness that so few of my salmon brothers return these days. Our people have noted the steady decline in the numbers of salmon returning each year. In the early 1900s, prior to the first dams being built, this once great river yielded hundreds of thousands of salmon and steelhead. More than a million came back to the river each year in their migration to their ancient spawning grounds upriver. Now, the return of salmon is measured in the tens of thousands. The salmon harvests on the river are so restricted we cannot meet the basic subsistence and commercial needs of our people. All North Coast sports and commercial fisheries have suffered along with us.

"Maybe I will quit catching fish for my family, I think, but this will not solve the problem. The threats to my salmon brothers must be removed. The water quality and streambed access for spawning salmon must be restored. The Yurok Tribe will protect our salmon brothers and we call upon all who love the earth and the river to join us, especially PacifiCorp and ScottishPower.

"Removal of these dams would be a historic step to restoring Klamath River fish populations. This is literally a once-in-a-lifetime opportunity that can save our salmon. Let's not allow this moment to pass and be lost along with the salmon forever."[164]

❨ ❨ ❨

Rivers, Indians, salmon, steelhead, sturgeon, bears, eagles, and so many others have suffered so much because of civilization. If far distant corporations that control dams on the Klamath and that profit from the murder of the river do not do the right thing and bring down the dams, perhaps it is time for those of us among the civilized who have awakened, who care, to begin to pay our debt for these crimes, to begin to make things right.

❨ ❨ ❨

I talked to someone who knows explosives. He's a friend of a friend, vouched for by the person who introduced us. Had we actually been going to *do* something, I would of course have required far more proof before I would have trusted him. But all we were going to do is talk—it seems that's all any of us ever do anyway—so this was good enough. We met at a baseball game. I'd like to say that had we been going to do more than talk we would have met somewhere more private, but it was at Olympic Stadium to watch the Montreal Expos, which I figured is about as lonely and private as you can get.

I thanked him for his offer to help me understand explosives, and asked him where he got his expertise.

He said, "Oh, it's my pleasure to help in any way I can. I've thought about this a lot, and there are two very important reasons I want to do this. First, it's crucial we take down civilization. If we're to save anything at all, we need to give everything we can to this struggle. I feel so much rage and despair, even moreso because of the things I know my skills could enable me to do. But I'm also trapped, because to act would put my family in jeopardy. I would have to leave them, and I'm not sure they would be able to make it. So day after day I sit here and the rage and despair burns and burns and feels as though it's eating up my soul. But I take some comfort in the sure and certain knowledge that soon, the sooner the better, the time will come to give this system the push that will topple it and free us all to begin to once again live as we should have all along."

I said, "You know, don't you, that we're going to win."

He nodded, then continued, "The second reason I want to help—and I know this might sound strange—is that the way I learned all this stuff also caused me to do many things that still eat at my soul, so if I am now able to put these skills to good use then I might feel a little less bad about that."

"I don't know what you mean."

"In 1981 I joined the Army National Guard in an artillery unit that was also capable of firing tactical nuclear weapons. While I had nothing to do with that, I had to have a security clearance just to be in the unit. After graduating in 1982 and running out of money and basically being homeless, I joined the marines because they were the only ones that could take me immediately. I was an infantryman. Since I already had a security clearance they stuck me in an experimental unit. We spent the first ten months training, and when all was said and done we were supposed to be 80 percent special forces qualified. My formal explosives training consisted of a three-week course taught by the army at Fort Bragg . . ."

"Three weeks? That doesn't seem very long."

"Oh, it's plenty long. To give you some perspective, that's longer than it took them to teach us sky and scuba diving together."

"What else did they teach you?"

"Besides sky and scuba diving there was mountain, desert, arctic, jungle, and temperate survival and warfare; rappelling; mountain climbing; sniper; booby traps; heavy weapons; guerilla tactics; intelligence gathering; counterintelligence operations; perimeter, interior and external security (including how to install and defeat animal, electronic, and mechanical security devices); long-range patrolling; reconnaissance; evasion and escape (this included things like hotwiring vehicles, breaking and entering, and so on); improvised weapons and explosive devices; battlefield first-aid, and finally the standard infantry tactical stuff.

"I have to say that while I have really conflicted feelings about the training I received and while I'm deeply disturbed, almost haunted, by the things I've done, I'm also glad I was taught these things that will ultimately help stop this madness that we live in."

"What did you do?"

"We were first deployed to Africa for almost a year: Angola, the Congo, and elsewhere. Our job was to teach counterinsurgency tactics. We were then deployed to Asia—Thailand and Cambodia—for a year, where we did the same thing. These were all places that had active wars going on, and although we were technically only advisors we still found far too many chances to use what we had learned. After I got out I read several books on more technical aspects they didn't touch in the Corps."

We all have skills, I thought, and no matter where we learned them, we need to use them in the service of our landbase. This man clearly has important skills to use.

《 《 《

When I first read the editorial by the Yurok man, there was one sentence that bothered me. It was, "We call upon PacifiCorp and ScottishPower's shareholders to take a bold, historic step forward in the preservation of a great species on a great river, the Klamath River, and remove the dams." To "call upon" the corporations sounded too much like the same old begging to me. And by now we all know that begging abusers isn't effective. Back to Bancroft, with just a few nouns changed: "You cannot get corporations and those who run them to change by begging or pleading. The only corporate (and political) decision makers who change are the ones who become willing to accept the consequences of their actions."[165] Because the entire system is set up to protect exploiters from these consequences, it becomes incumbent upon us to force consequences onto them. Again: "You cannot, I am sorry to say, get a corporate or political decision maker to work on himself by pleading, soothing, gently leading, getting friends to persuade him, or using any other nonconfrontational method. I have watched millions of protesters and activists attempt such an approach without success. The way you can help him change is to demand that he do so, and settle for nothing less."[166] And one more time: "The most important element in creating a context for change in one who is killing the planet is placing him in a situation where he has no other choice. Otherwise, it is highly unlikely that he will ever change his behavior."[167] Also, given the way corporations and governments lie and delay, and given that so often scientists study to death all possible negative effects of any action that might actually help the natural world before allowing it to move forward while applying a distinct lack of rigor to actions that do great harm (witness the lack of studies on the effects of (inevitable) dam failure on rivers, the lack of prior studies on nukes, on burning oil, on pesticides, and so on, *ad fucking nauseum*), I'm reasonably certain that a best-case scenario would see an absolute minimum of fifteen or twenty years passing before the dams actually come down. Look how long it's taking for the Elwha. And I don't think the Klamath salmon have fifteen or twenty years to spare.

But I realized that the Yurok man's statement is not necessarily begging. There is nothing wrong with explicitly calling for some action, even if to act in the way you are calling for would be out of an abuser's character. You might just get what you want. But what differentiates effective action from codependence are the actions one takes when the abuser fails to relent.

If, in this case, the corporations and governments which provide the muscle for them do not relent, or if they begin to take longer than the salmon have, what do we do then? At that point, where do our loyalties manifest in action? Are we more loyal to the laws and the corporations for which they are made, or to the fish? Are we more loyal to processes designed to maintain abusive power structures, or to the river? If it comes down to this—and you know that in so many cases, in so many places, it already has—which do you choose?

❨ ❨ ❨

The man said, "Onto the specifics. Buildings have points throughout them under relatively greater stress from gravity and other forces. This basic fact is well enough known and understood that when people design structures they take that into account and beef up stressed areas. Dams have two main points of stress that I'm aware of. The first is the farthermost point of the crescent, and the other is any place between sluice gates. These spots will be made stronger by making walls thicker or by reinforcing them with steel. Builders only reinforce as much as is necessary, and these places are jointed to normal sections of wall. Those joints are the places you want to attack, unless you have an infinite supply of explosives, in which case you always hit the points of greatest stress. The biggest thing here is that you have to know the details of the material used in construction, and if the builders used multiple types of material, you have to assume they used the strongest one throughout. This is all pretty basic stuff: demolition experts both within and without the U.S. military use this sort of analysis all the time."

I couldn't believe we were talking about this at a baseball game. The Expos were beating the Marlins 3–1, by the way.

He said, "To summarize, if someone were going to take out a dam that person would need: 1) Blueprints, or someone with enough related skills to be able to convey the same information: I've done enough of this that at this point for most dams I wouldn't need blueprints but could probably find the stress points by looking: the key is to train yourself to *see*, and to see things whole; 2) Knowledge of materials used in construction; 3) Knowledge of special terrain considerations such as being in an earthquake or hurricane zone that would cause the structure to be reinforced. All of this would tell you: 4) how much of what sort of explosive agent you need to accomplish your goal. Once you have this the next step is: 5) a plan, how to gain entry, who does what, when, and where, and what sort of detonator you'll use. Don't overlook this one: if the dam generates electricity it may interfere with an electric timer or a remote control device."

"Does it scare you to talk about this?" I asked. Not that it scared me. Not at all. Not in the slightest. I swear. And no, my voice didn't shake. Oh, okay, maybe the tiniest bit. All right, damn it, I was pretty fucking scared. But excited too.

He said, "This information isn't illegal. Like I said, this is Demolition 101, taught to tens of thousands of people at taxpayer expense in the military. If it's illegal for me to talk about it, why wasn't it illegal for them to teach me?"

"Because we're talking about using it in the service of people and the planet, not so the rich can accumulate more power. Because we're not talking about using it to hurt poor people all over the world."

"Damn," he said. "You're right. In that case it's probably illegal."

Silence.

Finally he said, "Great game, huh?"

"Yeah." It was 3–2 now.

He said, "You know, it just occurred to me: the military trains a *lot* of people in demolition, and a *lot* of these people understand how destructive civilization is."

I looked at him out of the corners of my eyes.

He had a dreamy look on his face. He shook it off and said, "Once you've got your plan, and once you make sure you've taken into account every possible accident and screw up, then you're ready for step six, which is that you need to practice doing not only your job but at least two others over and over until you and everyone else on your team is able to do it asleep. You don't want to fuck up. You want to do it right."

"Shit," I said. "I'm scared."

"Of course you are," he said. "It's scary stuff."

DAMS, PART IV

The river spirit is destroyed when you dam the river. Our people have lived here for thousands of years. When you destroy the spirit of the river, you destroy our culture. And when you destroy our culture, you destroy our people. If we are to live, the dam must go.

Milton Born With a Tooth[168]

THIS ISN'T A GAME. THE CONSEQUENCES ARE REAL. THOSE IN POWER BROOK no blasphemy, no breaking of the fourth or fifth premises of this book, no threat to their monopoly on violence, no threat to their property (which means no threat to their control), no threat to their power, no threat to their ability to destroy the world.

More people have been committing this blasphemy lately. More people have been recognizing that civilization is irredeemable. More people have come to know that the police will not protect us and the land we love from corporations, but instead that the police have as a primary purpose the facilitation of production, the facilitation of the destruction of everything we hold dear. More people have come to know that if the integrity of our own bodies and the integrity of the land we love is to be defended, we are the ones who must defend it. And we are recognizing that we must go on the offensive.

People are burning down ski resorts. They are torching logging trucks. They are putting sand or sugar into the gas tanks of fellerbunchers. They are burning sports utility vehicles, individually and collectively, at dealerships. They are burning down empty apartment complexes and empty luxury homes being built on sensitive habitat. They are pulling up genetically engineered crops. They are liberating animals from laboratories and factory farms. They are destroying research based on the torture of animals. They are burning down vivisection laboratories.

It's a start.

Not many of the people who have committed these actions have been caught. Among those who have been apprehended, the arrest has almost never come because of the sort of brilliant detective work we've come to expect by watching police propaganda: cop movies and television programs like *Forensic Files*. Instead arrests have almost always come because of careless mistakes on the parts of saboteurs. One animal liberator who later spent time in prison—and many other activists suffered jail time rather than testify against him—first came under suspicion because when he sent a communiqué to an organization that routinely does press work for animal liberators he wrote a phony account number on the FedEx envelope. Of course the letter never arrived, but went into the dead letter pile at FedEx, got opened, and revealed clues that could have been avoided had he paid the ten bucks to send it legitimately (or had he sent it for even cheaper through the Postal Service). Some other saboteurs burned a logging truck. The case broke because one of the activists bragged to his girlfriend, who was the daughter of a deputy sheriff. Another saboteur, now facing eighty years in prison if found guilty of arson, was on the run for nine months, until he was caught shoplifting boltcutters and identified. If I knew him at all, and had he asked, I would have given him the money to buy the damn boltcutters. One activist was sentenced to eight years in prison for burning an SUV dealership in southern California. How did the police come to suspect him? Soon after the arson, the police seized on the action as an opportunity to arrest a random activist and ransack his home. After the arrest, the alleged saboteur sent several emails to the *L.A. Times* insisting that police had arrested the wrong person. As if this weren't enough, the emails were sent from the library of the university the saboteur attended, and he was recorded on surveillance tapes entering the library minutes before each email was sent. He also bragged about the arson to his girlfriend and several friends, who said as much when questioned by police.

Let's be clear. He did not go to prison because he burned the SUV dealership. He went because he gave himself away.

This is bad enough. But he also gave the names of the people with whom he allegedly did the action, both to his friends and to the police.

Very rarely in my writing (or in my life) do I get prescriptive. I firmly believe people need to find and follow their own hearts. When people ask me what they should do to take down civilization, I steadfastly refuse to give any single answer, but tell them to listen to their own landbase and to their own heart, both of which will tell them what to do.

But here I will prescribe. Do not be stupid. Do not be careless. Real lives, including your own, are at stake. Some of us will be killed. Some of us will be imprisoned. We must of necessity take risks. But we must also of necessity make certain that the risks we take are worth the penalty if we are caught. It is one thing to shoplift boltcutters if you will serve a couple of months at most if caught. It is quite another to shoplift boltcutters if you're already wanted for crimes carrying sentences of decades.

If you are driving to a dam with liberation on your mind, do not speed. Do not have a broken taillight. Do not have an expired tab on your license plate. Do not tell your friends about it. Do not tell your girlfriend, the daughter of a deputy sheriff. Do not tell anyone about it. Do not tell anyone who does not need to know. Do not tell anyone who is not directly involved. Do not tell anyone except those you know will keep quiet even if it means they go to prison for forty years.

Do not tell anyone except those you trust with your very life, because that is exactly what you are doing.

Don't get me wrong. I have done many stupid things in my life, where the reward was in no way commensurate with the risk or cost. I have driven far too late into the night, and awakened to find myself speeding toward roadside reflectors. I've read my mail while driving and looked up to see the same. I nearly killed myself—ended my life—because I was in a hurry to open my telephone bill. When I was a kid I started a fire in a field by setting a match to a paper airplane, then throwing it. I had wanted to pretend it was a fighter plane that had been shot down. But it didn't look as cool as in movies. It looked like a piece of lined three-ring binder paper that some kid had set on fire and thrown into a bunch of dry weeds. It also looked like my friend—much smarter than I—leaping into those weeds to stomp out the fire before it could spread. I've wasted too many hours of my one and only life sitting motionless in front of a television watching some stupid program that has nothing to do with my life or with anything that is important to me. I have wasted too many hours of my one and only life sitting motionless in front of a computer playing some game that has nothing to do with my life or with anything that is important to me. I have at times ignored my health or physical well-being.

All of these are but the merest tip of the massive iceberg called Derrick's Stupid Actions. I am no better than any of these people. In fact they are better than I, because at least they have *done* something, taken the offensive. I am merely cautioning others to not make their same mistakes.

We are at war. War was declared against the world many thousands of years ago. War was declared against women. War was declared against children. War was declared against the wild. War was declared against the indigenous. The other side—the side that is killing the planet—right now has more guns than we do. There are not so many of us yet who are willing to fight back that we can afford to lose allies through careless mistakes.

I think sometimes about a former student I taught at the prison. One of his favorite crimes was to rob drug dealers. He said the adrenalin was unbeatable, because if caught you're lucky if they only kill you. Now, clearly we're already running into the high risk/low reward problem I've been talking about, but for him the rush evidently made it worth it. The point is that he never got caught. His plans were meticulous, with every contingency accounted for. He practiced till all actions were automatic. He made sure he was clean and sober. And then he acted. Within the context of this admittedly risky venture he was smart about his specific actions. So, you could ask, why was he in prison? Because one day

high on drugs he thought it would be a lark to steal a car. He got sloppy. He got stupid. He got caught.

Don't get sloppy. Don't tell anyone who doesn't need to know. Don't get caught.

❨ ❨ ❨

The Marlins had come back. It looked like they were going to win. I was glad for that. I told the man about the removal of the Embry Dam on the Rappahannock, and how it took two tries. I asked if six hundred pounds of explosives was a lot.

He asked if the newspaper account told what kind of explosive it was.

"No."

"Most of the time," he said, "the reporting is generic because of concerns about enabling the wrong (or right, in my opinion) people to get the information."

"A kid in southern California had a web page calling for revolution," I said. "He also had a page where visitors to his site could add links. Some rich right-wing kid added a link to a page describing how to fabricate explosives. The link wasn't even all that explicit, containing nothing you couldn't find at sites like howthingswork.com, loompanics.com, bombshock.com, totse.com, or even amazon.com. But because of that link the revolutionary kid got sentenced to a year in prison."

"I don't get it."

"It was on his website."

"But it was just a link. If they're going to arrest someone, why didn't they arrest the person who posted it? Why, then, don't the people at loompanics get popped? And what, as always, about the U.S. military?"

"Well, first, the feds were looking for an excuse to pop this guy anyway. So they took whatever they could find. Were it not that, they would have nailed him for jaywalking. Second, and more important, they didn't arrest the original poster, and they don't arrest the people at loompanics, because they don't talk about revolution. You can talk all you want about violence, as long as you don't mention social change. Witness training for the police and military. Similarly, you can talk all you want about social change, so long as you never mention violence. But you must *never* put them together."

"Why not?"

"It's premise four, man. That's *exactly* why it's okay for the military to teach so many people how to make and use explosives, and why it's okay for the military to blow people up all over the world. That's sending violence down the hierarchy. That's why it's okay for corporations to teach people how to make and use explosives to put in a mine and destroy a mountain. That's sending violence down the hierarchy. But if you mention explosives and the possibility of using them to go not down but up the hierarchy, you must be punished."

The Marlins had runners on first and third, one out.

He said, "Generally explosives are given in relation to TNT, so if C4 is two and a half times as powerful as TNT, then 600 pounds would actually mean 240 pounds or 30 blocks of C4, or composite C. I'm guessing that to blow Embry Dam they used commercial grade C4, plastic explosives. It not going off the first time probably meant one of the connections got wet and shorted out."

"Is that a large amount?"

"Six hundred pounds of TNT isn't that much. If I wanted to I could probably get a hold of it."

"How? Black market arms dealers?"

"Oh, god no. When you step into that world your risk increases dramatically, unless your name is Oliver North. If I were going to do it I'd probably take it from a mining operation or something."

"Seriously?"

"Lots of mining operations have the stuff lying around. It depends on the site. Some places secure it more carefully than others. But the thing about stored explosives is that because of their volatile nature they are *always* isolated from other buildings. This is a good thing because that makes those storage sites vulnerable."

"This is a world I know nothing about."

"*Anybody* could find explosives. All you have to do is pay attention to who uses it. Is it used to punch roads into a forest about to be destroyed? How about mining operations, either above or below ground? You're smart. You should be able to figure this stuff out."

I was thinking how glad I am that I'm a writer.

He continued, "One word here about theft. The authorities get really crazy when significant amounts of explosives disappear. It's possible, however, to steal it in ways that aren't so obvious."

I recalled that Georg Elser, one of the people who attempted to kill Hitler, had not stolen 120 pounds of explosives in one pop, but rather in small quantities from different containers over time so the thefts wouldn't be noticed.

He continued, "These are tactical questions that need careful consideration. Remember, also, that most security guards are very poorly paid thug types who couldn't or wouldn't qualify for the police."

I went back to thinking about how glad I am that I'm a writer.

He said, "Or maybe I would just make the explosives at home. It's amazingly easy to do. I've done it a lot. The weakest and easiest explosive is just ammonium nitrate fertilizer and diesel fuel mixed into a clay-like substance. If you throw in a little aluminum or some other oxidizer you increase the power by about 60 percent."

"You've done this a lot? I don't think I've ever met anyone who's made explosives before, at least not since the chemistry nerds in high school who used to set off bombs in the boys room."

He responded, "Every time I hear of a new recipe for an ammonium nitrate explosive, I make a batch that's very small but big enough to let me know it works. Trying out the recipe in a controlled environment is the only way to really know."

Did I mention that I'm glad I'm a writer?

☾ ☾ ☾

I asked the man one more question. "You said you could tell by looking where to put explosives. Could you also tell how much to use?"

"Oh, sure," he said. "How big was the Embry Dam?"

"They blew a hundred foot section that was twenty-two feet tall and at least eighteen inches thick."

"Do the math. Twenty-two feet times a hundred is 2,200 square feet. Divide that by six hundred pounds, and you get about four pounds per square foot, at eighteen inches thick. You could make similar calculations for whatever you want. Or you can just find yourself a chart."

☾ ☾ ☾

I give a talk. Afterwards someone asks, "I've heard you several times, and you always talk about the need to take out dams. Why don't you just shut up and do it?"

"That's a good question," I respond, "and one I ask myself all the time. I have three answers, all of which ring hollower every day than the day before. The first is that I'm scared."

"Of what?"

"Getting caught. I don't want to go to prison. I don't want to get killed. But in some ways those fears are secondary to the fact that I'm not quite ready to leave loved ones behind." It's like the person I mentioned whose people were people of the salmon, and who would be willing to take down dams once his children were a little older. I said, "I used to know a very good activist who said she'd go underground only after the death of her 23-year-old cat, whom she had rescued as a kitten from a vivisection lab. She couldn't leave her behind. I knew another who was going to do that work only after her elderly father died. She couldn't leave him behind. There are those I'm not yet ready to leave behind. I'm not particularly proud of this, nor am I ashamed. It just is. But I have to also tell you that the rapidity with which the world is being destroyed is making this fear increasingly pallid. And when this culture is poisoning our own bodies, the fear is also increasingly moot.

"The second answer has to do with proclivities. Some people love to play with explosives. They do it for fun. Some people love to write. I do it for fun. It seems kind of silly for me to learn all those skills that I don't enjoy and that I'm not very good at—my only D in college was in my chemistry quantitative analysis lab class—and make all of those mistakes of inexperience that are the *only* way we learn when there are already people out there who not only know this stuff but get off on it, probably the same way I get off on writing.[169] But if the people with these skills don't soon step forward, the murder of the world will necessitate me learning some new skills.

"The third reason—and this is the one that really stops me—is that I feel like the work I'm doing now is important, and I don't see anyone else doing it. I don't see enough people explicitly calling for us to bring down civilization, and making the sorts of comprehensive and comprehensible analyses I try to do. If there were enough other people doing this work, I'd sign up at the local YMCA for classes on bomb making. But there aren't, so I keep doing it. It's a matter of leverage. I still believe that, given my gifts and proclivities, writing continues to be the best way to multiply my efforts. Of course if or when I find a better lever I will, fears and proclivities aside, pick it up and use it. Likewise if it comes clear that I've overestimated the leverage provided by writing—if I find it's not helping enough—I'll do what's necessary to save salmon. It's the same old story of needing it all. Tecumseh not only fought against the civilized, he also spent a lot of time gathering an army.

"And while Tecumseh was able to conduct raids against the civilized at the same time he was recruiting, this was because the territory in which he operated was not yet completely occupied. By now the entire country, indeed most of the planet, has been overrun by the civilized. This means that there needs to be an absolute firewall between any public face and any underground activities.

Otherwise everyone might as well just go down to the police station for recreational mug shots. I am not blowing up dams precisely because I am talking about it publicly. I am talking about it publicly because someone needs to do that.

"The short answer is that for now this feels like the best thing for me to do. If that changes I'll do something else.

"The point is to accomplish something. After all, only the entire world is at stake."

❨ ❨ ❨

I do another talk. Afterwards someone—I found out later he's an old Dutch/American Indian farmer/hunter/gatherer/activist—says that several years before there had been big conflicts in his region between Indians and whites over fish. "Those conflicts made clear to me," he says, "that a primary difference between Indian warriors and most activists is that Indian warriors were willing to die for the fish, and I haven't seen many forest activists, for example, willing to die for trees." He pauses, then continues, "You talk about the salmon. My question for you is: would you die for those fish?"

I do not hesitate. I say, "In a fucking heartbeat."

❨ ❨ ❨

I do yet another talk. People ask about dams. Many people say many intelligent things. Afterwards I sign books. When I'm done I find my backpack, which I had dropped on the floor a few feet behind where I was sitting. I put my books and papers inside, say good night and thank you to the people who brought me, and head to my hotel. At the hotel I take my books and papers out of my pack to put into my suitcase. I'll head to the next town when I get up in the morning. But in my backpack I find a legal-sized envelope, with my first and last name typed in all caps on a piece of paper that is glued to the front. The envelope is sealed. I'm guessing that the glue on the flap was moistened with tap water, not saliva, and I'm guessing that if I searched for prints, the envelope would be clean. I open it. It contains a single sheet of paper, typewritten, single-spaced, then copied. No name, of course. I'm guessing the sheet also would not have prints.

It read, "I don't know if you are serious about blowing up a dam. When I was younger I blew up a few small dams, a couple of holes in the walls of vivisection labs, and a lot of logging and road building equipment. It was important work, and a whole lot of fun, but unfortunately it wasn't fun for the entire family as my husband has requested I quit.

"The best place I've found for how to make explosives is at a site called roguesci.org, under their explosives section. Sadly their domain name got revoked (ain't free speech a wonderful thing?) but I hope they'll be mirrored and back up soon. The site contains everything you'd ever want to know about explosives, including safety-conscious instructions on how to make them, and chemical supply companies that sell to the general public. You may already know that semtex now has a scent added to it for bomb sniffing dogs, and each batch is bar-coded and records kept of shipment. So it's no longer used as much by do-it-yourselfers. However, it's not hard to make PETN or RDX with a plasticizer. ANFO is always popular since ammonium nitrate and fuel oil are both easily available.

"If you can't find good information on explosives anywhere else, you can always get your hands on any number of United States military (especially Navy SEAL) manuals. I've always found it very interesting that the same information that is censored or even illegal elsewhere is put out at taxpayer expense by the U.S. military. That just about says it all, doesn't it? These training manuals are available at amazon.com or at bookstores.

"A couple of book recommendations. One is *Home Workshop Explosives (2nd Edition)*, by a gentleman who goes by the name Uncle Fester. It's $14, and well worth the money. He's very safety-conscious, which is of paramount importance. Explosives by their nature are dangerous. By day he's an industrial chemist and knows what he's talking about. He also has a book on biological weapons. If you're serious about this, here are a couple of other good texts: *Introduction to the Technology of Explosives*, by Cooper and Kurowski, and *The Chemistry of Powder and Explosives*, by Tenney L. Davis. All of these books are available through amazon.com or can be ordered from your friendly neighborhood independent bookstore. Make sure the independent bookstore opposes the Patriot Act and doesn't keep records on who orders books. Order them under a false name anyway and pay with cash. And order some other books at the same time, so your order might look like *The Red Pony, The Complete Poems of Robert Frost, Raising the Homestead Hog, Home Workshop Explosives (2nd Edition), Three Films by Ingmar Bergman, Maybe You Should Write A Book,* and *Winning Low Limit Texas Hold 'Em.*

"One final word: avoid *The Anarchist Cookbook*. The recipes are often wrong and dangerous to the maker.

"But who knows? You might not need explosives. In 1998, a twelve-year-old

hacker was playing around and broke into the computer system that runs Roo-
sevelt Dam in Arizona. Although he evidently neither knew nor cared, federal
authorities later said he had taken complete control of the system that controls
the dam's floodgates.

"Good luck."

I typed the note into my laptop. Had I been at home I would have burned the
original. Because I was in a hotel I tore it and the envelope into little pieces, and
I flushed them down the toilet. Then I flushed again, and again, and again.

☾ ☾ ☾

I cannot tell you how glad I am that I am a writer.

☾ ☾ ☾

Another town, another talk. Another Q and A session. This night someone asks,
"There are a lot of incredibly destructive dams going up right now in China. I
would love to go over there and take them down. But to do so will wipe out villages downstream and kill a lot of people. What do I do about that?"

There have been times when I've gotten questions like this that I have
responded as I wish I had to that pacifist who said that he wouldn't hurt a single human being to save an entire run of salmon, which is that the belief that
any human life, including my own, is worth more than the health of the landbase is precisely the problem. But for whatever reason I didn't feel like saying that
this night. I said, "First, we don't need to go to China. There are plenty of dams
right here that are very destructive. Second, we kill as surely by inaction as by
action: the dams wipe out villages and kill people, too, but those costs never
seem to be counted. And who gains from the dams? It's always those in power.
There's a very important third point, too, which is that there are hundreds of
thousands of dams that can be removed at absolutely no risk to human life. I
don't see us taking those out, which makes me think that this fear of killing people is just a smokescreen, another way to rationalize our inaction. If that really
is a concern, why don't we take out these smaller dams now, and then face the
question of the bigger dams when the time comes?"

☾ ☾ ☾

I'm going to take out a dam this coming year. I won't have to learn anything about explosives. I might not have to break the law.[170] It's very possible that no humans will even notice, although I guarantee the nonhumans who live there will. And many will be glad.

Of the two million dams in the United States, nearly all of them are less than six feet tall. Many of these are old. Many of these no longer serve any economic function. Many of these were illegal in the first place, and continue to be illegal. Many of them continue to strangle tiny streams only because no one has yet bothered to take them out. Those of us who care have no excuse, really, to not remove them.

☾ ☾ ☾

I recently spent a week on tour. In that time I saw six dams I could remove easily with no special knowledge. Each was no more than two or three feet tall, no wider than five or six feet, and at most six inches thick. Using a pick, sledge, and maybe a wedge, I could knock a notch in such a dam (which is really all that fish need)[171] in maybe fifteen minutes. Since a pick won't fit in my suitcase, and since I didn't feel like scrabbling at concrete with bare hands, I decided to try to find some dams closer to home.

It's time for me to take some nice afternoon walks along some beautiful streams, and then to take some other, more purposeful walks, late at night.

TOO MUCH TO LOSE
(SHORT TERM LOSS, LONG TERM GAIN)

I watched my father die because there was no money in our village to buy him medicine for his stomach. That's why I went
with the Zapatistas. . . . I decided to fight because if we're all
going to die it might as well be for something."

Raul Hernandez, seventeen-year-old
Zapatista prisoner of the Mexican state[172]

I DON'T LIKE TO KILL. IT'S REALLY DIFFICULT FOR ME. NO, I MEAN *REALLY* difficult. I usually catch mosquitoes and carry them outside, which is extreme even among my animal rights friends. I'm a terrible gardener because I hate killing weeds, and when the time comes I hate killing the vegetables themselves. I eat the pods of radishes so I don't have to kill the plant (but the pods taste just as good, and actually provide more food than the roots anyway). When I used to go fishing I was glad my fishing partner didn't mind killing.

That said, I will kill when I have to, and it doesn't scar me at all. I just don't like doing it. I've killed a lot. Of course we all kill, whether or not we choose to acknowledge it.

We don't need a philosopher to tell us that life feeds off life. It only takes a quick walk outside, or even a walk to the refrigerator. This failure or refusal or even inability to acknowledge the necessity of killing is a luxury born of our separation from life. It's a cliché by now to note that many of the civilized believe food comes from the grocery store, not from the flesh of living breathing plants and animals.

❲ ❲ ❲

This disconnection from killing is related to a disbelief in our own deaths, and even moreso to a disbelief in the rightness of our own deaths and a belief on the other hand that death is an enemy. As such it ties back to premise four of this book. Death is violence and violence *cannot* happen to us. We cannot die. We will not die. We are immortal. This delusion is based on the linear/historical view of the world we discussed so very long ago in this book, where life is not a circle where I feed off you who feeds off someone else who feeds off someone else who feeds off someone else who feeds off me (or put another way, where I feed someone else who feeds someone else who feeds someone else who feeds someone else who feeds me). Rather *we* are exempt from this cycle. We are at the top of a pyramid. We are consumers. I feed off you and I feed off someone else and I feed off someone else and I feed off someone else and I feed off someone else. No one feeds off me. I will never die.

❨ ❨ ❨

Have you ever eaten smoked salmon? I hadn't until a couple of years ago, but when I did I suddenly understood how the Indians of the Northwest (and earlier the Indians of the Northeast, and much earlier the indigenous of Europe) could eat salmon every day. It's some of the best-tasting food I've ever eaten. Every summer now I buy as much as I can afford, then ration it to make it last all year. The food is also about as politically correct as you can get: wild salmon caught on the Klamath River by a Yurok man, after which it is smoked by one of his tribal elders.[173] Maybe the PC quotient of the salmon partially expiates the definitely un-PC time I've spent playing Doom 2 on the computer.

I get up from in front of the computer, where I'm writing about taking down dams. I walk to the refrigerator, open the door, reach inside to pull a strip of salmon out of a bag. I peel some meat away from the skin. The salmon flesh speaks to me. It says, "Remember the bargain."

I eat the meat. I close the refrigerator door. I don't really want to think about it. I just want to eat.

I get back to work. But I want some more. It tastes so good. I return to the refrigerator, open it, pull out more salmon. It says the same thing: "Remember the bargain." I eat the food. Close the door. Try not to think about it.

I go back again. I remember the predator-prey bargain: If you consume the flesh of another, you take responsibility for the continuation of its community. I open the refrigerator. Eat more.

This time the salmon says something else to me: "I know you don't like killing. If you help take out the dams that will help us survive. Then you can kill and eat all the salmon you'd like. We will even jump out of the water and right to where you are waiting. You won't feel bad about killing us, because you have helped our community. We will gladly do this for you, if you will help us survive."

❨ ❨ ❨

Now, I know what you dogmatic humanists are thinking: *This guy is crazy. He thinks salmon meat talks to him. You know, many cult leaders believe that animals speak to them. And some psychopaths say figures in their dreams tell them what to do!*

I have a couple of responses. The first is that for the vast majority of human existence on this planet, humans have listened to what their nonhuman neighbors have had to say. That is *the* fundamental difference between civilized and indigenous peoples (well, apart from that little thing about the civilized killing the planet). We've been over this and over it, and I know I still can't convince you. But that's okay, because my second response is that in this case it doesn't matter in the slightest whether or not the flesh of the salmon spoke to me—the reality remains that if you want to eat salmon[174] you're going to have to remove dams.

Somebody has to take on this responsibility and help the salmon.

❨ ❨ ❨

In related news, the California Department of Fish and Game today released a report revealing that the fish kill on the Klamath River that I wrote about early in this book, that took place two years ago, was probably twice as bad as previously suspected. If you recall, this was the one where the federal government decided that fish don't need water and gave the water instead to a few heavily subsidized farmers in the Klamath Basin in southern Oregon. Fish and Game placed blame for the kill on low river flows, caused by dams on the Klamath. Studies by the U.S. Fish and Wildlife Service and the Yurok tribe have reached the same conclusion.

A spokesperson for the Bureau of Reclamation, in charge of the dams, denies the new estimates, saying, "We're two years later. How do we know that?" He also denies that the Bureau favors irrigators over fish, and denies that the dams create, in the words of the regional director of a commercial fishing organization, "a perpetual drought on the lower river." He further denies that the massive kill of salmon coming up to spawn two years ago will cause their offspring to return in record low numbers, saying, "I guess we just need to wait until next year and see what Mother Nature [*sic*] does."[175]

This is the spokesperson for the organization that is killing the fish.

I, too, am part of "Mother Nature," and so are you. What are you going to do about it? What are you going to do about the dams?

❨ ❨ ❨

We need to bring down civilization now. We need not hesitate any longer. The planet is collapsing before our eyes, and we do nothing. We hold our little protests, we make our little signs, we write our little letters and our big books, and the world burns.

Here is an article in a British newspaper, today's *Independent*, entitled "Disaster at sea: global warming hits UK birds." It reads, "Hundreds of thousands of

Scottish seabirds have failed to breed this summer in a wildlife catastrophe which is being linked by scientists directly to global warming.

"The massive unprecedented collapse of nesting attempts by several seabird species in Orkney and Shetland is likely to prove the first major impact of climate change on Britain.

"In what could be a sub-plot from the recent disaster movie, *The Day After Tomorrow*, a rise in sea temperature is believed to have led to the mysterious disappearance of a key part of the marine food chain—the sand eel, the small fish whose great teeming shoals have hitherto sustained larger fish, marine mammals and seabirds in their millions.

"In Orkney and Shetland, the sand eel stocks have been shrinking for several years, and this summer they have disappeared: the result for seabirds has been mass starvation. The figures for breeding failure, for Shetland in particular, almost defy belief.

"More than 172,000 breeding pairs of guillemots were recorded in the islands in the last national census, Seabird 2000, whose results were published this year; this summer the birds have produced almost no young, according to Peter Ellis, Shetland area manager for the Royal Society for the Protection of Birds (RSPB).

"Martin Heubeck of Aberdeen University, who has monitored Shetland seabirds for 30 years, said: 'The breeding failure of the guillemots is unprecedented in Europe.' More than 6,800 pairs of great skuas were recorded in Shetland in the same census; this year they have produced a handful of chicks—perhaps fewer than 10—while the arctic skuas (1,120 pairs in the census) have failed to produce any surviving young.

"The 24,000 pairs of arctic terns, and the 16,700 pairs of Shetland kittiwakes—small gulls—have 'probably suffered complete failure,' said Mr. Ellis.

"In Orkney the picture is very similar, although detailed figures are not yet available. 'It looks very bad,' said the RSPB's warden on Orkney mainland, Andy Knight. 'Very few of the birds have raised any chicks at all.'

"The counting and monitoring is still going on and the figures are by no means complete: it is likely that puffins, for example, will also have suffered massive breeding failure but because they nest deep in burrows, this is not immediately obvious.

"But the astonishing scale of what has taken place is already clear—and the link to climate change is being openly made by scientists. It is believed that the microscopic plankton on which tiny sand eel larvae feed are moving northwards as the sea water warms, leaving the baby fish with nothing to feed on.

"This is being seen in the North Sea in particular, where the water temperature has risen by 2C in the past 20 years, and where the whole ecosystem is thought to be undergoing a 'regime shift,' or a fundamental alteration in the interaction of its component species. 'Think of the North Sea as an engine, and plankton as the fuel driving it,' said Euan Dunn of the RSPB, one of the world's leading experts on the interaction of fish and seabirds. 'The fuel mix has changed so radically in the past 20 years, as a result of climate change, that the whole engine is now spluttering and starting to malfunction. All of the animals in the food web above the plankton, first the sand eels, then the larger fish like cod, and ultimately the seabirds, are starting to be affected.'

"Research last year clearly showed that the higher the temperature, the less sand eels could maintain their population level, said Dr. Dunn. 'The young sand eels are simply not surviving.'

"Although over-fishing of sand eels has caused breeding failures in the past, the present situation could not be blamed on fishing, he said. The Shetland sand eel fishery was catching so few fish that it was closed as a precautionary measure earlier this year. 'Climate change is a far more likely explanation.'

"The spectacular seabird populations of the Northern Isles have a double importance. They are of great value scientifically, holding, for example, the world's biggest populations of great skuas. And they are of enormous value to Orkney and Shetland tourism, being the principal draw for many visitors. The national and international significance of what has happened is only just beginning to dawn on the wider political and scientific community, but some leading figures are already taking it on board.

"'This is an incredible event,' said Tony Juniper, director of Friends of the Earth. 'The catastrophe [of these] seabirds is just a foretaste of what lies ahead.

"'It shows that climate change is happening now, [with] devastating consequences here in Britain, and it shows that reducing the pollution causing changes to the earth's climate should now be the global number one political priority.'"[176]

Remind me again, what are we waiting for?

❨ ❨ ❨

In other news of the day, the U.S. stock market rose slightly in heavy trading.

❨ ❨ ❨

A couple of days after the above article appeared in the *Independent*, a far shorter version of it—three column inches—appeared on page D10 (the bottom back of the sports section) of the *San Francisco Chronicle*. The front page contained, among many other things, a four column inch teaser to a full-page article celebrating the life and mourning the death of funk legend and violent misogynist Rick James.

⟨ ⟨ ⟨

I hate this culture.

⟨ ⟨ ⟨

I drive through the small town where I live. I get stuck behind a van at a stoplight. The van has four bumper stickers. The first states: "I got a gun for my wife: it was the best trade I ever made." The second: "I miss my ex, but my aim is getting better." The third: "My ideal girlfriend is a nympho liquor store owner." Above the others, pretty much summing them up, is a bumper sticker with an American flag and the caption: "God Bless America."

⟨ ⟨ ⟨

I really do hate this culture.

⟨ ⟨ ⟨

For the past several months, ever since I started speaking to the fisheries biologists, I've been haunted by a specific phrase used by more than one of them: Catastrophic dam failure always involves short term habitat loss and long term habitat gain.

Short term habitat loss, long term habitat gain. Short term loss, long term gain. What was a primary reason my mother stayed so long with my father? The fear of short term loss outweighed the prospect of long term gain. Why does anyone stay in any abusive relationship? Chances are good it comes down to a fear of the short term loss being greater than the perceived possibility of long term gain. Why do people stay in any self-destructive relationships? Why do people stay at jobs they hate? Why do addicts stay addicted? Why don't people take out dams? Why don't people get rid of civilization? Short term loss, long term gain.

Why is it, as Zygmund Bauman wrote, that "rational people will go quietly, meekly, joyously into a gas chamber, if only they are allowed to believe it is a bath-room"?[177] Why is it that so many of us today do not resist?

I think the fisheries biologists gave us part of the answer.

We will accept short term loss—even murder, both personal and planetary—rather than take the risks that would lead to long term gain.

❨ ❨ ❨

Of course it's not quite so simple. Is not Christianity based on teaching us to give up the ever-so-small short term loss of being in our bodies on a beautiful planet in exchange for the long term gain of heaven? Is not technological civilization based on teaching us to give up the short term loss of the natural world in exchange for the technotopia that awaits us around the corner? Isn't capitalism based on teaching us to give up the short term loss of our daily happiness to work jobs we hate so we can eventually retire rich (on a dying planet)?

What's the difference?

That last question stumped me for a couple of days, till I took a long walk in the forest and suddenly got the answer. Domination. The important question is not whether someone is more repelled by the short term losses than attracted by the long term gains, but rather who gains by this stasis. Who gains by a woman staying with her abuser? Who is exploiting whom? Who gains by some-one staying in a job he hates? Who is exploiting whom? Who gains as we give our lives away here in hopes of a better life in heaven? Who is exploiting whom?

❨ ❨ ❨

Just ten seconds ago I received the following note from a fishing guide on the Klamath River, forwarded to concerned parties: "I was fishing down near Blue Creek a couple days ago and found this fish dying in the water. I netted it so I could look at it. The water temp. right below Blue Creek was 74 degrees. The steel-head are stacking up in the cooler water along the edge of the river. (Just like a couple years ago.) Is there any way to get some more water? Please do something."

❨ ❨ ❨

Pretend you are not civilized. Pretend you love the land where you live. Pretend you were never taught to value economic production over life, or pretend you unlearned this. Pretend you were never taught that everyone else is here for you to use, or pretend you unlearned this. Pretend you do not feel entitled to take from those around you. Pretend you know that someday you will die. Pretend you are not separate from your landbase, but a part of it, as it is a part of you. Pretend you were taught to take care of your landbase as though your life depends on it. Pretend that's what you do.

❨ ❨ ❨

Those who hold others captive are captives themselves. Freedom feeds freedom, and captive keeps captive. This is true of those who hold humans captive. This is true of those who hold the world captive. It is true of those who hold rivers captive. Not only the wild water on one side of a dam is held captive, but entire communities on all sides of every dam. Freedom feeds freedom, and a river no longer captive feeds the freedom it finds to everyone and everything downstream which was once held captive by the captivity of the river.

Have you ever listened to a plant as it drinks up water? You can hear the water drawing up its roots, moving through stems, moistening (ever so slightly) the surfaces of leaves. When a dam breaks it is not only water but freedom that is absorbed and expired, exuded and extended, because much of the community downstream from a dam is, in a very real sense, locked up or extremely limited in terms of its proper movement until the medium of motion—wild water— is restored.

I want to be there when river after river finds freedom and feeds it to everyone else. This will be the most beautiful and appropriate kind of feeding frenzy.

Those who really know freedom will not and cannot want to keep this freedom from others anymore than they can want to lose it themselves. Only terrified, helpless, stupid slaves—including the civilized—could and would protect their own prisons, *be* their own bars.[178]

❨ ❨ ❨

Pretend you are not civilized. Pretend you are not a slave. Pretend you are a free human being. Pretend you were taught to value freedom, your own and others, enough that you will fight for it. Pretend that is what you do.

PSYCHOPATHOLOGY

How is it conceivable that all our lauded technological progress—our very Civilization—is like the axe in the hand of the pathological criminal?

Albert Einstein

JUST AS CIVILIZATION FIRST CONVERTS THE LIVING TO THE DEAD
psychologically and socially (including perceptually), and then physically—perceiving trees as dollar bills makes deforestation inevitable—so, too, civilization
needs to be reversed in the same order, first psychologically and socially (including perceptually), and then physically. If we could help people to not see trees
as dollar bills, we would not have to fight so hard to stop deforestation. We
might not have to fight at all. The same is true of course of rapists and other
abusers: if we could help them to see women as women and not as objects to be
exploited, not nearly so many women would get raped. This is true of all who
perceive themselves entitled, and to all forms of exploitation by all exploiters.
If we can change their ways of perceiving, their behavior will change.

But there are two problems with this. The first is that we don't have time.
Spotted owls and marbled murrelets are being extirpated right now, and we
don't have the time to change the way Charles Hurwitz perceives the world.
Even if we did, we'd then have to change the way his replacement perceives the
world, as Hurwitz would very soon be sacked for no longer maximizing profits. While I know that changing hearts and minds is desperately important, and
while I know that this is especially true for the young (which is why I wrote a
book on education[179]), I also know that if we wait on this, much of the world will
be dead before these children are adults (presuming that somehow we are powerful, persuasive, and pervasive enough to overcome the effects of child abuse,
industrial education, advertising, etc. on all of these children, if we were powerful and pervasive enough to bring about this change right now, why wouldn't
we just go ahead and bring down the dams ourselves?).

The second problem is that even if we did have the time, not everyone could
be converted. Remember the monkeys made permanently insane by their isolated upbringing. Remember the rates of recidivism for abusers. What's worse
is that the psychopaths who run this culture are socially rewarded for making
antisocial, indeed psychopathic decisions. Charles Hurwitz is rewarded very
well for destroying the redwood forests of northern California. Within the context of this culture he would be a fool to change his behavior. Of course this is
true for every abuser.

❨ ❨ ❨

I just used the term *psychopath* to describe Charles Hurwitz and others who guide this culture. I don't use this term spuriously.[180] A psychopath can be defined as one who willfully does damage without remorse: "Such individuals are impulsive, insensitive to other's needs, and unable to anticipate the consequences of their behavior, to follow long-term goals, or to tolerate frustration. The psychopathic individual is characterized by absence of the guilt feelings and anxiety that normally accompany an antisocial act."[181] Dr. Robert Hare, who has long studied psychopaths, makes clear that "among the most devastating features of psychopathy are a callous disregard for the rights of others and a propensity for predatory and violent behaviors. Without remorse, psychopaths charm and exploit others for their own gain. They lack empathy and a sense of responsibility, and they manipulate, lie and con others with no regard for anyone's feelings."[182] Hare also states, "Too many people hold the idea that psychopaths are essentially killers or convicts. The general public hasn't been educated to see beyond the social stereotypes to understand that psychopaths can be entrepreneurs, politicians, CEOs and other successful individuals who may never see the inside of a prison."[183]

You can take your pick between these definitions. Both work for Hurwitz, both work for corporations and those who run them, and both work for the culture at large.

❨ ❨ ❨

Speaking of psychopaths, today *The New York Times* ran an article entitled "Bad news (and good) on Arctic warming." The article begins, "The first thorough assessment of a decades-long Arctic warming trend shows the region is undergoing profound changes, including sharp retreats of glaciers and sea ice, thawing of permafrost, and shifts in ocean and atmospheric conditions that are likely to harm native communities, wildlife and economic activities, while offering some benefits."

Benefits? Here they are: "The potential benefits of the changes include projected growth in marine fish stocks and improved prospects for agriculture and timber harvests in some regions, as well as expanded access to Arctic waters.[184] There, sea-bed deposits of oil and gas that have until now been cloaked in thick shifting crusts of sea ice could soon be exploitable, and ice-free trade routes over Siberia could significantly cut shipping distances between Europe and Asia in the summer."[185]

To speak of "benefits" such as these at the expense of the natural world is insane: it is to be out of touch with physical reality.

((((

Global warming may be far worse than any of us fear. I can do no better in describing this than John Atcheson, in the *Baltimore Sun*: "There are enormous quantities of naturally occurring greenhouse gasses trapped in ice-like structures in the cold northern muds and at the bottom of the seas. These ices, called clathrates, contain 3,000 times as much methane as is in the atmosphere. Methane is more than 20 times as strong a greenhouse gas as carbon dioxide.

"Now here's the scary part. A temperature increase of merely a few degrees would cause these gases to volatilize and 'burp' into the atmosphere, which would further raise temperatures, which would release yet more methane, heating the Earth and seas further, and so on. There's 400 gigatons of methane locked in the frozen arctic tundra—enough to start this chain reaction—and the kind of warming the Arctic Council [a group of world-class climatologists convened to discuss the state-of-the-art understanding of global warming] predicts is sufficient to melt the clathrates and release these greenhouse gases into the atmosphere.

"Once triggered, this cycle could result in runaway global warming the likes of which even the most pessimistic doomsayers aren't talking about.

"An apocalyptic fantasy concocted by hysterical environmentalists?

"Unfortunately, no. Strong geologic evidence suggests something similar has happened at least twice before.

"The most recent of these catastrophes occurred about 55 million years ago in what geologists call the Paleocene-Eocene Thermal Maximum (PETM), when methane burps caused rapid warming and massive die-offs, disrupting the climate for more than 100,000 years.

"The granddaddy of these catastrophes occurred 251 million years ago, at the end of the Permian period, when a series of methane burps came close to wiping out all life on Earth. More than 94 percent of the marine species present in the fossil record disappeared suddenly as oxygen levels plummeted and life teetered on the verge of extinction. Over the ensuing 500,000 years, a few species struggled to gain a foothold in the hostile environment. It took 20 million to 30 million years for even rudimentary coral reefs to re-establish themselves and for forests to regrow. In some areas, it took more than 100 million years for ecosystems to reach their former healthy diversity."[186]

Remind me again of the benefits of global warming.

Remind me again why we do not use any and all means necessary—and I mean any and all means necessary—to stop this from happening.

⟨ ⟨ ⟨

How many times do we need to say it: Global warming may be far worse than any of us fear. Another article: "Researchers have found a new and potentially devastating danger from the huge volumes of carbon dioxide (CO_2) produced by industry and transport, already threatening the planet with climate change.

"Now, they warn, it is also rapidly turning the world's oceans to acid as it is dissolved in seawater, and putting an enormous array of marine life at risk. Ocean acidification [as dissolved carbon dioxide reacts with water to form carbonic acid] may wipe out much of the microscopic plankton at the base of the marine food web, and have a knock-on fatal effect up through shellfish to major human food species such as cod. It is already having a serious impact on organisms such as coral, and putting a question mark against the future of coral reefs."[187]

Remind me one more time of the benefits of global warming.

Remind me again why we do not use any and all means necessary—and I mean any and all means necessary—to stop this from happening.

⟨ ⟨ ⟨

Psychopathology in action.

Scientists have finally realized that birds are intelligent. According to a *New York Times* article: "Today, in the journal *Nature Neuroscience Reviews*, an international group of avian experts is issuing what amounts to a manifesto. Nearly everything written in anatomy textbooks about the brains of birds is wrong, they say.[188] The avian brain is as complex, flexible and inventive as any mammalian brain, they argue, and it is time to adopt a more accurate nomenclature that reflects a new understanding of the anatomies of bird and mammal brains."

So far so good. Of course we don't need scientists to tell us that birds are intelligent: birds do a fine job of that, if only we pay attention.

This realization that birds are intelligent is part of a revolution, according to Dr. Peter Marler. Once again, so far so good. But his revolution is not the same as mine. Here is his: "I think that birds are going to replace the white rat as the favored subject for studying functional neuroanatomy."[189]

Let's be clear: after all these years scientists realize what they should have known all along, which is that birds are actually sentient, intelligent, complex creatures, and their first inclination is to torture them.

This is the essence of psychopathology. This is the essence of civilization.

❧ ❧ ❧

Just as earlier I went through the characteristics of abusers and showed they were applicable to this culture as whole, I'd like to do the same, only far more briefly, for characteristics of psychopaths. The characteristics come from the ICD-10 Classification of Mental and Behavioural Disorders, World Health Organization, Geneva, 1992, section F60.2 on Dissocial (Antisocial) Personality Disorder:

A: Callous unconcern for the feelings of others
B: Gross and persistent attitude of irresponsibility and disregard for social norms, rules, and obligations
C: Incapacity to maintain enduring relationships, though having no difficulty in establishing them
D: Very low tolerance to frustration and a low threshold for discharge of aggression, including violence
E: Incapacity to experience guilt and to profit from experience, particularly punishment
F: Marked proneness to blame others, or to offer plausible rationalizations, for the behaviour that has brought the patient into conflict with society[190]

I'm sure you can see how these apply to the culture at large, and to those who run it. But let's quickly go through them.

Callous unconcern for the feelings of others. What measure would you like? Have the civilized ever cared for the feelings of the indigenous whose land they've stolen? How about the half-million Iraqi children whose deaths Madame Albright said were a price those in power were willing to pay?

How many times have we heard that emotion must be removed from all scientific study? How many times have we been told that emotions must never interfere with decisions having to do with cold hard cash?

Not only are the feelings of those to be destroyed by this culture given, at best, token attention, in most cases they're "scientifically" denied. Do chickens in slaughterhouses have feelings? How about pigs? How about monkeys in vivisection labs? How about trees? How about rivers? How about stones? This culture not only has "callous unconcern" for the feelings of these others, but denies the feelings even exist.

Next, gross and persistent attitude of irresponsibility and disregard for social norms, rules, and obligations. Obviously the civilized are persistently irresponsible: I don't believe it's possible to be more irresponsible than to kill the planet. So far as the latter part of this characteristic, because the society we're talking about is the entity suffering from this disorder, we cannot use its norms, rules, and obligations as the standard by which we judge. That would be akin to asking whether Ted Bundy was acting according to his own norms, rules, and obligations when he raped and killed women. And what are the norms, rules, and obligations of this culture? Norms: The rape of women, the abuse of children, the destruction of landbases. Rules: Laws made by the powerful, for the powerful, to maintain their power. Obligations: To amass as much power as possible, to never deviate from premise four of this book; to always—*always*—protect abusers and abusive social structures.

But let's move to a larger scale, and talk about the norms, rules, and obligations of living sustainably on this planet, such as the fundamental predator-prey relationship,[191] giving at least as much to your landbase as you take, and living in long-term intrahuman and interspecies cooperative arrangements. I have read scores of accounts of the indigenous saying they do not understand the civilized, because the civilized violate every rule of living. Recall the words of the Sauk Makataimeshiekiakiak (Black Hawk),[192] "An Indian who is as bad as the white men could not live in our nation; he would be put to death, and eat up by the wolves. The white men are bad schoolmasters; they carry false looks, and deal in false actions; they smile in the face of the poor Indian to cheat him; they shake them by the hand to gain their confidence, to make them drunk, to deceive them, to ruin our wives. We told them to let us alone, and keep away from us; but they followed on, and beset our paths, and they coiled among us, like the snake. They poisoned us by their touch. We were not safe. We lived in danger. We were becoming like them, hypocrites and liars, adulterers, lazy drones, all talkers, and no workers."[193]

Next, psychopaths have an incapacity to maintain enduring relationships, though no difficulty in establishing them. How long has this culture been on this continent? I live on Tolowa land, and the Tolowa lived here at least twelve thousand years (if you believe science, or since the beginning of time if you believe Tolowa myths instead). They had enduring relationships with their human and nonhuman neighbors. We do not. It's hard to maintain enduring relationships with those you exploit.

Psychopaths have a very low tolerance to frustration and a low threshold for discharge of aggression, including violence. How many times has the United States invaded other countries? How many of the indigenous have been slaughtered by the civilized? What slight excuses are always made for this aggression?

The next characteristic is an incapacity to experience guilt and to profit from experience, particularly punishment. How much guilt does Charles Hurwitz experience over the destruction of ancient redwood forests? After deforesting the Middle East, the Mediterranean, Western Europe, Britain, Ireland, much of North and South America, Africa, Oceania, and Asia, and with damage from this deforestation increasing daily, can we honestly say that those who run this culture are learning from prior experience? And how much do those in this culture profit from the repeated experience of failure as technology after introduced technology is promised to be clean yet inevitably leads to widespread destruction? How much are they learning from the experience of changing the climate through burning coal, oil, and natural gas? Will that stop them from pursuing genetic engineering? How about nanotech? Will they learn from their pursuit of nukes? Pesticides? Dams? Of course not.

Finally, there's a marked proneness to blame others or to offer plausible rationalizations for the behavior that has brought the psychopath into conflict with society. How much responsibility has Hurwitz taken for his violence? George W. Bush for his? The typical rapist for his? Bush blames forest fires for his urge to deforest. Clinton and company blamed beetles. It's the same old psychopathic story. And I'm sick of it.

I've long used the simile that sharing our finite planet with the dominant culture is like being locked in a room with a psychopath.[194] There's no way out, and although the psychopath may choose other targets first, eventually it will be our turn. Eventually we'll have to fight. There's no way around it. And the sooner we fight back—the sooner we kill this psychopath—the more life will remain.

PACIFISM, PART I

The people in power will not disappear voluntarily; giving flowers to the cops just isn't going to work. This thinking is fostered by the establishment; they like nothing better than love and nonviolence. The only way I like to see cops given flowers is in a flower pot from a high window.

William S. Burroughs[195]

MANY HUNDREDS OF PAGES AGO, AND NOW FOR ME MANY YEARS AGO, I wrote that this book was originally going to be an exploration of when counterviolence is an appropriate response to the violence of the system. In fact what has become this book was supposed to be nothing more than a pamphlet in which I took the main arguments normally presented by pacifists and examined them to see if they make any sense. Here now is that pamphlet.

Here are some standard lines thrown out by pacifists. I'm sure you, too, have heard them enough that if we had a bouncing red ball we could all sing along. Love leads to pacifism, and any use of violence implies a failure to love. You can't use the master's tools to dismantle the master's house. It's far easier to make war than to make peace. We must visualize world peace. To even talk about winning and losing (much less to talk about violence, much, much less to actually do it) perpetuates the destructive dominator mindset that is killing the planet. If we just visualize peace hard enough, we may find it, because, as Johann Christoph Friedrich von Schiller tells us, "Peace is rarely denied to the peaceful." Ends never justify means, which leads to Erasmus saying, and pacifists quoting, "The most disadvantageous peace is better than the most just war." Gandhi gives us some absolutism, as well as absolution for our inability to stop oppressors, when he says, "Mankind has to get out of violence only through non-violence. Hatred can be overcome only by love." Gandhi again, with more magical thinking, "When I despair, I remember that all through history the way of truth and love has always won. There have been tyrants and murderers and for a time they seem invincible but in the end, they always fall—Think of it, ALWAYS."[196] Violence only begets violence. Gandhi again, "We must be the change we wish to see." If you use violence against exploiters, you become like they are. Related to that is the notion that violence destroys your soul. If violence is used, the mass media will distort our message. Every act of violence sets back the movement ten years. If we commit an act of violence, the state will come down hard on us. Because the state has more capacity to inflict violence than we do, we can never win using that tactic, and so must never use it. And finally, violence never accomplishes anything.

Let's take these one by one. Love leads to pacifism, and any use of violence implies a failure to love. If we love we cannot ever consider violence, even to protect those we love. Well, we dealt with this several hundred pages ago, and I'm not sure mother grizzly bears would agree that love implies pacifism, nor mother moose, nor many other mothers I've known.

You can't use the master's tools to dismantle the master's house. I can't tell you how many people have said this to me. I can, however, tell you with reasonable certainty that none of these people have ever read the essay from which the line comes: "The Master's Tools Will Never Dismantle The Master's House," by Audre Lorde (certainly no pacifist herself). The essay has nothing to do with pacifism, but with the exclusion of marginalized voices from discourse ostensibly having to do with social change. If any of these pacifists had read her essay, they would undoubtedly have been horrified, because she is, reasonably enough, suggesting a multivaried approach to the multivarious problems we face. She says, "As women, we have been taught either to ignore our differences, or to view them as causes for separation and suspicion rather than as forces for change. Without community there is no liberation, only the most vulnerable and temporary armistice between an individual and her oppression. But community must not mean a shedding of our differences, nor the pathetic pretense that these differences do not exist."[197] We can say the same for unarmed versus armed resistance, that activists have been taught to view our differences as causes for separation and suspicion, rather than as forces for change. That's a fatal error. She continues, "[Survival] is learning how to take our differences and make them strengths. *For the master's tools will never dismantle the master's house.*"[198]

It has always seemed clear to me that violent and nonviolent approaches to social change are complementary. No one I know who advocates the possibility of armed resistance to the dominant culture's degradation and exploitation rejects nonviolent resistance. Many of us routinely participate in nonviolent resistance and support those for whom this is their only mode of opposition. Just last night I and two other non-pacifists wasted two hours sitting at a county fair tabling for a local environmental organization and watching the—how do I say this politely?—supersized passersby wearing too-small Bush/Cheney 2004 T-shirts and carrying chocolate-covered bananas. We received many scowls. We did this nonviolent work, although we accomplished precisely nothing. But many dogmatic pacifists refuse to grant the same respect the other way. It is not an exaggeration to say that many of the dogmatic pacifists I've encountered have been fundamentalists, perceiving violence as a form of blasphemy (which it is within this culture if it flows up the hierarchy, and these particular fundamentalists have never been too picky about reaping the fiscal fruits of this culture's routine violence down the hierarchy), and refusing to allow any mention of violence in their presence. It's ironic, then, that they end up turning Audre Lorde's comment on its head.

Our survival really does depend on us learning how to "take our differences"—including violent and nonviolent approaches to stopping civilization from killing the planet—"and make them strengths." Yet these fundamentalists attempt to eradicate this difference, to disallow it, to force all discourse and all action into only one path: theirs. That's incredibly harmful, and of course serves those in power. The master's house will never be dismantled using only one tool, whether that tool is discourse, hammers, or high explosives.

I have many other problems with the pacifist use of the idea that force is solely the dominion of those in power. It's certainly true that the master uses the tool of violence, but that doesn't mean he owns it. Those in power have effectively convinced us they own land, which is to say they've convinced us to give up our inalienable right to access our own landbases. They've effectively convinced us they own conflict resolution methods (which they call *laws*), which is to say they've convinced us to give up our inalienable right to resolve our own conflicts (which they call *taking the law into your own hands*). They've convinced us they own water. They've convinced us they own the wild (the government could not offer "timber sales" unless we all agreed it owned the trees in the first place). They're in the process of convincing us they own the air. The state has for millennia been trying to convince us it owns a monopoly on violence, and abusers have been trying to convince us for far longer than that. Pacifists are more than willing to grant them that, and to shout down anyone who disagrees.

Well, I disagree. Violence does not belong exclusively to those at the top of the hierarchy, no matter how much abusers and their allies try to convince us. They have never convinced wild animals, including wild humans, and they will never convince me.

And who is it who says we should not use the master's tools? Often it is Christians, Buddhists, or other adherents of civilized religions. It is routinely people who wish us to vote our way to justice or shop our way to sustainability. But civilized religions are tools used by the master as surely as is violence. So is voting. So is shopping. If we cannot use tools used by the master, what tools, precisely, can we use? How about writing? No, sorry. As I cited Stanley Diamond much earlier, writing has long been a tool used by the master. So I guess we can't use that. Well, how about discourse in general? Yes, those in power own the means of industrial discourse production, and those in power misuse discourse. Does that mean they own all discourse—all discourse is one of the master's tools—

and we can never use it? Of course not. They also own the means of industrial religion production, and they misuse religion. Does that mean they own all religion—all religion is one of the master's tools—and we can never use it? Of course not. They own the means of industrial violence production, and they misuse violence. Does that mean they own all violence—all violence is one of the master's tools—and we can never use it? Of course not.

But I have yet another problem with the statement that the master's tools will never dismantle the master's house, which is that it's a terrible metaphor. It just doesn't work. The first and most necessary condition for a metaphor is that it make sense in the real world. This doesn't.

You can use a hammer to build a house, and you can use a hammer to take it down.

It doesn't matter whose hammer it is.

I'm guessing that Audre Lord, for all of her wonderful capabilities as a writer, thinker, activist, and human being never in her entire life dismantled a house. Had she done that, she could never have made up this metaphor, because you sure as hell can use the master's tools to dismantle his house.[199] And you can use the master's high explosives to dismantle the master's dam.

ℂ ℂ ℂ

There's an even bigger problem with the metaphor. What is perhaps its most fundamental premise? That the house belongs to the master. But there is no master, and there is no master's house. There are no master's tools. There is a person who believes himself a master. There is a house he claims is his. There are tools he claims as well. And there are those who still believe he is the master.

But there are others who do not buy into this delusion. There are those of us who see a man, a house, and tools. No more and no less.[200]

ℂ ℂ ℂ

Those in power are responsible for their choices, and I am responsible for mine. But I need to emphasize that I'm not responsible for the way my choices have been framed. If someone puts a gun to my head and gives me the choice of taking a bullet to the brain now or watching twelve straight hours of Dennis Miller, I don't think I could be held entirely responsible for taking the easy way out and telling the person to pull the trigger.

That's a joke (sort of), but the point is a serious one. I want to be clear: I am responsible for the choices I make. I am also responsible for attempting to break the confines which narrowly limit my choices, whenever and wherever possible.

❆ ❆ ❆

The next argument I've often heard for pacifism is that it's much easier to make war than to make peace. I have to admit that the first ten or fifteen times I heard this I didn't understand it at all: whether war or peace is harder is irrelevant. It's easier to catch a fly with your bare hand than with your mouth, but does that mean it's somehow better or more moral to do the latter? It's easier to take out a dam with a sledgehammer than a toothpick, but doing the latter wouldn't make me a better person. An action's difficulty is entirely independent of its quality or morality.

The next ten or fifteen times I heard this phrase it seemed to be an argument for violent resistance. If I want to live in a world with wild salmon, and if I'm all for doing this the easiest way possible, they're telling me I should make war. Certainly we have enough difficulties ahead of us in stopping those who are killing the planet without adding difficulties just for the hell of it.

The next ten or fifteen times I heard it I started going all psychotherapeutic on those who said it, wondering what it is about these pacifists that causes them to believe struggle for struggle's sake is good. Sounds like a martyr complex to me. Or maybe misplaced Calvinism. I don't know.

But after I heard it another ten or fifteen times I decided I just don't care. The argument is nonsensical, and I don't want to waste time on it that I could put to better use, like working to bring down civilization.

If all they're saying, by the way, is that oftentimes creativity can make violence unnecessary, I wish they would just say that. I would have no problem with that, so long as we emphasize the word *oftentimes*.

❆ ❆ ❆

It's tricky, though. Not many people take responsibility for their actions. Instead of recognizing that the framing conditions constrain their options and choosing from there, many instead *blame* the framing conditions for their choices. To take a patriarchal cliché of an unhappy family the miserable husband does not choose to have an affair, but is forced to by his wife's recalcitrance around sex: *I don't want to leave the marriage, nor do I want to lead a sexless life, so what am I going to do?* The miserable wife does not choose to live a sexless life, but is forced to by her husband's unwillingness or inability to communicate on any but a physical level: *I do not want to leave the marriage, nor do I want sex without intimacy, so what am I going to do?*

Similarly, I've never heard of an abuser saying, "I hit you because I wanted to terrorize you into submission." Instead he might say, "I wouldn't have hit you if you wouldn't have kept yelling and yelling and yelling at me about coming in so late." The framing conditions caused his violence. If we move this to a larger scale, how often have we heard politicians speak of the necessity of preemptive attacks on other countries (which just happen to sit atop coveted resources)? They rarely say, "I choose to invade this country." Instead they say they've been forced into this regrettable action by those they are about to subjugate, er, liberate. The Nazis played this same card—*everybody* plays this same card—they only invaded Poland because they had no choice; they only invaded the Soviet Union because they had no choice; they only killed *untermenschen* because they had no choice. Sigh. It's a terribly dirty job, but somebody's got to do it.

CEOs follow the same logic. If it were up to them, they would keep factories open (not that this is a good thing from the perspective of the planet, but within the confines of this culture most people consider it good), pay workers livable wages, maintain solid retirement programs, and so on. But you know how things are. They have no choice but to lay off workers and move the factories to Bangladesh, where they have no choice but to pay Bangladeshis eight cents per hour (as they themselves pull down a cool million per year, which converts to about five hundred dollars per hour, or more in one minute than they pay a Bangladeshi for a hundred hours). And if the Bangladeshis complain, the CEOs will have no choice but to move the factory on to Vietnam. Market pressures, you know. And these same market pressures force them to pollute, to clearcut, to overfish.

I'm sorry, each and every one of us can say, *we have no choice but to destroy the planet. It's really not our fault.*

Bullshit.

We may as well acknowledge that our entire culture—from top to bottom, inside out, personally and socially—is founded on, motivated by, and requires a systematic and absolute avoidance of responsibility. This is true both for our actions and our failures to act. What, ultimately, is environmental degradation? Any and all environmental degradation is a manifestation and a consequence of avoidance of responsibility. What is pollution? It is a manifestation and a consequence of avoidance of responsibility. What is overfishing? Deforestation? They are manifestations and consequences of avoidance of responsibility.[201]

And what is our failure to stop each of these things? It's just as much an avoidance of responsibility.

(((

We must, we are told, visualize world peace. My first thought on hearing this is always that the abused spouse is so often told that if she can just love her husband enough, he might change. Meanwhile her daughter may very well be wishing she gets a pony for Christmas, but that isn't going to happen either. My second thought on hearing this is always that visualizing world peace is essentially the semi-secular new age equivalent of praying.

All that said, I have to admit that I actually *am* a huge fan of visualization. I just normally call it daydreaming. When I was a high jumper in college, I used to more or less constantly picture myself floating over the bar. I'd do this in the shower, driving, walking to classes, certainly all through my classes. Later when I coached high jumping I used to guide my students through visualizations as a routine part of our practice. Now I constantly daydream about my writing. And more importantly I visualize people fighting back. I visualize people knocking down dams. I visualize them taking down the oil and electrical infrastructures. I visualize wild salmon returning in greater numbers every year. I visualize migratory songbirds coming back. I even visualize passenger pigeons returning. So I guess I don't have a problem with visualizing world peace, so long as people are also working for it. Except that as I made clear early on, civilization requires the importation of resources, which means it requires the use of force to maintain itself. This means that if these folks who are visualizing world peace really are interested in actualizing world peace, they should also be visualizing industrial collapse. And bringing it about.

But I don't think most of the people with "Visualize World Peace" bumper stickers on their old Saabs are interested in doing the work to take down civilization. It's too messy. I keep thinking about that line by Gandhi, "We want freedom for our country, but not at the expense or exploitation of others." I've also had this line crammed down my throat more times than I want to consider—often phrased as "You keep saying that in this struggle for the planet that you want to win, but if someone wins, doesn't that mean someone has to lose, and isn't that just perpetuating the same old dominator mindset?"—and I've always found it both intellectually dishonest and poorly thought-out.

A man tries to rape a woman. She runs away. Her freedom from being raped just came at his expense: he wasn't able to rape her. Does this mean she exploited him? Of course not. Now let's do this again. He tries to rape her. She can't get away. She tries to stop him nonviolently. It doesn't work. She pulls a gun and shoots him in the head. Obviously her freedom from being raped came at the expense of his life. Did she exploit him? Of course not. It all comes back to what I wrote earlier in this book: defensive rights *always* trump offensive rights. My right to freedom always trumps your right to exploit me, and if you do try to exploit me, I have the right to stop you, even at some expense to you.

All of this leads us to the fuzzy thinking. Anybody's freedom from being exploited will *always* come at the expense of the oppressor's ability to exploit. The freedom of salmon (and rivers) to survive will come at the expense of those who profit from dams. The freedom of ancient redwood forests to survive will come at the expense of Charles Hurwitz's bank account. The freedom of the world to survive global warming will come at the expense of those whose lifestyles are based on the burning of oil. It is magical thinking to pretend otherwise.

❨ ❨ ❨

Every choice carries with it costs. If you want air conditioning, you (and many others) are going to have to pay for it. If you want automobiles, you (and many others) are going to have to pay for them. If you want industrial civilization, you (and many others) are going to have to pay for it.

If you want freedom, you will have to fight for it and those who are exploiting you are going to have to pay for it. If you want a livable planet, at this point you will have to fight for it and those who are killing the planet are going to have to pay for it.

❨ ❨ ❨

Schiller's line, too, that "Peace is rarely denied to the peaceful," is more magical thinking, and the people who spout it really should be ashamed of themselves. What about the Arawaks, Semay, Mbuti, Hopi? Peace has been denied them. What about the peaceful women who are raped? What about the peaceful children who are abused? What about salmon? What about rivers? What about redwood trees? What about bison? What about prairie dogs? What about passenger pigeons? I hate to steal a line from someone so odious as John Stossel, but give me a break.

❆ ❆ ❆

Sometimes this book scares me. I'm calling for people to bring down civilization. This will not be bloodless. This will not be welcomed by most of the civilized. But I do not see any other realistic options. I cannot stand by while the world is destroyed. And I see no hope for reform. This is true whether we talk about the lack of realistic possibility of psychological or social reform, or whether we talk about the structural impossibilities of civilization (which requires the importation of resources) ever being sustainable. And really, think about it for a moment: this culture is changing the climate—*changing the climate*—and those in power are doing nothing to stop it. In fact they're burning more oil each year than the year before. If changing the earth's climate is not enough to make them change their ways, nothing will. Nothing. Not petitions, not letters, not votes, not the purchase of hemp hackysacks. Not visualizations. Not sending them love. Nothing. They will not change. They must be stopped. Through any means necessary. We are talking about the life of the planet. They must be stopped.

This scares me.

I sent a note saying all this to my publisher, who wrote me back, "Nothing could be scarier than this culture. I dare you to scare me."

Back to work.

❆ ❆ ❆

The next pacifist argument is that the ends never justify the means. While adding the word *almost* just before the word *never* makes this true for many trivial ends—I would not, for example, be willing to destroy a landbase so I can magnify my bank account—it's nonsense when it comes to self-defense. Are the people who spout this line saying that the ends of not being raped never justify the means of killing one's assailant? Are they saying that the ends of saving salmon—who have survived for millions of years—and sturgeon—who have survived since the time of the dinosaurs—never justify the means of removing dams without waiting for approval from those who are saying they wish salmon would go extinct so we can get on with living [*sic*]? Are they saying that the ends of children free from pesticide-induced cancer and mental retardation are not worth whatever means may be necessary? If so, their sentiments are obscene. We're not playing some theoretical, spiritual, or philosophical game. We're talking about survival. We're talking about poisoned children. We're talking about a planet being killed. I will do whatever is necessary to defend those I love.

Those who say that ends *never* justify means are of necessity either sloppy thinkers, hypocrites, or just plain wrong. If ends *never* justify means, can these people ride in a car? They are by their actions showing that their ends of getting from one place to another justify the means of driving, which means the costs of using oil, with all the evils carried with it. The same is true for the use of any metal, wood, or cloth products, and so on. You could make the argument that the same is true for the act of eating. After all, the ends of keeping yourself alive through eating evidently justify the means of taking the lives of those you eat. Even if you eat nothing but berries, you are depriving others—from birds to bacteria—of the possibility of eating those particular berries.

You could say I'm reducing this argument to absurdity, but I'm not the one who made the claim that ends *never* justify means. If they want to back off the word *never*, we can leave the realm of dogma and begin a reasonable discussion of what ends we feel justify what means. I suspect, however, that this would soon lead to another impasse, because my experience of "conversations" with pacifists is that beneath the use of this phrase oftentimes is an unwillingness to take responsibility for one's own actions coupled with the same old hubris that declares that humans are separate from and better than the rest of the planet. Witness the pacifist who said to me that he would not harm a single human to save an entire run of salmon. He explicitly states—and probably consciously believes—that ends never justify means, but what he really means is that no humans must be harmed by anyone trying to help a landbase or otherwise bringing about social change.

I sometimes get accused of hypocrisy because I use high technology as a tool to try to dismantle technological civilization. While there are certainly ways I'm a hypocrite, that's not one of them, because I have never claimed that the ends never justify the means. I have stated repeatedly that I'll do whatever's necessary to save salmon. That's not code language for blowing up dams. Whatever's necessary for me includes writing, giving talks, using computers, rehabilitating streams, singing songs to the salmon, *and* whatever else may be appropriate.

Setting rhetoric aside, there is simply no factual support for the statement that ends don't justify means, because it's a statement of values disguised as a statement of morals. A person who says ends don't justify means is simply saying: I value process more than outcome. Someone who says ends do justify means is merely saying: I value outcome more than process. Looked at this way, it becomes absurd to make absolute statements about it. There are some ends that justify some means, and there are some ends that do not. Similarly, the same means may be justified by some people for some ends and not justified by or for others (I would, for example, kill someone who attempted to kill those I love, and I would not kill someone who tried to cut me off on the interstate). It is my joy, responsibility, and honor as a sentient being to make those distinctions, and I pity those who do not consider themselves worthy or capable of making them themselves, and who must rely on slogans instead to guide their actions.

❨ ❨ ❨

It's pretty clear to me that our horror of violence is actually a deep terror of responsibility. We don't have issues with someone being killed. We have issues about unmediated killing, about doing it ourselves. And of course we have issues with violence flowing the wrong way up the hierarchy.

❨ ❨ ❨

Erasmus's statement, "The most disadvantageous peace is better than the most just war," used to strike me as insane and cowardly (not that this was true of all Erasmus's work). Now I just say I disagree.

Gandhi came out with a different version of this when he said, "My marriage to non-violence is such an absolute thing that I would rather commit suicide than be deflected from my position." I guess there are ways I can understand this, in that there are things I would kill myself rather than do. But this statement seems inflexible to the point of insanity. Is he saying that if he had the opportunity to stop a rape/murder, but could do so only through physically stopping the assailant, he would kill himself (and let the other person be raped/murdered) rather than break his sacred vow to non-violence? Is he saying that if he had the opportunity to stop the murder of the planet, but could do so only through physically stopping the assailants, he would kill himself (and let the planet be murdered) rather than violate his sacred vow to non-violence?

Unfortunately, he does seem to be saying these things. Now it's true that Gandhi perceived cowardice as worse even than violence (and please note that while I'm accusing Gandhi of fuzzy thinking, naïveté, and, as you'll see in a while, misogyny, never would I accuse him of cowardice: the man was stone cold brave), saying, for example, "Where the choice is between only violence and cowardice, I would advise violence," and "To take the name of non-violence when there is a sword in your heart is not only hypocritical and dishonest but cowardly." Even more to the point—and if all of Gandhi's words were this great he'd certainly be *my* hero—he said, "Though violence is not lawful, when it is offered in self-defence or for the defence of the defenceless, it is an act of bravery far better than cowardly submission. The latter befits neither man nor woman. Under violence, there are many stages and varieties of bravery. Every man must judge this for himself. No other person can or has the right." And here's one I like even more: "I have been repeating over and over again that he who cannot protect himself or his nearest and dearest or their honour by nonviolently facing death may and ought to do so by violently dealing with the oppressor. He who can do neither of the two is a burden. He has no business to be the head of a family. He must either hide himself, or must rest content to live forever in helplessness and be prepared to crawl like a worm at the bidding of a bully."

But damn if he doesn't follow this up with more of that old time pacifist religion. His very next paragraph is: "The strength to kill is not essential for self-defence; one ought to have the strength to die. When a man is fully ready to die, he will not even desire to offer violence. Indeed, I may put it down as a self-evident proposition that the desire to kill is in inverse proportion to the desire to die. And history is replete with instances of men who, by dying with courage and compassion on their lips, converted the hearts of their violent opponents."

Let's do a little exegesis. Sentence one: "The strength to kill is not essential for self-defence; one ought to have the strength to die." Problem: Although this makes a good sound bite, it also makes no sense. The first clause is a statement of faith (why does this not surprise me?), logically and factually unsupported and insupportable yet presented as a statement of fact. The same is true for the second. Perhaps worse, if one of the purposes of self-defense is to actually defend oneself (to keep oneself from harm, even from death), then saying that self-defense requires the strength to die becomes exactly the sort of Orwellian absurdity we've all by now become far too familiar with from pacifists: *self-defense requires the strength to allow self-destruction*, and *self-destruction requires strength* take their fine place alongside *freedom is slavery, war is peace*, and *ignorance is strength*. His sentence would imply that the Jews who walked into the showers or laid down so they could be shot in the nape of the neck by members of *einsatzgruppen* were actually acting in their own self-defense. Nonsense. Now sentence two: "When a man is fully ready to die, he will not even desire to offer violence." Once again, a statement of faith, logically and factually unsupported and insupportable yet presented as a statement of fact. I have read hundreds of accounts of soldiers and others (including mothers) who were fully prepared to die who sold their lives as dearly as possible. Sentence three: "Indeed, I may put it down as a self-evident proposition that the desire to kill is in inverse proportion to the desire to die." This is actually a pretty cheap rhetorical trick on his part. Any writer knows that if you label something as self-evident people are less likely to examine it, or even if they do and find themselves disagreeing with it, they're prone to feeling kind of stupid: *If it's so self-evident, how stupid must I be to not see it the same way?* A far more sophisticated and accurate examination of the relationship between a desire to kill and a desire to die was provided earlier in this book by Luis Rodriguez. Oftentimes a desire to kill *springs from* a desire to die. It's certainly true that the dominant culture—I've heard it called a thanatocracy—manifests a collective desire to kill self and other. But there is something far deeper and far more creepy going on with this sentence. Read it again: "Indeed, I may put it down as a self-evident proposition that the desire to kill is in inverse proportion to the desire to die." Let's pretend it's true. It is Gospel. You have never in your life read anything so true as this. Now let's ask ourselves whether Gandhi had a desire to kill. The answer is pretty obviously absolutely not. He said as much many times. What, then, does that mean Gandhi had a desire to do? If we take him at his word, it means he had a correspondingly absolute desire to die. He has an absolute death wish. Suddenly I understand why he would rather kill himself than break his marriage to nonviolence. Suddenly I understand his more or less constant rhetoric of self-sacrifice. Suddenly I understand his body hatred (we'll get to this in a moment). Suddenly I understand why Gandhi—and by extension so many other pacifists who are drawn to his teachings—was so often so little concerned with actual physical change in the real physical world. Pacifism as death wish. And don't blame me for this one, folks: it's nothing more than a strict literal interpretation of Gandhi's own text. Gandhi repeatedly stated his absolute desire to not kill, and stated here explicitly: "the desire to kill is in inverse proportion to the desire to die."

But that isn't even what bothered me most about his paragraph. Sentence four horrified and appalled me: "And history is replete with instances of men who, by dying with courage and compassion on their lips, converted the hearts of their violent opponents." If Gandhi's statement contained a shred of evidence to support it, the Nazis would have quickly stopped, domestic violence would cease,[202] the civilized would not kill the indigenous, factory farms would not exist, vivisection labs would be torn down brick by brick. Worse, by saying this, Gandhi joins the long list of allies of abusers by subtly blaming victims for perpetrators' further atrocities: *Damn, if only I could have died courageously and compassionately enough, I could have converted my murderer and kept him from killing again. It's all my fault.* Nonsense. Many killers—and nearly all exploiters—would vastly prefer intended victims not resist. The overwhelming preponderance of evidence just doesn't support Gandhi's position.

And his position leads him into (even more) grotesque absurdity. During World War II, as Japan invaded Myanmar (then called Burma), Gandhi recommended that if India were invaded, the Japanese be allowed to take as much as they want. The most effective way for the Indians to resist the Japanese, he said, would be to "make them feel that they are not wanted."[203] I am not making this up. Nor am I choosing one out-of-character statement. Gandhi urged the British to surrender to the Nazis, and recommended that instead of fighting back, both Czechs and Jews should have committed mass suicide (death wish, anyone?). In 1946, with full knowledge of the extent of the Holocaust, Gandhi told his biographer Louis Fisher, "The Jews should have offered themselves to the butcher's knife. They should have thrown themselves into the sea from cliffs."[204]

This is—and all you pacifists can get your gasps out of the way right now—both despicable and insane.

The insanity continues. If you recall, Gandhi said, "Mankind has to get out of violence only through non-violence. Hatred can be overcome only by love."[205] By now you should be able to spot the premises that, like any good propagandist, he's trying to slide by you. Violence is something humankind "has to get out of." Nonviolence is the only way to accomplish this. Hatred is something that needs to be overcome. Love is the only way to accomplish that.

These premises are statements of faith. They are utterly unsupported and unsupportable in the real world, and they are extremely harmful. Let's go back to the same basic example we've been using. A man breaks into a woman's home. He pulls out a knife. He is going to rape and kill her. She has a gun. Perhaps if she just shows him by shining example the beauty of nonviolence, perhaps if she dies with courage and compassion on her lips—or if she offers herself to the butcher's knife or throws herself into the sea from a cliff—she will convert his heart and he will realize the error of his ways and repent, to go and rape no more. Perhaps not. If she guesses wrong, she dies. And so do the rapist's next victims.

Gandhi's statement reveals an almost total lack of understanding of both abusive and psychopathological dynamics. His comment is one of the worst things you can say to anyone in an abusive situation, and one of the things abusers most want to hear. As I mentioned earlier, among the most powerful allies of abusers are those who say to victims, "You should show him some compassion even if he has done bad things. Don't forget that he is a human, too."[206] As Lundy Bancroft commented, "To suggest to her that his need for compassion should come before her right to live free from abuse is consistent with the abuser's outlook. I have repeatedly seen the tendency among friends and acquaintances of an abused woman to feel that it is their responsibility to make sure that she realizes *what a good person he really is inside*—in other words, to stay focused on his needs rather than her own, which is a mistake."[207] I want to underscore that Gandhi's perspective is, following Bancroft, "consistent with the abuser's outlook."

Too often pacifists have said to me, "When you look at a CEO, you are looking at yourself. He's a part of you, and you're a part of him. If you ever hope to reach him, you must recognize the CEO in your own heart, and you must reach out with compassion to this CEO in your heart, and to the CEO in the boardroom." It's revealing that none of these pacifists have ever said to me, "When you look at a clearcut, you are looking at yourself. It is a part of you, and you are a part of it. If you ever hope to help it, you must recognize the clearcut in your own heart, and you must reach out with compassion to this clearcut in your heart, and to the clearcut on the ground." The same is true for tuna, rivers, mountainsides. It's remarkable that pacifists tell me to look at the killer and see myself, while never telling me to look at the victim and see myself: they are telling me to identify with the killer, not the victim. This happens so consistently that I have come to understand it's no accident, but reveals with whom the people who say it do and do not themselves identify (and fear).

So far as psychopaths, Gandhi ignores their first characteristic: a "callous unconcern for the feelings of others." Far worse, he fails to understand that some people are unreachable. He wrote Hitler a letter requesting he change his ways, and was evidently surprised when Hitler didn't listen to him.

His statement also ignores the role of entitlement in atrocity. I can love Charles Hurwitz all I want, I can nonviolently write letters and nonviolently sit in trees, and so long as he feels entitled to destroy forests to pad his bank account, and so long as he is backed by the full power of the state, within this social structure, none of that will cause him to change in the slightest. Nor, and this is the point, will it help the forests. Similarly, so long as men feel entitled to control women, loving them won't change them, nor will it help women.

There's yet another problem with Gandhi's statement, which is that he has made the same old unwarranted conflation of love and nonviolence on one hand, and hatred and violence on the other.

There is a sense in which the last sentence—and only the last sentence—of his statement could be true, with some significant modifications. Instead of saying, "Hatred can be overcome only by love," we could say, "If someone hates you, your best and most appropriate and most powerful responses will come out of a sense of self-love." I like that infinitely better. It's far more accurate, intellectually honest, useful, flexible, and applicable across a wide range of circumstances. But there's the key right there, isn't it? Within this culture we're all taught to hate ourselves (and to identify with our oppressors, who hate us, too, and call it love).

(((

This leads to the next line by Gandhi often tossed around by pacifists: "When I despair, I remember that all through history the way of truth and love has always won. There have been tyrants and murderers and for a time they seem invincible but in the end, they always fall—Think of it, ALWAYS."[208]

You know how there are some people whose work you're supposed to respect because everyone else seems to? And you know how at least with some of these people your respect fades over time, slowly, with each new piece of information that you gain? And you know how sometimes you feel you must be crazy, or a bad person, or you must be missing something, because everyone keeps telling you how great this person is, and you just don't get it? And you know how you keep fighting to maintain your respect for this person, but the information keeps coming in, until at long last you just can't do it anymore? That's how it was with me and Gandhi. I lost a lot of respect when I learned some of the comments I've mentioned here. I lost more when I learned that because he opposed Western medicine, he didn't want his wife to take penicillin, even at risk to her life, because it would be administered with a hypodermic needle; yet this opposition did not extend to himself: he took quinine and was even operated on for appendicitis. I lost yet more when I learned that he was so judgmental of his sons that he disowned his son Harilal (who later became an alcoholic) because he disapproved of the woman Harilal chose to marry. When his other son, Manilal, loaned money to Harilal, Gandhi disowned him, too. When Manilal had an affair with a married woman, Gandhi went public and pushed for the woman to have her head shaved. I lost more respect when I learned of Gandhi's body hatred (but with his fixation on purity, hatred of human (read animal) emotions, and death wish this shouldn't have surprised me), and even more that he refused to have sex with his wife for the last thirty-eight years of their marriage (in fact he felt that people should have sex only three or four times in their lives). I lost even more when I found out how upset he was when he had a nocturnal emission. I lost even more when I found out that in order to test his commitment to celibacy, he had beautiful young women lie next to him naked through the night: evidently his wife—whom he described as looking like a "meek cow"— was no longer desirable enough be a solid test.[209] All these destroyed more respect for Gandhi (although I do recognize it's possible for someone to be a shitheel and still say good things, just as it's possible for nice people to give really awful advice). But the final push was provided by this comment attributed to him: "When I despair, I remember that all through history the way of truth and love has always won. There have been tyrants and murderers and for a time they seem invincible but in the end, they always fall—Think of it, ALWAYS." This is as dismissive as his treatment of his wife and sons. It's as objectifying as his treatment of the young women he used as tests. It's as false as his advice to Jews, Czechs, and Britons. The last 6,000 years have seen a juggernaut of destruction roll across the planet. Thousands of cultures have been eradicated. Species are disappearing by the hour. I do not know what planet he is describing, nor what history. Not ours. This statement—one of those rallying cries thrown out consistently by pacifists—is wrong. It is dismissive. It is literally and by definition insane, by which I mean not in touch with the real physical world.[210]

Further, even if it were accurate—which it absolutely isn't, except in the cosmic sense of everything eventually failing—it's irrelevant. So what if the tyrant eventually falls? What about the damage done in the meantime? That's like saying that because a rapist will eventually die anyway we need not stop him now.

RESPONSIBILITY

Action springs not from thought, but from a readiness for responsibility.

Dietrich Bonhoeffer, killed by the Nazis
for his role in the resistance

WHAT DOES IT MEAN TO BE RESPONSIBLE? HOW CAN ONE *BECOME* responsible?

Maybe it will help to know what the word *means*. Let's take a walk through a dictionary. "Responsible: 1) liable to be called upon to answer; liable to be called to account as the primary cause, motive, or agent; being the cause or explanation; 2) able to answer for one's conduct or obligations; able to choose for oneself between right and wrong."[211]

Please note especially this final phrase: "able to choose for oneself between right and wrong." Gandhi doesn't choose for us. Cops don't choose for us. The lobbyists and politicians who write laws don't choose for us. Those who write books about taking down civilization don't choose for us. I choose for myself. You choose for yourself. It's an awesome and delightful and often scary task. But that's life.

Now, let's follow back the etymology. "Responsible: 1599, 'answerable (to another, for something),' from Fr. *responsable,* from L. *responsus,* pp. of *respondere* 'to respond' (see *respond*). Meaning 'morally accountable for one's actions' is attested from 1836. Retains the sense of 'obligation' in the Latin root word."

Let's keep going back. "Respond: c.1300, *respound,* from O.Fr. *respondere* 'respond, correspond,' from L. *respondere* 'respond, answer to, promise in return,' from *re-* 'back' + *spondere* 'to pledge.' Modern spelling and pronunciation is from c.1600."[212]

To be responsible is to promise in return. The questions become: To whom is this promise made? And in return for what?

This goes to the heart of everything we've been exploring in this book. It is, in some ways, the thread that binds everything together, from the discussion of morality over a glass of water; to the distinctions between civilized and land-based religions; to the conversation at the post office with the clerk who has forgotten he is an animal, and who bought a gun so he can kill himself when civilization falls; to questions of whom or what you most closely identify with; to the understanding that within abusive social dynamics everything is set up to serve the abuser; to what the predator pledges to the prey in return for the sustenance of its flesh.

Questions.

Who feeds you?

What is the source of your life?

To whom do you owe your life?

If your experience—far deeper than belief or perception—is that your food comes from the grocery store (and your water from the tap), from the economic system, from the social system we call civilization, it is to this you will pledge back your life. If you experience this social system as the source of your life, you will be responsible to this social system. You will defend this social system to your very death.

If your experience—far deeper than belief or perception—is that food and water come from your landbase, or more broadly from the living earth, you will make and keep promises to your landbase in exchange for this food. You will honor and keep and participate in the fundamental predator/prey relationship. You will be responsible to the community that supplies you with food and water. You will defend this community to your very death.

When the social system into which you've been enculturated is destroying the landbases on which all life depends, that question of who you are responsible to—to whom you make and keep your promises—makes all the difference in the world.

☾ ☾ ☾

Here are some more questions. To whom will you be called upon to answer? By whom do you *wish* to be called upon to answer?

With every word I write—especially when what I write scares me—I think about these questions. And here are the answers I come to every day. I write for the salmon, and for the trees, and for the soil beneath my feet. I write for the bees, frogs, and salamanders. I write for bats and owls. I write for sharks and grizzly bears. When I find myself wanting to not tell the truth as I understand it to be—when I find the truth too scary, too threatening—I think of them, and I think of what I owe them: my life. I will not—cannot—disappoint them.

And I consider myself answerable to—responsible to—the humans who will come after, who will inherit the wreckage our generation is leaving to them. When I want to lie, to turn my face away from the horrors, to understate the magnitude of what we must do and what we must unmake, to give answers that are not as deep and clear and real as I can possibly comprehend and articulate, I picture myself standing before humans a hundred years from now, and I picture myself answering to them for my actions and inactions. Them, too, I will not—cannot—disappoint.

I can sometimes lie to myself. I could probably even lie to you. But to them—to all of those to whom I hold myself responsible—I could never lie. To them, and for them, I give my brightest, deepest truth.

PACIFISM, PART II

The whole history of the progress of human liberty shows that all concessions yet made to her august claims, have been born of earnest struggle. The conflict has been exciting, agitating, all-absorbing, and for the time being, putting all other tumults to silence. It must do this or it does nothing. If there is no struggle there is no progress. Those who profess to favor freedom and yet depreciate agitation, are men who want crops without plowing up the ground, they want rain without thunder and lightning. They want the ocean without the awful roar of its many waters.

This struggle may be a moral one, or it may be a physical one, and it may be both moral and physical, but it must be a struggle. Power concedes nothing without a demand. It never did and it never will. Find out just what any people will quietly submit to and you have found out the exact measure of injustice and wrong which will be imposed upon them, and these will continue till they are resisted with either words or blows, or with both. The limits of tyrants are prescribed by the endurance of those whom they oppress. In the light of these ideas, Negroes will be hunted at the North, and held and flogged at the South so long as they submit to those devilish outrages, and make no resistance, either moral or physical. Men may not get all they pay for in this world, but they must certainly pay for all they get. If we ever get free from the oppressions and wrongs heaped upon us, we must pay for their removal. We must do this by labor, by suffering, by sacrifice, and if needs be, by our lives and the lives of others.

Frederick Douglass[213]

I'VE HEARD TOO MANY PACIFISTS SAY THAT VIOLENCE ONLY BEGETS VIOLENCE. This is manifestly not true. Violence can beget many things. Violence can beget submission, as when a master beats a slave (some slaves will eventually fight back, in which case this violence will beget more violence; but some slaves will submit for the rest of their lives, as we see; and some will even create a religion or spirituality that attempts to make a virtue of their submission, as we also see; some will write and others repeat that the most disadvantageous peace is better than the most just war; some will speak of the need to love their oppressors; and some will say that the meek shall inherit what's left of the earth). Violence can beget material wealth, as when a robber or a capitalist[214] steals from someone. Violence can beget violence, as when someone attacks someone who fights back. Violence can beget a cessation of violence, as when someone fights off or kills an assailant (it's utterly nonsensical as well as insulting to say that a woman who kills a rapist is begetting more violence).

Back to Gandhi: "We must be the change we wish to see." This ultimately meaningless statement manifests the magical thinking and narcissism we've come to expect from dogmatic pacifists. I can change myself all I want, and if dams still stand, salmon still die. If global warming proceeds apace, birds still starve. If factory trawlers still run, oceans still suffer. If factory farms still pollute, dead zones still grow. If vivisection labs still remain, animals are still tortured.

I have worked very hard to become emotionally healthy, to heal from this culture, my childhood, and my schooling. I'm a genuinely nice guy. But I don't do that emotional work to try to help salmon. I do it to make life better for myself and those around me. My emotional health doesn't help salmon one bit, except insofar as that health leads me to dismantle that which is killing them. This is not cognitively challenging at all.

Next: If you use violence against exploiters, you become like they are. This cliché is, once again, absurd, with no relation to the real world. It is based on the flawed notion that all violence is the same.[215] It is obscene to suggest that a woman who kills a man attempting to rape her becomes like a rapist. It is obscene to suggest that by fighting back Tecumseh became like those who were stealing his people's land. It is obscene to suggest that the Jews at who fought back against their exterminators at Auschwitz/Birkenau, Treblinka, and Sobibór became like the Nazis. It is obscene to suggest that a tiger who kills a human at a zoo becomes like one of her captors.

Related to that is the notion that committing an act of violence destroys your soul. A couple of years ago I shared a stage with another dogmatic pacifist. He said, "To harm another human being irretrievably damages your very core."

I didn't think Tecumseh would have agreed. I asked, "How do you know?"

He shook his head. "I don't know what you're asking."

"How do you know that violence irretrievably damages your very core?"

He looked at me as though I had just asked him how he knows that gravity exists.

I asked, "Have you ever killed anyone?"

"Of course not."

"So you don't know this by direct experience. Have any of your friends ever killed anyone?"

Disgust crossed his face. "Of course not."

"Have you ever even spoken with anyone who has killed someone?"

"No."

"So your statement is an article of faith, unsupported, based not on direct experience or conversations with anyone who would know."

He said, "It's self-evident."

Nice rhetorical trick, I thought. I said, "I have friends at the prison who've killed people, and I'm acquaintances with many others who've done the same. Because I've heard so many pacifists make this claim before, I asked these men if killing really changed them."

He didn't look at me. He certainly didn't ask about their answers.

I told him anyway. "The answers are unpredictable, and as varied as the people themselves. A few were devastated, just as you suggest. Not many, but a few. A bunch said it didn't fundamentally change anything. They were still the exact same people they were before. One said he'd been stunned by how easy it is, physically, to take someone's life, and that made him realize how easily he, too, could be killed. The act of killing made him feel very frightened, he said. Another said it made him feel incredibly powerful, and it felt really, really good. Another said the first time was hard, but after that it quickly became easy."

The pacifist looked like he was going to throw up.

I thought, *This is just reality, man. Reality is a lot more complex than any dogma could ever be. That's one of the problems with abstract principles: they're always smaller and simpler than life, and the only way to make life fit your abstractions is to cut off great parts of it.* I said, "A few told me their answers depended entirely on who they were killing: they regretted some of their murders, but wouldn't take back others even if it meant they could get out of prison. One man, for example, overheard a rapist bragging how he'd made his victim tell him she liked it, and made her beg for more so he wouldn't kill her. The man I spoke with invited the rapist into his cell for a friendly game of chess, and strangled him to death because of what he did to that woman. That murder had felt right at the time, he said, and he knew it would feel right for the next fifteen years till he got out. And one man told me that the thing he was most proud of in his entire life was that he killed three people."

The pacifist shook his head. "That's really sick," he said.

"Let me tell you the story," I responded. "He was a migrant farm worker, from a large Mexican family. He was fifteen. One day he didn't go to the fields but to town. That day three men killed his father. Soon there was a family meeting, and he violated family tradition by interrupting his elders. He insisted that because he was the youngest, the only one without a family relying on him, that he be the one to avenge their father. For the next few years he worked hard to establish a business that would support his mother later on, and when the time came he killed the three men who had killed his father. The next day he went to the police station and turned himself in. He's now serving life."

"He should have let the law handle it."

"I cannot blame him for his actions. They were human." I paused a moment, then said, "And I have known others who killed because they were human. I have known women who killed their abusers. They had no regrets. Not one. Not ever."

"You cannot sway me," he said. "They should let the law handle it."

"The law," I replied. "The law. Let me tell you another story. A woman killed her mother's boyfriend, who had battered her mother for years and finally murdered her mother. And—surprise of all surprises—the district attorney refused to charge him with murder. I suppose this was because women aren't people whose lives actually count. So the woman did a sit-in at the DA's office. For three days, she just kept saying over and over 'You're going to call it murder.' The DA finally had her arrested for trespassing. Having gotten no satisfaction from the system, she bought a gun, tracked the boyfriend down and shot him dead. Because of her sit-in stunt, the lawyers were able to argue temporary insanity. She served two years in prison and didn't regret a single day of it."[216]

The pacifists who say that fighting back against those who are exploiting you or those you love destroys your soul have it all backwards. It is just as wrong and just as harmful to not fight back when one should as it is to fight when one should not. In fact in some cases it may be far more harmful. The Indians who spoke of fighting, killing, and dying—and who fought, killed, and died—to protect not only their land but their dignity from theft by the civilized understood this. So did Zapata. So did the Jews who rose up against the Nazis. Of those who rose up against their exterminators at Auschwitz/Birkenau, and who were able to kill seventy SS, destroy one crematoria, and severely damage another, concentration camp survivor Bruno Bettelheim[217] wrote that "they did only what we would expect all human beings to do: to use their death, if they could not save their lives, to weaken or hinder the enemy as much as possible; to use even their doomed selves for making extermination harder, or maybe impossible, not a smooth running process. . . . If they could do it, so could others. Why didn't they? Why did they throw their lives away instead of making things hard for the enemy? Why did they make a present of their very being to the SS instead of to their families, their friends, even to fellow prisoners; this is the haunting question."[218] Bettelheim also wrote, this specifically of Anne Frank's family, "There is little doubt that the Franks, who were able to provide themselves with so much, could have provided themselves with a gun or two had they wished. They could have shot down one or two of the SS men who came for them. There was no surplus of SS men. The loss of an SS with every Jew arrested would have noticeably hindered the functioning of the police state."[219] Bettelheim—and he is joined by many in this—states explicitly that such actions could most likely have slowed the extermination process. Ward Churchill responds, "It should be noted that similar revolts in Sobibór and Treblinka in 1943 were even more effective than the one at Auschwitz/Birkenau a few months later; Sobibór had to be closed altogether, a reality that amplifies and reinforces Bettelheim's rather obvious point."[220]

Bettelheim comments, in words he could have written about us as we watch our TVs and wait for the end of the world, "The persecution of the Jews was aggravated, slow step by slow step, when no violent fighting back occurred. It may have been Jewish acceptance, without retaliatory fight, of ever harsher discrimination and degradation that first gave the SS the idea that they could be gotten to the point where they would walk into the gas chambers on their own. Most Jews who did not believe in business-as-usual survived the Second World War. As the Germans approached, they left everything behind and fled to Russia, much as many of them distrusted the Soviet system. . . . Those who stayed on to continue business-as-usual moved toward their own destruction and perished. Thus in the deepest sense the walk to the gas chamber was only the last consequence of a philosophy of business-as-usual."[221]

Bettelheim also writes, in words that are just as applicable, "Rebellion could only have saved either the life they were going to lose anyway, or the lives of others."[222] And, "Inertia it was that led millions of Jews into the ghettos the SS had created for them. It was inertia that made hundreds of thousands of Jews sit home, waiting for their executioners."[223]

Ward Churchill sums up Bettelheim's description of this inertia, which Bettelheim "considers the basis for Jewish passivity in the face of genocide, as being grounded in a profound desire for 'business as usual,' the following of rules, the need to not accept reality or to act upon it. Manifested in the irrational belief that in remaining 'reasonable and responsible,' unobtrusively resisting by continuing 'normal' day-to-day activities proscribed by the nazis through the Nuremberg Laws and other infamous legislation, and 'not alienating anyone,' this attitude implied that a more-or-less humane Jewish policy might be morally imposed upon the nazi state by Jewish pacifism itself."[224]

Bettelheim observes that "we all wish to subscribe to this business-as-usual philosophy, and forget that it hastens our own destruction," and that we have a "wish to forget the gas chambers and to glorify the attitude of going on with business as usual, even in a holocaust."[225]

But remember, the Jews who participated in the Warsaw Ghetto uprising, even those who went on what they thought were suicide missions, had a higher rate of survival than those who did not fight back. Never forget that.

Instead of saying, "If we fight back, we run the risk of becoming like they are. If we fight back, we run the risk of destroying our souls," we must say, "If we do not fight, we run the risk of not just acting like but *becoming* slaves. If we do not fight back, we run the risk of destroying our souls and our dignity. If we do not fight back, we run the risk of allowing those who are exterminating the world to move ever faster."

WHAT IT MEANS TO BE HUMAN

We must all share the burden of responsibility. . . . I could never look the wives and children of the fallen in the eye if I did not do something to stop this senseless slaughter.

Count Claus von Stauffenberg, killed by the Nazis
for his part in the resistance[226]

I THINK THAT ABOVE I UNDERSTATED THE PROBLEM. THE WORLD is being killed. It is not true that if we do not fight back we run the *risk* of destroying our own souls and our dignity, and we run the *risk* of allowing those who are exterminating the world to move ever faster. At this point, with all excuses long since exhausted, if we do not fight back we *destroy* our own souls and our own dignity, and far worse, we do *allow* those who are exterminating the world to move ever faster.[227] No risk about it.

☾ ☾ ☾

I give a talk. Afterwards four of us—myself and three women—go get something to eat. We sit at a dark pizza place—one of the few places still open at this time of night—and I marvel at how much cheaper pizzas are in the Midwest than in California.

We drink water while we wait for our food. We talk. A lull in the conversation turns into a longer silence. We can all tell that one woman is thinking, and we don't want to interrupt her. Finally she says, "I used to know a professor where I went to college who vivisected rhesus monkeys. He's a Jewish man whose family fled to the United States in the 1930s to get out of Nazi Germany. Throughout the 1960s he was a strong supporter of civil rights, and worked for all sorts of humanitarian causes. At the same time he was an extremely successful and respected researcher at the university. In this research he did horrible things to the monkeys. I remember him telling me, 'A terrible enemy of mine once accused me of being the Dr. Mengele of the monkeys. How could he say such a ghastly thing?' Even more than me remembering what he said I remember the utter disbelief on his face. He had no idea what this 'enemy' was talking about."

One of the women responds, "Do you believe that?"

The first woman says immediately, "No, not for a second. He knew exactly what the other person meant. He was bullshitting me to make himself feel better."

The second woman responds again, "Why did you say he had no idea, then?"

Just as quickly the first woman says, "I was bullshitting myself to make myself feel better for not killing him for what he did to those monkeys."

None of us say anything.

She continues, "The more I think about it, the more I believe his disbelief was actually over the fact that this other person had the gall to compare him to Mengele out loud, not that it wasn't the truth."

The third woman says, "His disbelief was that someone had broken the silence."

The second: "We're all taught very well, aren't we?"

The first again: "This particular researcher was always upset by the things he 'had' to do: the horrible surgeries, and the gross results that followed. For instance, he and his colleagues would do brain surgeries, even though they were not surgeons—they thought they were so entitled. And sometimes, he admitted out loud, the surgical instruments would slip, just a little here and there, and sometimes the monkeys would lose the use of their limbs because a portion of their brains had been destroyed."

The second: "Someone should slice his brain, see how he likes it."

The first: "The stories were atrocious. I hated that he told them to me, and he knew it, but he did it anyway. I don't think he could help himself. He talked constantly, day in and day out, about what he was doing, as well as what the Jews went through, all of the atrocities, weaving the stories in and out of each other. It was amazing. He was a smart man, so surely he knew full well the glaring connection. I think what he couldn't believe was that someone who knew him, this enemy who knew his own Jewish past, not to mention all the 'good' work he had done over the years, had the awesome nerve to accuse him of what was so obviously a truth he wanted others to keep silent about, even though in his own stories, he constantly exposed them. Unlike some abusers who can't ever admit what they've done, I always thought he knew. His stories gave him away."

The second woman says, "I think he was telling those stories again and again because he wanted everyone's approval. Each time he said it and no one stopped him or even spoke to him, he received implicit permission to continue."

We all nod. No one says anything for a moment. I take a sip of water, then say, "If this were a novel, and the monkeys he tortured were figments of some writer's imagination, I suppose the knowledge deep in his soul of the crimes he has committed would somehow be enough punishment, but . . ."

The second woman: "No, it's not. It's not nearly enough."

We all know what the others are thinking.

The first woman continues, "Here's another thing I remember. One day he went slumping through the department muttering under his breath about how awful one of the female monkeys he'd just been 'working with' had been because she'd thrown her shit at him. He couldn't understand that either."

She pauses, and this time no one speaks.

She says, "He knew he horrified me. One day he came tromping through the halls waving a bloody lab jacket and saw me standing there. He barked at me, 'You'll be happy to know I'm no longer using rhesus monkeys. The program is *over*.' He marched into his office and slammed the door."

What can we say?

She says, "I'll bet the dead monkeys were glad the 'program' was over, too."

I look away.

She says, "And by the way, in the college's old physiology/anatomy department before they built their new buildings, you could smell the burning of the research animals twice a day—at 10 a.m. and 3:00 p.m.—and that included the rhesus monkeys."

Finally the second woman says, "That doctor has to be stopped before he begins another program. He cannot be allowed to continue."

The third woman begins to speak quickly, "Along with every other research scientist in that department and all of the graduate research students who so desperately kiss ass and do whatever is needed to assure themselves of a good career continuing that behavior, and along with every department that's like that in this country and any country, and all of the research corporations that do that. But instead they get rewards and money and have things named after them."

The second woman: "For now."

"What?" responds the third. "Oh, yes. I understand."

The first woman says, softly, "Last I heard he had cancer. But he's probably getting pain killers, which is more than he managed to do for his monkeys."

❮ ❮ ❮

When I think of resilience, I think of a stream near my home where tiny fry of coho salmon swim above a bottom clogged by sandy sediment from logging. I think of the pond outside my home where the black eggs of northern red-legged frogs—disappearing, too—hang suspended in jelly clinging to underwater branches, and I think of the tadpoles who survive UV from ozone depletion, survive pesticides, survive predators to hop, tiny as dimes, onto the shore and into the forest. I think of aromatic Port Orford cedars—disappearing like the rest— fighting against an introduced disease (and even moreso against an introduced culture, introduced timber corporations, and introduced chainsaws). And I think of American chestnuts, whose crowns once grew one hundred feet across, felled also by an introduced illness: young trees rise up, die, then sprout again from the roots. Where does that pool of strength come from—for chestnuts, for all of them? What is that rootstock of resilience from which, given a chance, these others regenerate?

When I think of resilience, I remember the determination I once saw in the eyes and in the set jaw of a child who'd vowed when he grew up he wouldn't strike his son or daughter as his father had struck him. I think of the open tears of fright from a grown woman taken back by an innocent gesture to a time in her childhood when her father could and would have killed her had she not slipped from his grasp, and I think of how she has successfully fashioned a creative life from the wreckage of her childhood. I think of the pride with which another woman—this one beaten and raped by her father as a child—states that she has never struck nor even shouted at her sons.

When I think of resilience I wonder where all of this strength comes from, and I wonder how people so violated—stabbed in the arms and chest with a steak knife, or beaten with ropes, or starved, or forced by fist to finish plate after plate after plate of unwanted food (and these are just people I know personally)—can sometimes grow up to live lives marked by grace and compassion.

My own first experience of resilience—or rather of conditions that called it forth, then shaped it to my body and emotions, made it necessary—came early, from the physical and sexual violence my father inflicted upon us.

One of the ways I survived was by pretending nothing was happening, nothing was amiss. I had a deal with my unconscious: because I was spared the beatings, I made myself believe that if I didn't consciously acknowledge the abuse, it wouldn't be visited directly upon me. My father's first visit to my bedroom didn't abrogate the deal. It couldn't, because without the deal I couldn't have survived. In order to maintain the illusion of control in an uncontrollably painful situation, that is, in order to stay alive, the events in my bedroom necessarily didn't happen. His body behind mine, his penis between my legs, these images slipped in and out of my mind as easily and quickly as he slipped into and out of my room.[228]

Of course it's simply not possible to survive such trauma. The pain was too strong, the pressure too deforming, for me to bear. I repeatedly erected psychological and emotional walls to keep out this relationship too terrifying to tolerate, and just as repeatedly these walls were smashed down in the next wave of violence, only to be re-erected by a child desperate to keep some parts of himself safe, separate from the violence, and thus untainted by terror.

❨ ❨ ❨

When I was a child, I used to climb out my bedroom window at night to lie beneath the stars. The tiny points would get bigger and bigger as they rushed closer to me, or I to them, and soon I would hear their voices. They would say to me that none of this was my fault, that none of this was right, that things were not supposed to be this way. They told me they loved me. Had they not told me all of this, I would have died.

❨ ❨ ❨

My childhood, while dramatic, wasn't unusual. We've all seen the numbers. According to the United States Centers for Disease Control, just within this country a half million children are killed or seriously injured by their parents or guardians each year.[229] Studies elsewhere show that nearly one in three girls and one in six boys are sexually abused by the time they're eighteen.[230]

❨ ❨ ❨

I often spent afternoons by myself in the irrigation ditch that ran behind our house. I'd catch crawdads and garter snakes, or climb up the banks to lie on my belly and watch the comings and goings of ants in their hills. I got to know and love the songs of meadowlarks and robins, and the song of the water in the ditch, its sighs and whispers and gloops as it slid around branches and across reeds. Sometimes I came with friends, sometimes with siblings. But my father never came here, nor did I bring him with me.

❨ ❨ ❨

There are those who pass on to others the abuse they received—I know many people like this, as I'm sure you do—but there are those also who do not. Despite the seeming impossibility of survival, there are children—and adults— who do not accept, wear, and pass on this mantle given to them by those who would initiate them into this lineage of abuse. In fact it happens all the time. I've come to know many people who've survived the unsurvivable, and whose lives are now full of joy. Indeed, because many of them have had to struggle so hard to find, allow, and realize love in their lives, their appreciation of this is far more profound, layered, and textured than it might be for many who have never been forced to feel the dreadful and grinding ache of terror deep in the marrow of their bones. When and if those formed in such a crucible do achieve some form of

hard-won emotional connectedness—with other humans, nonhumans, the natural world, music, art, writing, or even with every breath they take—they often
find themselves then able to feel passion more acutely, and to savor those connections with a strength as unfathomable to those for whom these connections
are first nature—that is, transparent—as are the original traumas themselves.

Given the near-ubiquity of abuse within our culture—and I'm talking not only
about the deformations of child abuse, but of coercive schooling, the wage economy requiring people to waste lives working jobs they'd rather not do, the
trauma of living in a world being destroyed before our eyes—the question
becomes, what helps some people to open out after having been subject to
abuse, and what causes others to shut down? In other words, what causes or
allows resilience?

❰ ❰ ❰

I often walk through the forests where I live. *Walk* might not be the best word,
because the forests are so thick I crawl along game trails, snaking my way
between branches and beneath clinging vines. The forest rewards me. Last week
I saw a red-legged frog the size of a small dinner plate, and this week the biggest
pile of bearshit I've ever seen, dark blue and signaling a diet of berries, as well
as once again answering in the affirmative the age-old question of whether the
bear does in fact shit in the woods. Once, I stumbled across a spot where the bear
beds down, and saw tufts of black hair twining with grasses flattened outward
beneath a big downed log. I was far from any roads, and lost beyond all hope.
This is where she sleeps, I thought. This is her place of refuge.

❰ ❰ ❰

All things need places where they are allowed to be who they are, places where
they can—like the roots of the chestnut trees—derive sustenance and strength
from their surroundings. Terror and exploitation do not engender growth, and
it is especially true that those normally subject to these need refuges where they
can regenerate in peace.

I knew all of this as a child. Everyone does. Thus my relationship to the stars.
Thus my relationship to the creatures in the irrigation ditch. Thus—and this may
seem odd, but I'd wager this is true for many others thus violated—my relationship to places within my own body that remained safe, places my father
could not touch.

It is possible to look back on one's history, no matter how horrible, and find places of relative safety, where fear was never allowed to permeate. Those places can teach us, if we let them, that as well as knowing fear we can know—as I learned from the ditch, from the stars—safety, peace. We can know what it feels like to not have our guard up, to experience a world where the strong do not exploit the weak, where dogs do not eat dogs. This allows us not only to breathe, but to learn that openness feels different from defendedness, that relationships can be pleasing and beneficial. The key, then, to resilience, is to find or remember those places of refuge, and build out from there. Because I knew that peace exists, and because I experienced the difference between peace and abuse, I was able to migrate, slowly, toward openness, at first only toward the creatures in the ditch, and toward the stars, and then toward others equally nonthreatening, and then toward other people.

Perhaps even more important than providing me a template, those places provided me with the understanding that the pain I suffered was neither natural nor inevitable, that there are other ways to be. This understanding is crucial to resilience, and in fact to the continuation of life, because if all of life consisted of abuse and exploitation, what would be the use in going on?

❨ ❨ ❨

We are living in the time of industrial capitalism's greatest ascendancy. One can buy a Big Mac and a Coke ("the real thing") in nearly every nation of the world. Even more telling of our way of living's temporary stranglehold on how humans live is the fact that the world has even been carved into nations in the first place. And even more telling than this is that we do not find this startling. All of this means that there are few places anymore (inside or out) safe from civilization's reach. In the north, polar bear fat is contaminated with dioxin, and their fate is sealed by global warming: wild populations will probably be gone within another couple of generations. In the south, ice caps melt quickly enough to make the most stolid of scientists who study them weep. Trawlers capable of "handling" three hundred and forty-four tons of fish per day spread their nets more than a mile long, scraping the sea floor, destroying all life—fish, birds, other animals—in their paths, tossing much of it—called by-catch—back overboard, dead. Trident submarines patrol the oceans, too, first-strike weapons capable of launching twenty-four missiles simultaneously, each missile containing up to seventeen independently targeted nuclear warheads, each warhead ten times more powerful than the bomb that incinerated Nagasaki, each

warhead capable of traveling 7,000 miles, meaning that just one of these subs—and the United States has twenty-two—could effectively eliminate 408 cities across an entire hemisphere. Coral reefs will soon be dead. Glaciers melt around the world. Mount Everest is littered with tons of trash. Ninety-seven percent of North America's native forests have been cut. Human languages disappear as quickly as so many dreams, as culture after culture is consumed by civilization's voracious way of "living." Where is safety?

The future resides in these places of refuge, these places of freedom, small as the inside of our hearts and minds and bodies, and big as the deepest bottom of the oceans where trawlers' nets cannot reach. Without freedom, without these places that are free of terror and exploitation where we can develop comfortable and nurturing relationships—to streams, to ponds, to pieces of ground, to stars, to human beings, to art, to pets, to music, to *ourselves*—there can be no resilience. For resilience *is* relationship, to other and to self, and grows naturally where relationships are allowed to flourish. Salmon in cold streams free of sediment grow to reinhabit other streams. Port Orford Cedars free of the disease grow as well to reinhabit their former territories. And even parts within us that we can by any means keep free of the taint of terror can provide reservoirs of resilience and help us remember what it means to be human.

❨ ❨ ❨

To reach the middle of the ocean, those in power must have oil to run their ships and metals to build them. To deforest, they must have gasoline to power their chainsaws and metals to build the saws. No oil, and the ships have no capability to reach the center of the ocean, which means that the oceans can begin to live again. No oil, and chainsaws sputter—actually they don't even do that—and forests must only contend with local use. If those in power have no oil, they cannot rebuild the dams we remove. Part of taking down civilization is the destruction of the oil economy.

Of course in the longer run we must remember how to live in place with what the land willingly gives, but before we can even seriously think about doing that we must remove this threat to the entire planet. To do otherwise is the equivalent of trying to decide how we shall live next summer as we ignore the upraised butcher's knife in front of us.

The first step in taking down civilization is to realize in our own hearts and minds that the dictionaries lied to us, that civilization is not "a high stage of social and cultural development,"[231] or "a developed or advanced state of human society."[232] I am not talking about convincing some hypothetical mass movement of people, which will not happen within this culture. As I said earlier, when fathers are raping daughters, when lovers are beating those they purport to love, there is no hope for the salmon. I am talking about me realizing this in my own heart, and you realizing it in yours.

The next step in taking down civilization is finding a few other people who feel the same. It is hard enough to take on this entire abusive social structure—where everything is set up to protect the abusers—without having to fight our friends as well. It can be lifesaving to have friends who will say, and mean, with courage, love, and determination glistening in their eyes, "Yes, it is unacceptable to me that salmon be exterminated from this river. I will do what it takes to save them." I am talking about small groups of people—small enough to know and trust each other with your very lives—coming to this understanding, and beginning to act upon it.

Next, taking down civilization means understanding that we are in the midst of a war, that war was long ago declared on the natural world, including on humans, and that we must fight back. I am not speaking metaphorically.

Next, taking down civilization means understanding that very few wars are won on actual battlefields. Economic production allows governments to win wars. And the destruction of the means of economic production causes them to lose them. Recall the U.S. military analysis that determined that World War II attacks on German railroads were "the most important single cause of Germany's ultimate economic collapse."[233] This is not to belittle the sacrifices of the soldiers who beat back the Nazis at Stalingrad and elsewhere, but to remind people of the truism that an army fights on its stomach. This includes an army of consumers.

Taking down civilization means acting. It means committing ourselves to defending our landbases, which means committing ourselves to removing the economic and transportation infrastructures, which means committing ourselves to hitting them, and hitting them again, and again, and again. This may be, as we shall see in a few pages, easier than it seems.

Once the economic and transportation infrastructures have been taken down, our fights over how to live sustainably in our own landbases will be local, and face to face, which means they will be human, which means they are eminently winnable, through discourse or violence or some other means.

❨ ❨ ❨

Bringing down civilization means depriving the rich of their ability to steal from the poor, and it means depriving the powerful of their ability to destroy the planet.

❨ ❨ ❨

I feel kind of silly trying to reduce my articulation of what it means to bring down civilization to a page. It is what this entire book is about. Indeed, it is the sum (and more) of all of my books.

❨ ❨ ❨

I said this before. It bears repeating. I have no interest in spiritual purity. I want to live in a world with wild salmon and old growth forests and oceans full of wild fish and mothers who do not have dioxin in their breast milk, and I will do whatever it takes to get there.

PACIFISM, PART III

The primary purpose of everything we do must be to make this society increasingly unmanageable. That's key. The more unmanageable the society becomes, the more of its resources the state must expend in efforts to maintain order "at home." The more this is true, the less the state's capacity to project itself outwardly, both geographically and temporally. Eventually, a point of stasis will be reached, and, in a system such as this one, anchored as it is in the notion of perpetual growth, this amounts to a sort of "Doomsday Scenario" because, from there, things start moving in the other direction—"falling apart," as it were—and that creates the conditions of flux in which alternative social forms can really begin to take root and flourish.

Ward Churchill[234]

THE NEXT ARGUMENT THROWN OUT BY PACIFISTS IS THAT WE MUST never use violence, because if we do the mass media will distort our message.

I'm presuming that the people who say this have never actually *read* a newspaper or watched the news, because an unstated and unfounded premise here is that the purpose of the capitalist media is to tell something that resembles the truth. It's not. Well, maybe I should correct myself. The purpose of the capitalist media *is* to tell something that *resembles*—is a toxic mimic of—the truth. The truth itself? That comes only often enough to keep us guessing. The mass media distorts our message anyway. What do you think the mass media is *for?*

Two examples. The other day a translator who works for the U.S. occupying troops in Iraq was killed. This happens often. But this time the *San Francisco Chronicle* featured an account of the woman's life and death on the front page. The article began: "Rwaida's death hit everyone hard. Partly because she was a buddy they had known for a long time. Mostly, because she was an innocent, brutally slain for the simple reason that she worked for the Americans." We learn later in the article that she was more than an interpreter to the occupiers. She once pushed a U.S. soldier out of the path of a bullet. She was a "buddy," said one soldier. Another said she was "one of us." Soldiers flirted with her, asked her to marry them, to which she responded, in true American fashion, "How much money you got, eh?" The occupiers have now hired another interpreter, a twenty-two year old woman the soldiers call Nadia. Nadia says, "I am not afraid. I know this is my duty and that I should do that." The article states that she appreciates the money the occupiers give her, but more important than money is the fact that "Morals should be the most important thing for everyone." The reporter for the U.S. corporate press lets her mouth the real moral of this story: "I have great respect for the Christian people. They respect God and I love God. Too many Muslims hurt each other."[235]

What just happened? I can't speak for you, but I feel like I just got mindfucked. Of course, that's what the capitalist press is supposed to do. There are three main avenues by which this writer just attacked us. The first is that he chose to write an article about this woman, while ignoring the many thousands of women killed by U.S. troops. To give each of these women the same attention would take more space than the newspaper has. The second is that the journalist opens his piece by telling us that this woman was "innocent," then goes on to describe the actions of a collaborator. Imagine the Iraqi military invading the United States.[236] Imagine them occupying this country. Imagine a woman from Phoenix, Arizona who goes to work for these occupying soldiers, who flirts with them, who is their buddy. She would receive the treatment that collaborators of all times have received, which is the treatment this woman received. And the occupiers would call her an innocent hero. The third avenue of attack is the final sentence: this woman was not killed because she was a collaborator, but instead because Muslims have a nasty habit of hurting each other.

My point? All writers are propagandists. And lest you think that the story above merely makes the pacifists' point, that if the Iraqis would simply lay down their arms the press would tell their story fairly, let's take a look at an editorial in the following Sunday's *Chronicle*. The editorial is entitled "Biggest pests of all—agricultural biotech opponents," and says that those who oppose genetic engineering have an "anti-social agenda," and that "they should be held accountable." Check out the author's first sentence: "California is under attack by terrorists, six-legged ones: glassy-winged sharpshooters, which are leaf-hopping insects that are among the state's most insidious agricultural pests."

Okay, so what just happened this time? Mind-fuck again, of course, but what did we expect? I have been told by pacifists that I need to watch my rhetoric or environmentalists will be labeled terrorists. I have news for them: if insects are labeled terrorists then it doesn't much matter what we say or do. If we oppose economic production, even by sucking the sap from grapevines, we are going to be called names.

"The press is the hired agent of a monied system," wrote Henry Adams, "set up for no other purpose than to tell lies where the interests are concerned."

They're going to lie about us no matter what we do, so we may as well do what we want.

❨ ❨ ❨

Another night, another talk. Another pacifist plagues me like a biting black fly. He says, "Every act of violence sets back the movement ten years."

I respond, "How do you know that?"

He stares at me, this time as though I've asked him to prove not the existence of gravity, but of air. He shakes his head.

"What evidence do you have?"

Still shaking his head, "I don't . . ."

"It's an article of faith. You can't have any evidence to support your position because no environmentalists or animal rights activists have yet committed any acts of violence against a human being, which means they *can't* have set the movement back."[237]

"They've burned SUVs."

"That's not violence."

"It still sets the cause back."

"How?"

"It harms public opinion."

"Okay," I say. "Let me know if this is how it goes. So long as activists behave themselves and follow the rules—set up by those in power—then some theoretical mass of people will be willing to listen to them, maybe even agree with them, and possibly even send them money."

"Let's leave money out of this."

I continue, "But if someone breaks the rules—set up by those in power—then the great mass of fence-sitters will write good activists like you off as lunatics. Then you'll have to be good for another ten years to make up for the lost goodwill, right?"

"I don't like how you're spinning it, but it's okay."

I keep going. "We have to follow the rules of polite discourse in order to be heard. But why do these rules apply only to us? Why is it that when the people and companies and institutions we're opposing commit violence or otherwise break the rules of polite discourse it doesn't set them back ten years? Further, if we only act in ways that are acceptable to those who are benefiting from the exploitation in the first place, we will never be able to stop the exploitation."

There were plenty of other questions that night, so I moved on, but had I more time I would have said more. I would have said for the thousandth time that all life is circumstantial, and that some acts of violence may set some movements back some number of years, and that some acts of violence may move them forward. Some acts of non-violence may set some movements back, and some may move them forward. Some failures to act at the right time with the right tactic (violent or nonviolent) may set movements back or move them forward. The trick is knowing when and how to act. Well, that's the first trick. The *real* trick is kicking aside our fear and acting on what we already know (because, truly, we depend on those around us, and they are dying because they depend on us, too).

I would have talked about resistance movements in Latin America, Asia, and Africa where violence helped throw off overt colonialism. I would have talked about resistance by indigenous peoples. I would have talked about violence by abolitionists, and I would have mentioned that Harriet Tubman carried opiates with her, and she carried a gun. The opiates were to drug the people she was transporting in case they got too frightened, and the gun was to shoot them if they wouldn't stop screaming. Did Harriet Tubman set "the movement" back ten years?

I would also have said that the notion that some act could set some "movement" back implies that the "movement" is actually accomplishing something in the first place. That's a doubtful proposition, at best.

Next, I would have recalled where I've previously heard this sentiment of fearing that more militant actions will threaten one's own resistance, which is in accounts of discussions between death camp inmates about whether or not they should try to escape. There are those who wish to make things as comfortable as they can within the confines of the razor wire and electrified fences, and those who want to break away entirely. Of course those who want to break away will "set things back" for those whose goals are limited to gaining a sliver of soap and an extra potato in their broth.

Trauma expert Judith Herman describes the "constriction in initiative and planning" that often takes place among captives: "Prisoners who have not been entirely 'broken' do not give up the capacity for active engagement with their environment. On the contrary, they often approach the small daily tasks of survival with extraordinary ingenuity and determination. But the field of initiative is increasingly narrowed within confines dictated by the perpetrator. The prisoner no longer thinks of how to escape, but rather how to stay alive, or how to make captivity more bearable. A concentration camp inmate schemes to obtain a pair of shoes, a spoon, or a blanket; a group of political prisoners conspires to grow a few vegetables; a prostitute maneuvers to hide some money from her pimp; a battered woman teaches her children to hide when an attack is imminent."[238] And environmentalists work as hard as they can to (temporarily) save some scrap of wilderness.

Now, I certainly have nothing but respect for those environmentalists working to save scraps of wilderness (something I've done myself) and the same is true for others of the abused as they try to hide their children, hide some money, grow vegetables, or sneak a spoon, and given the choice I'd prefer to be slightly more comfortable as a prisoner rather than less. But I'd rather not be a prisoner at all.

An act of violence will set the movement back ten years? Good, we only have another several thousand years to go, then.[239] The existence of an environmental movement at all is an acknowledgement that something is desperately wrong with the culture. A healthy culture would have no need, any more than it would need battered women's shelters or drug and alcohol rehabilitation centers, or those sanctuaries I mentioned that are refuges from atrocity. And ultimately I don't give a shit about the health of the movement, any movement, anyway. I care about the health of the landbase.

Reading Herman's passage suddenly helped me understand the desperate vehemence of some dogmatic pacifists. I certainly understand and have an appreciation for differences of opinion, and I've repeatedly described my support for and participation in nonviolent resistance, yet so often when I have spoken with pacifists I have encountered an absolute refusal to even enter into reasonable discussion about the use of violence. Recall the argument of the pacifist on stage: "Violence schmiolence." This man was not stupid, as this comment makes him seem.

But now I understand it. And I understand, also, a primary reason we are so terribly ineffective in our attempts at resistance. It is because we are captives of this culture who have not been entirely "broken," but have been traumatized to the point that our "field of initiative" has been "increasingly narrowed within confines dictated by the perpetrator." Judith Herman describes this process of narrowing in words that will surely resonate with many of us: "The constriction in the capacities for active engagement with the world, which is common even after a single trauma, becomes most pronounced in chronically traumatized people, who are often described as passive or helpless. Some theorists have mistakenly applied the concepts of 'learned helplessness' to the situation of battered women and other chronically traumatized people. Such concepts tend to portray the victim as simply defeated or apathetic, whereas in fact a much more complex inner struggle is usually taking place. In most cases the victim has not given up. But she has learned that every action will be watched, that most actions will be thwarted, and that she will pay dearly for failure. To the extent that the perpetrator has succeeded in enforcing his demand for total submission, she will perceive any exercise of her own initiative as insubordination."[240]

Now reread this passage, substituting the word *activist* for *victim*. Consider especially the sentences, "In most cases the activist has not given up. But she has learned that every action will be watched, that most actions will be thwarted, and that she will pay dearly for failure." There we have the psychology of most environmental activism in two sentences.

And consider Herman's next passage in light of another complaint of pacifists, that if we commit an act of violence, the state will come down hard on us. She states, "Prolonged captivity [and several thousand years of civilization certainly counts as prolonged] undermines or destroys the ordinary sense of a relatively safe sphere of initiative, in which there is some tolerance for trial and error. To the chronically traumatized person [and to the civilized, to the slave] any action has potentially dire consequences. There is no room for mistakes. Rosencof describes his constant expectation of punishment: 'I'm in a perpetual cringe. I'm constantly stopping to let whoever is behind me pass: my body keeps expecting a blow.'"[241]

And the blows will come, and keep coming, till civilization is no more. It has shown itself to be insatiable, implacable, the demand for submission ultimately total. Indigenous Quichua in the village of Sarayacu, Ecuador recently refused an offer of $60,000 for an oil company to drill on their land. A spokesperson said, "We are fighting not only for Sarayacu, but for all Amazon communities. Petroleum development has been a disaster in Ecuador, generating environmental, social and cultural crises, and ultimately causing the extinction of indigenous peoples. We want to maintain our way of living, free of contamination, in harmony with nature." The response by the Ecuadoran Minister of Energy Eduardo Lopez was, as one reporter put it, to announce a "total opening of the southern Amazon to oil exploitation and to describe organizations that oppose the policy as undesirable. He also said he preferred to come to an agreement with Sarayacu 'before employing force.'"[242]

Here is the pattern, as clear as it is every other time. If you let us destroy your community and your landbase, we will give you money. If you do not accept the money, we will destroy you as well.

❲ ❲ ❲

WHY CIVILIZATION IS KILLING THE WORLD, TAKE TWENTY-FOUR. It's not just oil companies. It's the whole damned culture. Here is a headline from today's *San Francisco Chronicle*: "Ecuador free-for-all threatens tribes, trees: Weak government lets loggers prevail." You can guess the content. In case you can't, one sentence is all it takes to make it clear: "The Ecuadoran rain forest has long attracted rubber-tappers, oil companies and timber concerns backed by a federal government eager to exploit the natural riches of the Amazon."[243] The government is strong when it comes to backing corporations, and weak when it comes to stopping them. If corporations are going to be stopped from destroying the world, it is up to us to stop them.

❲ ❲ ❲

A story, and then a study.

The story, unfortunately true, is told here in the words of an extraordinary forest activist named Remedy: "Humboldt County, California: Five Mattole Forest Defense activists were arrested early Wednesday morning after serving a subpoena to Pacific Lumber's head of Security. Carl Anderson, an ex-sheriff's deputy, who has led the timber corporation's face-to-face opposition to activists in the woods for over a decade, was served a subpoena to appear at the infamous pepper spray trial in San Francisco, which starts September 7. The case dates back to 1997, when non-violent forest activists were subjected to torture, in the form of pepper spray swabbed in their eyes [by Humboldt County Sheriffs, the same department with the near-perfect zero percent rate of going after rapists. The Sheriffs videotaped themselves swabbing pepper directly onto the eyeballs of activists who had locked down in the office of a Congressman deeply beholden to Pacific Lumber.] The pepper spray victims, who are the plaintiffs in the case, subpoenaed Anderson to testify at trial.

"Both the service of the subpoena, and the subsequent assault, took place on state park land near the entrance to PL property. The activists documented proper service of the subpoena with a video camera, as they have learned the hard way that PL representatives have ignored legal subpoenas in the past. Activists have been threatened and unlawfully detained during previous attempts to serve legal documents in another pending case involving PL.

"Shortly after the subpoena was served, activists were met by a truck from Columbia Helicopters, which is contracted by PL to stack logs from clear-cuts. Activists reported the truck driver was aggressive with his driving, pushing into activists' bodies to get through them. When another truck appeared, this one a personal red pickup truck, the driver jumped out and assaulted the woman with the camera, which was held around her neck by a strap. After he wrestled with the woman, throwing her to the ground, he began choking her with the strap as he attempted the take the camera from her. She tried to protect the camera by wrapping her body around it, but he was determined to take it. He pulled out a knife and eventually cut the strap, but not without cutting the woman in the process.

"The sheriffs arrived sometime thereafter, but refused to take reports from activists. They did, however, take notes on the report offered by Carl Anderson. The activists attempted to notify the deputies that they had been assaulted, to which Anderson reportedly joked to the Officer Carla Bolton, 'What are you going to do, arrest me?' Given the number of times activists have witnessed Anderson giving orders to the sheriffs, opening the doors to their trucks and helping himself to their vehicle phones and equipment,[244] the answer wasn't hard to guess. Five activists were arrested, and the videotape and mangled camera confiscated along with the proof of service of the subpoena.

"As of late Friday morning, the activists are still in jail, waiting arraignment."[245]

This is one way violence routinely plays out in this culture.

Now, the study, in the words of Brian Martin, author of *Nonviolence Versus Capitalism*: "In the early 1970s, a group of researchers investigated attitudes to violence by surveying over 1,000 U.S. men. Among their revealing findings were that more than half the men thought that burning draft cards was violence and more than half thought that police shooting looters was not violence. The researchers concluded that 'American men tend to define acts of dissent as "violence" when they perceived the dissenters as undesirable people.' In other words, many of the U.S. men used the label 'violent' when they thought something was bad and 'nonviolent' when they thought it was good."[246]

This will come as no surprise to anyone who has paid any attention to premise four of this book. It will come as no surprise to anyone who has paid any attention to this culture.

Now the letter. I sent Remedy's article to a friend, who wrote me a letter back. It read: "You know, the people who always insist on 'letting the system work' are the ones who have never tried to actually do anything. I was thinking how I could, in a matter of minutes, come up with a long list of men—easily into triple digits—who have raped, battered, molested, stalked, and tortured women and girls. And I can't think of one who has ever gone to jail. Hell, I can't think of one who's even gone to trial.

"My friend Mari was in a feminist theory class once where the professor asked one of those ridiculous new age questions about 'What would you do if you only had thirty days to live?' And of course everybody comes back with 'I'd go to the ocean' and 'I'd sit outside and smell the flowers' or whatever. Except Mari who said, 'I'd make a list of all the men who have raped women—just the women I know personally—and I would get a gun and I would take out as many as I could until I got caught.' The class was horrified. Her response was, 'And what else is going to stop them?'

"I think there's a tremendous psychological barrier here. People really want to believe that the world is fair—fair enough that even if injustice happens, it will eventually be righted by the rule of law. 'The system may need some change, but it's essentially sound.' Because otherwise it's just unbearable. And you are then faced with your own agency and responsibility, and acting on that will literally make you an outlaw. Better to keep buying recycled toner cartridges and taking your kids to multicultural story hour at the library and vaguely believing it'll all come out right in the end.

"These people always have lovely anecdotes to back them up. I remember once at a weekend event I attended, there was a man there I knew *for a fact* was a child molester. And there were children present. There was a point where the women and men split into separate groups so I said to the women's group what I knew, and how I knew it, and asked if anyone wanted to help me make him leave. No, of course nobody wanted to do that: 'He needs healing!' 'He needs community!' 'That's what we're here for!' I could see there were others in the group looking down and not looking at me, obviously confused and afraid to buck the new age articles of faith. Eventually the whole debate turned on two points. One, I was too angry and also needed healing, and because I wouldn't 'admit' that then no one had to listen to me. And two, everything was getting so much better in the world, everything. Proof? Somebody had seen a man in a pickup truck with right-wing bumper stickers get out of his truck and pick up some garbage on the side of the road. I'm like, huh? So fucking what? Lots of people hate litter. The KKK adopted a highway in Missouri to keep clean. What does that have to do with sexual abuse being basic socialization in patriarchy?

"And of course, when it was all said and done, two of the women came up to me and desperately wanted to know who the man was so they could keep their kids away from him. They already had him pegged as a creep because of how he'd been behaving. And 'Thanks for speaking up, sorry I didn't help you, you're so brave . . .' I'm not that goddamn brave. If I was that brave, that man would have come to serious bodily harm. All I was asking for was the nonviolent approach—eject him from the event, let him know some people were onto him and maybe were watching him. Protect the kids. I mean if we aren't going to protect children, what are we willing to do? Is the answer really nothing?"[247]

Indeed, if we are not going to protect children, what *are* we willing to do?

❨ ❨ ❨

I asked a friend what he thought is meant by the phrase, "Every act of violence sets back the movement ten years."

He responded, "I think it's a cop out mostly driven by fear. That's certainly a cop out that too often I take myself. More often than not, before I say anything radical or militant at all in any sort of public forum, I wonder who is taking in my words. And I wonder what will be the consequences if I say something that may threaten the worldview of those in power. Jumper cables hooked up to my testicles are one of my biggest fears. Another fear that runs through my mind is that some members of the Black Panthers haven't seen the light of day since the seventies. There's a reason those in power do these things: they work. And when the fear of these forms of retribution take control of our hearts or minds, pacifism can grow to seem a viable option. I know from my experience as a former pacifist that pacifism won't piss off too many people. Judging by my own fears, and my reaction to those fears, saying that violence will set back a cause may just be bowing to the consequences of pissing off those in power."

He paused, then continued, "I think identity has a lot to do with resistance to violent acts. It's pretty apparent to us all at a very early age that you're absolutely forbidden by the master to use the 'tools of the master to destroy the master's house.' Imagine a child who is routinely beaten with a two-by-four, who one day picks it up and fights back. Imagine especially what happens to this child if he's not yet big enough to *effectively* fight back, to win. Not good. On the larger scale I don't think many people are willing to identify themselves with these types of acts or with anyone willing to commit these types of acts simply because it is forbidden by those in power and therefore to be feared.

"And as much as I'd hate to have my testicles electrified, I don't think the fear is even primarily physical, but instead is something even deeper. We are social creatures, and our biggest fear is to not be accepted. Unfortunately, a lot of people want to be accepted, and to be liked, by those at the top of the hierarchy. I sometimes think back to our social groups in high school. You could be hanging out with a friend, and when someone who is a bit more popular joins the group your friend's loyalty might change real fast. Your friend wants to be accepted by the more popular person. Sometimes your friend won't want to identify with you anymore. If that means making you feel inferior to get a chuckle out of the popular person, well, that's what will happen. We've all seen this. This type of dynamic is played out not only in high school, but also in society in general on a daily basis."

Another short pause, and then he concluded, "The way I see it, the phrase about setting the movement back is coming from a place of fear. It surely can't be coming from the perspective of successful pacifist resistance to the machine. If it did, we wouldn't be here discussing how to stop the atrocities committed by this culture."[248]

❬ ❬ ❬

The landbase is not only primary, it is everything. It is the source of all life. After all is said and done—and usually more is said than done—the reality is that our landbases are being killed. We can be as spiritually groovy as we want, and it won't matter. We can be as full of love as we want and it won't matter. We can be as energy efficient as we want and it won't matter. We can recycle as much as we want, and it won't matter. We can be as pacifistic as we want, and it won't matter. We can be as violent as we want, and it won't matter. We can blow up dams or we can not blow up dams, and it won't matter. We can write or read books, and it won't matter. None of this will matter except insofar as it helps stop the murder of our landbases. It really is that simple. The health of our landbases is the gauge by which those who come after will measure us. It is the gauge by which every one of our actions must be measured.

$$\mathbb{C} \quad \mathbb{C} \quad \mathbb{C}$$

A few pages ago I referred to another oft-mentioned pacifist argument, that we must not commit an act of violence (or I would say counterviolence) because if we do, the state will respond with overwhelming violence back at us and at anyone else who happens to be in the area. After hearing all of the other arguments against violence that don't make any sense to me, I've always found this one refreshingly honest. There's no appeal to a faux higher moral ground, no failure of logic presented as moral imperative, no doublespeak. Nothing but good old-fashioned fear.

From the beginning the state has been founded on and supported by the threat of violence. Remember Stanley Diamond's famous opening line to his book *In Search of the Primitive*: "Civilization originates in conquest abroad and repression at home."[249] Or to bring this up to date, consider the following report from a "peaceful" protest in Miami: "No one should call what [Police Chief] Timoney runs in Miami a police force. It's a paramilitary group. Thousands of soldiers, dressed in khaki uniforms with full black body armor and gas masks, marching in unison through the streets, banging batons against their shields, chanting, 'back . . . back . . . back.' There were armored personnel carriers and helicopters. The forces fired indiscriminately into crowds of unarmed protesters. Scores of people were hit with skin-piercing rubber bullets; thousands were gassed with an array of chemicals. On several occasions, police fired loud concussion grenades into the crowds. Police shocked people with electric tasers. Demonstrators were shot in the back as they retreated. One young guy's apparent crime was holding his fingers in a peace sign in front of the troops. They shot him multiple times, including once in the stomach at point blank range."[250]

The motto of the police may be "to protect and to serve," but you and I both know what they are protecting and whom they are serving. It's not only in Miami that the police are a paramilitary organization protecting the interests of those in power. As Christian Parenti told me years ago: "We need to always remember that while the police do everything from getting kittens out of trees and enhancing public safety to killing strikers and framing radicals, the social control function has always been at the heart of what they do, even though most of what they do is not that."

I replied, "A couple of years ago I got burgled, and the first thing I did was call the cops."[251]

Christian said, "Most of us would do that. But the fact remains that it's an important distinction to see, that while most of what the police do is mundane sort of pseudo public safety functions, the heart of what they do, the most important social function, is to intervene at times of political crisis against rebels and to prevent such rebellion, too."

I just today received an email from a friend about this: "Whether a campaign is waged through violence or nonviolence, the oppressors are going to respond the moment the uprising gets serious. I'm going to name names: white middle-class Americans have their heads in the sand about this. The powerful will react to protect their power and since they're allowed to use violence, they will. Non-violent demonstrators will get shot. Arrestees will be threatened and tortured in jail. If anyone out there is serious about building a resistance movement, they are going to have to face what they're potentially risking: life and limb. The resistors' nonviolence does not in any way preclude the oppressor's use of violence. Quite the opposite, really. Because the more serious the opposition, the more serious the powerfuls' response. Whether we choose violence or nonviolence, it is *resistance* that challenges power, and that power will protect itself. Plenty of peaceful protestors have been killed in all kinds of struggles. To borrow another well-used Audre Lourde-ism: 'Your pacifism will not protect you.'"[252] Because the state is based on violence anyway, the best we can hope for, really, is that this violence isn't aimed at us.

I think for many people, pacifism comes from having been pacified.[253] I mean this in the sense of the U.S. military "pacifying," to use its term, villages by blowing them up and terrorizing residents into submission, and I mean it in the sense of giving a child a pacifier, a phony tit that shuts her up by providing artificial comfort; by getting her to attach herself to something she pretends is a source of life but that in reality gives her no nourishment at all.

We should for once be honest with ourselves that a great many of us in the center of civilization reap overwhelming material rewards in exchange for our compliance (which means in exchange for our dignity, humanity, animality, and any hint of moral high ground). Our cars, stereos, closets full of clothes, computers, vacations in Cancun or Acapulco are all giant pacifiers we eagerly place into our mouths and on which we greedily suck. But no matter how we suck, we never get what we need. And then we wonder why we are so (spiritually) hungry.

It's all carrots and sticks. Or rather plastic pacifiers in the shape of carrots and sticks. So long as we keep that plastic nipple (or is it a metal bit? I'm never quite sure) firmly between our teeth, so long as we keep sucking and sucking at nothing at all, and in so doing consume the entire world, gaining nothing of the nourishment from our landbase that would be our birthright to receive and our landbase's birthright to give (and receive in turn), those in power—the abusers, the exploiters—need not too often use the stick. But spit out the pacifier—spit out the bit—and they'll show you the stick. Ball up your fists and they'll raise it. Hit them hard, and they'll make you wish you'd never been born.

It's pretty effective, effective enough to cause us to stand by while the entire world is murdered.

If I didn't have to worry about going to prison, not a dam would stand anywhere I could reach.

I'm not sure, however, that I want to acknowledge that my compliance has been bought so cheaply as it has, for a bunch of cheap plastic consumables, perhaps my own thirty pieces of silvered plastic; (temporary) approval from those at the top of the pyramid; and them granting me the boon of not torturing— remember the CIA interrogation manuals—or killing me.[254] I don't want to acknowledge that fear—even very real fear—is the primary reason I'm failing to adequately protect those I so loudly proclaim I love. I don't at all like what that says about me.

I think that others, too, might not like what it says about them. Thus all the highfalutin but ultimately nonsensical moral arguments for pacifism. Thus the stridency with which many dogmatic pacifists disallow mention of violence, or dismiss it with absurdities: "Violence, schmiolence." A response by a pacifist to Helen Woodson provides a great example of that insane stridency. One of her "crimes" was to walk into a bank with a starter's pistol, tell everyone she was not going to hurt them, demand cash from the tellers, and burn $25,000 while delivering a statement on the evils of money. Now, check out the response by one pacifist online: "1. I'm curious how folks feel about this? When I heard about it, I was pretty shocked. First of all, that's a lot of $ to burn![255] Second, I don't necessarily think $ per se is the ROOT of all evil, whatever evil is (as you define it). And last, but not least, I believe it is a violent act to hold folks at gunpoint (even if it's a toy pistol) to make a point. There's got to be a better way. Now, if Helen Woodson wanted primarily to be locked up (which sounds apparent, anyway) she certainly achieved her mission.[256] But why involve innocent bank tellers and customers? (I'm curious . . . does anyone know the name/location of the bank?)[257] 2. I found the report quite disturbing and was puzzled about it being in this [pacifist] conference. What she did must have been terrifying for the people in the bank. If this is nonviolence, let me off the boat."

This response reveals many of the reasons I have so little respect for so many pacifists. If we leave saving the world up to people like this, there will be nothing left.

I think a central reason for their stridency has to do with the old Jack and Jill discussion we had from R. D. Laing. It does no good for Jack to forget that his refusal to take down those in power is based on his fear of the violence he has seen them do to others, if Jill keeps reminding him of it. He must make her forget as well, and make her feel morally inferior to boot.

This is not the legacy I wish to leave. I do not want to have to look into the eyes of those humans one hundred years from now—or look into the eyes of the salmon now, or any other wild beings in this beautiful world being destroyed, or the animals in the industrialized hell of factory farms or laboratories—and say, "I did not do what was necessary because I was too afraid." I do not want that.

God—land, universe, muse, spirit, whomever—grant me strength, and more courage than I have.

❆   ❆   ❆

I'm also not sure this is an argument against violence anyway, so much as it is an argument against getting caught. Which is an argument for being really smart.

And believe it or not, the odds are on our side. Study after study has shown that nearly all crimes go unpunished. Jessica Mitford, in her book *The American Prison Business*, writes, "The President's Commission on Causes and Prevention of Violence says that for an estimated nine million crimes committed in the United States in a recent year [this was forty years ago, but the statistics still generally hold], only 1 percent of the perpetrators were imprisoned. Carl Rauh, advisor to the deputy attorney general of Washington, D.C., describes the process: 'Of 100 major crimes [felonies], 50 are reported to the police. For fifty incidents reported, 12 people are arrested. Of the 12 arrested, 6 are convicted of anything—not necessarily of the offense reported. Of the 6 who are convicted, 1.5 go to prison or jail."[258] I think we would have to adjust the 1.5 number up a little bit, to account for the fact that selective law enforcement officers and the courts nearly always select laws to enforce that have to do with violence or sabotage going up the hierarchy and ignore laws that have to do with violence going down the hierarchy (witness the Humboldt County Sheriffs Department going to great lengths to take out tree-sitters and ignoring both environmental degradation and rape). But then we'd have to adjust it back down a bit to account for crimes in which the perpetrator is obvious as well as those committed with no planning by people who are drug- or alcohol-impaired (recall my student at the prison who was never caught robbing drug dealers, but who got caught stealing a car: I could tell you dozens of those stories, and my students could tell you far more). When you take the obvious, the foolish, and the damn unlucky (I want to leave myself an out in case I ever get sent to prison) into account, I'm not sure exactly who else is getting popped.

One of the striking implications of this is that those in power *must* rely on us to police ourselves. No matter how they try, they cannot be everywhere at once, unless they can get inside the hearts and minds of each and every one of us and convince us to do their work for them. This is one of the ways that many pacifists are powerful allies of those at the top of the hierarchy: it's not only scary, they say, but immoral to fight back. Whom does this position serve?

Near the end of our book *Welcome to the Machine: Science, Surveillance, and the Culture of Control*, George Draffan and I wrote, "A high-ranking security chief from South Africa's apartheid regime later told an interviewer what had been his greatest fear about the rebel group African National Congress (ANC). He had not so much feared the ANC's acts of sabotage or violence—even when these were costly to the rulers—as he had feared that the ANC would convince too many of the oppressed majority of Africans to disregard 'law and order.' Even the most powerful and highly trained 'security forces' in the world would not, he said, have been able to stem that threat."[259]

We continued, "As soon as we come to see that the edicts of those in power are no more than the edicts of those in power, that they carry no inherent moral or ethical weight, we become the free human beings we were born to be, capable of saying *yes* and capable of saying *no*."[260]

This is what those at the top of the hierarchy fear more than anything else in the world.

❨ ❨ ❨

There is no direct relationship between laws and morality. Some laws are moral, and some laws are immoral.

Let's return to the questions I asked earlier: To whom will you be called upon to answer? By whom do you *wish* to be called upon to answer?

There is a difference between being called upon to answer, and being punished. This is a difference that far too many activists and others forget. When Plowshares activists bang on a missile then wait to get arrested—or as Philip Berrigan put it in the interview I quoted early in this book, "And you take the heat. You stand by and wait for the arrest"[261]—they are forgetting that there is no moral reason to "take the heat," to "wait for the arrest." In fact doing so reinforces the mistaken and dangerous belief that governments have legitimacy beyond their capacity to impose punishment. It reinforces the mistaken and dangerous belief that the government is not a government of occupation. It reinforces the mistaken and dangerous belief that one should be responsible to—answerable to—the government, and not to one's landbase.

❨ ❨ ❨

The next argument often thrown out by pacifists is that because the state has more capacity to inflict violence than we do, we can never win using that tactic, and so must never use it. But if we can never use a tactic the state has more capacity to use then we do, we might as well hang it up right now. The state has more capacity to propagate discourse than we do: this logic would suggest we can never win using discourse, so we must never use it. The state has more capacity to raise funds and to use money than we do, which means we can never use fundraising either. We can say this for every possible tactic, except perhaps the tactic of sending pink bubbles of pure sweet love toward our enemies—oh, sorry, toward those wonderful souls who happen to be wounded in ways that are causing them to become CEOs, politicians, and police. We have the monopoly on this one.

The argument is just not true anyway. The United States had several orders of magnitude greater capacity to kill than the Vietnamese, yet the Vietnamese drove out the United States. I think often of Ho Chi Minh's famous line: "For every one of yours we kill, you will kill ten of ours. But in the end it is you who will grow tired." Now, you could argue that at this remove the Vietnamese may have won the war, but McDonald's and Nike have won the ensuing "peace," but we could say the same about the decolonization of India, that Gandhi may have won his peaceful revolution—which could not have happened, by the way, had Britain not already been bled white by World War II, nor without armed revolutionaries also fighting for freedom—but that Monsanto and Coca-Cola have won the "peace" that followed.

What I said a few hundred pages ago—the fact that those in power can always outspend us does not mean that we should never attempt to use money for good—applies here as well. Here is what I said, altered to fit the present subject: "But we must never forget that if we attempt to economically, rhetorically, or physically/violently go head-to-head with those who are destroying the planet, we will always be at a severe, systematic, inescapable, and functional disadvantage. Me not buying an airline ticket won't do squat. Me writing one book won't do squat. Me blowing one bridge won't do squat. But all is not lost. The questions, yet again: Where are the fulcrums? How do we magnify our power?"

FEWER THAN JESUS HAD APOSTLES

It could be that, in the future, people will look back on the American Empire, the economic empire and the military empire, and say, "They didn't realize that they were building their whole empire on a fragile base. They had changed that base from brick and mortar to bits and bytes, and they never fortified it. Therefore, some enemy some day was able to come around and knock the whole empire over." That's the fear.

Richard Clarke, head of the President's Critical Infrastructure Advisory Board[262]

I HAVE A CONFESSION TO MAKE. I'VE BEEN HOLDING OUT ON YOU.
Quite a while ago I had one of the most positive conversations I've ever had. It makes me think it really could be possible to speed up the process of bringing down civilization.

I talked with some hackers. I hope you'll forgive me if I don't tell you when or where we spoke, or the oddly satisfying circumstances under which we met. I also won't say their names or genders. Nor will I describe them. Presume they're men. Presume one of them looks like your bench partner from your high school advanced laboratory class,[263] and the other looks strangely familiar, too, like someone you saw once in the far corner of a library, surrounded by books on Nestor Makhno, Emiliano Zapata, August Spies, and Albert Parsons. Or maybe he looks like someone you saw standing at the very back of an auditorium as he listened to someone speak passionately about the necessity of taking down civilization *now*.

In any case, here I am sitting across a table (or at least you can presume I'm sitting across a table) from these folks, sharing a pitcher of water at a café. Let's presume we're in Asheville, North Carolina, and it's late, very late, on a hot summer night.

"Let's start small," I say. "Would it be possible to inflict serious economic damage on a major corporation by hacking into computer systems?"

"You're presuming," responds the first one, let's call him Brian, "that this doesn't happen already."

The other, let's call him Dean, nods. I look back and forth between the two.

Brian continues, "It's in the corporations' best interests to not let on that this stuff happens all the time."

"Why is that?" I ask.

"You think they want people to know how easy it is to hack into a system?" He winks, then pauses for effect. "And it's getting easier all the time. Take the use of wireless technologies that have come on strong these past few years. See that thermostat over there?"

He points to the far side of the room. I turn to look, then turn back when I hear him start talking again.

He says, "Those are oftentimes computerized, and send and receive signals through the air from a main system. The other day I hacked into the main computer of a major corporation through the thermostat."

My jaw drops.

He throws back his head and laughs, then says, "I didn't do anything. I was just trying to see if I could do it."

"But could you have done damage?"

"Oh, yes."

"How?"

"Name a nasty corporation," he says.

"Ha!" says Dean, "Name one that isn't."

"Freeport McMoRan is pretty nasty."

They both shake their heads.

"Most polluting company in the United States. Pollutes all over the world. Machine guns natives in West Papua. Imprisons others in shipping crates."

Dean looks at me intently before asking, "How does it make its money?"

"Mainly mining. Gold in Indonesia, sulfur in the Gulf of Mexico. Other minerals too."

"Okay," Dean says. "Piece of cake."

"What do you do?" I ask. "Mess with their bank accounts, pretend it's *Fight Club* and destroy their credit card accounts?"

Brian wrinkles his nose.

Dean says, "Shipping. All the shipping these days is computerized."

Brian interrupts to ask, "Did you know the U.S. economy almost ground to a halt last year?"

"What?" I exclaim.

"The dockworkers strike on the West Coast," Dean says. "The big companies couldn't get their raw materials and parts. They were within a day or two of running out. Do you know what happens then?"

It's clear the question is rhetorical.

He asks another question, "Do you know how much it would cost GM to shut down its assembly line?"

"I have no idea."

"Millions of dollars per minute."

"Jesus," I say.

"No," Brian responds. "Dockworkers."

"Or," Dean says, "Hackers. Let's say Freeport McMoRan ships through Singapore. Singapore is the most automated port in the world. What happens if you reroute canister after canister headed for New Orleans instead to Honduras, Belize, Turkey?"

"The people who work for these companies," Brian adds, "rely more on computers than common sense. They have to. The companies are so big, the movements of people and resources so complicated, that people can't keep track of it all. Last month I hacked into the security system of a major corporation and had the computer issue me an ID card. I went to the company headquarters, swiped my new card, and it okayed me to enter. I walked over to the security people and told them I was a hacker who had just breached their security. They refused to believe me. They said that the computer okayed me, so I should just quit joking and head on in."

"They believed the computer over their own ears."

"I tried to persuade them, but nothing I said convinced them to listen to me."

"Do you think," I ask, "that hackers could do more than just mess with a big corporation or two?"

Brian smiles. "You're presuming, again, that nobody is already doing this."

"No," I say. "Do you think they could bring it all down, could take down civilization?"

Brian nods, and so does Dean. Dean says, "I've spent the past twenty years studying *how* the economic *system* works. I don't mean economic theory, although I certainly understand that. But rather the nuts and bolts of it. Transport of raw materials like we were talking about with those canisters. And the thing that amazes me is that the system hasn't already collapsed. It's incredibly fragile. And incredibly vulnerable."

As Dean talks, Brian pulls what looks like a walkie-talkie from a holster on his belt. The walkie-talkie has a small LED screen. Suddenly the machine squawks, and a light turns green.

"Guess what," Brian says. "Somebody in Asheville is receiving a page. He pulls a hand calculator from another holster, and punches a few buttons. He shows me the screen. I read information about the page. He smiles, proudly, then says, "I made a few minor modifications . . ."

I ask, "Why do you do this?"

"It makes me giddy to figure things out. I love the rush when I suddenly understand something new."

I know the feeling. I feel what I'd imagine is the same rush whenever I suddenly *get* the relationship, for example, between pornography and science.[264]

I ask, "If you love computers, would you take down civilization?"

Dean says, "Yes."

Brian says, "In a heartbeat."

"Why?"

"Do you know where they put computer books in bookstores?" Brian asks. "In the business section."

"And?"

"Computers were supposed to set us free. That was the rhetoric. That's always the rhetoric. But they've just been used to further enslave us, to further enslave the poor, to further enslave the planet."

I become aware of the silence in the room. I take a sip of water.

Brian continues, "Let's say you have a soldering iron that you love to use. You love soldering pieces of metal together. You love burning beautiful designs into pieces of wood furniture. Now, what would you do if somebody started using that soldering iron to torture people? I can't speak for you or anyone else, but I would pull the plug on the iron. I'd do that," he repeats, "in a heartbeat."

A soft sound breaks the silence of the café. Across the room the lone employee has begun stacking chairs on tables.

Brian says, "I'm in love with figuring out how things work. And the existence or nonexistence of machines doesn't mean we can't figure things out. If I smash this calculator, that doesn't invalidate Ohm's Law. Ohm's Law is still there. Nature is still there, under all this concrete, under all these machines. And have you gone outside during a blackout? The lights are still there; they're up in the sky. And it's so quiet you can finally start to hear."

I ask again, "And you'd be willing to help bring it down?"

The woman is stacking chairs closer. We don't have much time.

They both laugh and say, "Of course."

"You've thought about this a lot."

Again, both laugh and say, "Of course."

I have to know. "If they were dedicated enough, and knew what they were doing, how many people do you think it would take to bring down civilization?"

Brian says, "It would take far fewer than Jesus had Apostles."

The woman has stacked all the chairs but ours.

Dean says, "Let's go."

I nod, then say, "It's late."

PACIFISM, PART IV

The West won the world not by the superiority of its ideas or values or religion (to which few members of other civilizations [*sic*] were converted) but rather by its superiority in applying organized violence. Westerners often forget this fact, non-Westerners never do.

Samuel Huntington[265]

THE FINAL ARGUMENT I'VE OFTEN HEARD FROM PACIFISTS IS THAT violence never accomplishes anything. This argument, even more than any of the others, reveals how completely, desperately, and arrogantly out of touch many dogmatic pacifists are with physical, emotional, and spiritual reality.

If violence accomplishes nothing, how do these people believe the civilized conquered North and South America and Africa, and before these Europe, and before that the Middle East, and since then the rest of the world? The indigenous did not and do not hand over their land because they recognize they're faced with "a high stage of social and cultural development." The land was (and is) seized and the people living there were (and are) slaughtered, terrorized, beaten into submission. The tens of millions of Africans killed in the slave trade would be surprised to learn their slavery was not the result of widespread violence. The same is true for the millions of women burned as witches in Europe. The same is true for the billions of passenger pigeons slaughtered to serve this economic system. The millions of prisoners stuck in gulags here in the U.S. and elsewhere would be astounded to discover that they can walk away anytime they want, that they are not in fact held there by force.

Do the pacifists who say this really believe that people all across the world hand over their resources to the wealthy because they enjoy being impoverished, enjoy seeing their lands and their lives stolen—sorry, I guess under this formulation they're not stolen but received gracefully as gifts—by those they evidently must perceive as more deserving? Do they believe women submit to rape just for the hell of it, and not because of the use or threat of violence?

One reason violence is used so often by those in power is because it works. It works dreadfully well.

And it can work for liberation as well as subjugation. To say that violence never accomplishes anything not only degrades the suffering of those harmed by violence but it also devalues the triumphs of those who have fought their way out of abusive or exploitative situations. Abused women or children have killed their abusers, and become free of his abuse. (Of course, often then the same selective law enforcement agencies and courts that failed to stop the original abuse now step in to imprison those who sent violence the wrong way up the hierarchy.) And there have been many indigenous and other armed struggles for liberation that have succeeded for shorter or longer periods.

In order to maintain their fantasies, dogmatic pacifists must ignore the harmful and helpful efficacy of violence.[266] Years ago I was asked by a publisher to review a book-length manuscript they had just received from a household-name pacifist activist. The document was a mess, and they said they might want me to help edit it. I was younger then, and far less assertive, so my comments were fairly minor throughout, until I came to a statement that made me curse and hurl my pen across the room, then get up and stalk outside for a long walk. The activist claimed that the American movement against the war in Vietnam was a triumph for pacifist resistance, and that it showed that if enough people were just dedicated enough to nonviolence they could bring about liberation in all parts of the globe. He mentioned the four dead at Kent State as martyrs to this nonviolent campaign, and also mentioned "our unfortunate soldiers who lost their lives fighting for this unjust cause," but never once mentioned the millions of Vietnamese who outfought, outdied, and outlasted the invaders. My point is not to disparage or ignore the importance of nonviolent protests in the United States and elsewhere, but rather to point out what the pacifist pointedly ignored: the antiwar movement didn't stop the U.S. invasion—it *helped* stop the invasion. The primary work—and primary suffering—was done by the Vietnamese.

Oddly enough, the publisher didn't hire me to edit it.

I am just being honest when I say that I have talked to hundreds of people who are ready to bring the war home. I've talked to those who went down to assist the Zapatistas but were told, "If you really want to help, go home and start the same thing there." I've talked to family farmers, prisoners, gang members, environmentalists, animal rights activists, hackers, former members of the military who have had their fill of their own enslavement and the destruction of all they love, and who are ready at long last to begin to fight back. I have spoken to Indians who have said their people are ready to bring back out the ceremonial war clubs they have now kept buried or hidden for so long. I have spoken to students and other men and women in their teens, twenties, thirties, forties, fifties, sixties, seventies, eighties who know the world is being killed, and are ready to fight and to kill and if necessary to die to stop this destruction, who, like me, are not willing to stand by while the world is destroyed.

I give a talk. Afterwards someone asks, "How do we hold CEOs accountable for their actions?"

I look hard at the person, but before I can piece together my answer, I hear a voice from the back of the room, "A bullet to the brain does wonders."

I don't say anything. I am surprised, I have to admit, at the number of people I see nodding solemnly. At least half.

The person again shouts out, "What other accountability is there?"

Finally I speak, "There is no legal accountability: when was the last time you saw a CEO put in prison for murder (or for anything, really)? When was the last time you saw a war criminal who won put in prison? Can you say *Henry Kissinger?* Put in the name of your favorite politician. And there is no moral accountability. A lot of these jokers think they're going to heaven. They all have their claims to virtue, and many of them probably believe them. And there is no communal accountability. These people are, like Hitler, admired. What's left?"

The same person shouts out, "Flesh. They're mortal. They die as surely as do the people they've killed."

It's a big hall, and it's dark in the back. I can't see who it is. It doesn't matter. Many people have expressed these same thoughts to me, only in private. I cannot tell you how many times I have thought them myself, only once again in private.

Someone else calls out, "But they'll just get replaced."

And a third person, "Take them out, too. And the next and the next. Eventually they'll get the message."

I feel certain this is what Tecumseh would have done.

The second person again, "Violence never acts as a deterrent."

A sharp laugh from the back and someone says, "Ted Bundy."

"What?"

"The state's violence deterred him from killing again."

"He didn't have to be killed."

"He was kept in prison by force."

A woman in the front says, "And the violence of men against women is a huge deterrent. Why do you think I don't walk alone at night? I have been deterred by violence. Don't tell me that violence is not a deterrent."

"Why do you think it is," someone else chimes in, "that we don't all rise up right now to overthrow this horrid system? We're afraid of getting killed or sent to prison. Violence works great as a deterrent. It's just *we* don't use it."

"Someone show me," I said, "a peaceful way we can make those in power stop killing the world, and I will be on board faster than you would think possible. But I just don't see it. I just don't see it."

❰ ❰ ❰

Ward Churchill puts it well: "There is not a petition campaign that you can construct that is going to cause the power and the status quo to dissipate. There is not a legal action that you can take; you can't go into the court of the conqueror and have the conqueror announce the conquest to be illegitimate and to be repealed; you cannot vote in an alternative, you cannot hold a prayer vigil, you cannot burn the right scented candle at the prayer vigil, you cannot have the right folk song, you cannot have the right fashion statement, you cannot adopt a different diet, build a better bike path. You have to say it squarely: the fact that this power, this force, this entity, this monstrosity called the State maintains itself by physical force, and can be countered only in terms that it itself dictates and therefore understands. That's a deep breath time; that's a real deep breath time.

"It will not be a painless process, but, hey, newsflash: it's not a process that is painless now. If you feel a relative absence of pain, that is testimony only to your position of privilege within the Statist structure. Those who are on the receiving end, whether they are in Iraq, they are in Palestine, they are in Haiti, they are in American Indian reserves inside the United States, whether they are in the migrant stream or the inner city, those who are 'othered' and of color in particular but poor people more generally, know the difference between the painlessness of acquiescence on the one hand and the painfulness of maintaining the existing order on the other. Ultimately, there is no alternative that has found itself in reform; there is only an alternative that founds itself—not in that fanciful word of revolution—but in the devolution, that is to say the dismantlement of Empire from the inside out."[267]

❰ ❰ ❰

I'm really angry that I had to spend the last couple of months deconstructing pacifist arguments that don't make any sense anyway. I'm angry that I've had to spend the last three years writing this book to show conclusions that should be pretty damn obvious. Newsflash: Civilization is killing the planet. (I've often heard that pattern recognition is one sign of intelligence. Let's see if we can spot this pattern in less than six thousand years. When you think of the landscape of Iraq, where civilization began, do you normally think of cedar forests so thick sunlight never reaches the ground? That's how it was prior to civilization. How about the Arabian peninsula? Do you think of oak savannah? That's how it was prior to civilization. When you think of Lebanon, do you think of cedars? At least they have one on their flag. Prior to the arrival of civilization, it was heavily forested, as were Greece, Italy, North Africa, France, Britain, Ireland. How long will it take you to see this pattern? How long will it take you to do something about it?) Newsflash: Civilization is based on violence. Newsflash: The system is psychopathological. Newsflash: This entire culture requires our disconnection from each other and especially from our landbases. Newsflash: This entire culture inculcates us into irresponsibility and would not survive were we to gain even a shred of responsibility.

I just received an email from a friend: "There are so many people who fear making decisions and taking responsibility. Kids are trained and adults are encouraged not to make decisions and take responsibility. Or more accurately they are trained to engage only in false choices. Whenever I think about the culture and all the horrors it perpetrates and we allow, and whenever I consider our typical response to being faced with difficult choices, it seems clear to me that everything in the culture leads us to 'choose' rigid, controlled, unresponsive 'responses' over fluidity, real choice, and personal responsibility for and to those choices. Every time. Every single time.

"Pacifism is but one example of this. Pacifism is of course less multifaceted in its denial and delusions than some aspects of the culture (in other words, more obvious in its stupidity), but it's all part of the same thing: control and denial of relationship and responsibility on one hand versus making choices and taking responsibility in particular circumstances on the other. A pacifist eliminates choice and responsibility by labeling great swaths of possibility off-limits for action and even for discussion. 'See how pure I am for making no wrong choices?' they can say, while in reality facing no choices at all. And of course they actually are making choices. Choosing inaction—or ineffective action—in the face of exploitation or abuse is about as impure an action as anyone can conceptualize. But these ineffective actions can provide the illusion of effectiveness: no matter what else can be said about pacifism, even with the gigantic problems we face, pacifism and other responses that do not threaten the larger concentration camp status quo are certainly achievable. That's something, I guess. But it all reminds me of those who go to therapists to create the illusion that they're doing something, rather than the few who actually work to face their fears and patterns and take an active role in transformation.

"Pacifism is a toxic mimic of love, isn't it? Because it actually has nothing to do with loving another. Could it be said that toxic mimics are toxic in part because they ignore responsibility, they ignore relationship, they ignore presence, they substitute control for fluidity and choice? Toxic mimics are of course products and causes of insanity. Could it be said that a lack of responsibility, relationship, and presence, and the substitution of control for fluidity and choice are causes and products of insanity?"

GET THERE FIRST WITH THE MOST

The art of war is simple enough. Find out where your enemy is. Get at him as soon as you can. Strike at him as hard as you can, and keep moving on.

Ulysses S. Grant[268]

WE ARE GOING TO WIN. DAY BY DAY CIVILIZATION BECOMES MORE brittle, more top-heavy. Day by day it becomes more clear to ever more of us that we must make a choice between civilization and the planet, and day by day ever more of us are choosing the planet. Day by day it becomes more clear that the earth itself and its wild members are beginning to fight back, and day by day they more strongly beckon us to join them.

☾ ☾ ☾

How do you win? Someone once asked the Confederate General Nathan Bedford Forrest how he won so many battles. His response summed up the essence of military strategy in six words: "Get there first with the most."[269]

Let's break it down. *Get there.*

You choose where you fight. The person or force who chooses the battlefield has a better chance of winning. Indeed, much military strategy consists of attempting to get your enemy to attack you where you're strong and to not attack you where you're weak, while simultaneously probing for your enemy's weak spots to attack. This is true on battlefields, it is true in antagonistic discourse,[270] it is true in all areas of conflict.

The Battle of Fredericksburg during the Civil War is a great example. Confederate general Robert E. Lee had the Army of Northern Virginia entrench behind a stone wall, up a hill, behind a river. Federal general Ambrose Burnside sent his Army of the Potomac across the river and into a series of frontal assaults up the hill. His troops were slaughtered. On hearing this news a few days later, another Confederate general, Joseph E. Johnston, commented peevishly, "What luck some people have. Nobody will ever come to attack me in such a place."[271] And that, once again, is what you want. You want your enemy to attack you where you are strong, and to not attack you where you are weak. You want insofar as possible to control *where* and *over what* you fight—the *terms* and *terrain* of the battle.

The same is true in discourse. We are all familiar with the infamous line from the attorney, "When did you stop beating your wife?" The field of battle has shifted from *whether* to *when.* A few years ago I wrote about an exemplary case: a representative of the capitalist press was moderating a "debate" between two capitalists running to head the Washington State Department of Natural Resources. His first question: "Do you think environmental regulations work, or do they go too far?" Notice how he framed the field of battle, what he included and excluded by his framing. A similar thing happened a few pages ago with the article about the woman in Iraq who was shot for working as a translator for the U.S. military. Calling her "innocent" puts us on one discourse battlefield, and calling her a "collaborator" puts us on a whole different one. The same is true in discussions of pacifism. This is one reason pacifists so often try to claim the "moral high ground," military language if I've ever heard any. Allowing them that moral high ground gives them an advantage similar to allowing soldiers to shoot at you from above. Shifting the field of discourse such that what they claimed was moral high ground is now a plain or valley or swamp,[272] or shifting to discussions of efficacy, or as in the case with this book, shifting the field of discourse to one of being present to one's circumstance and valuing context and relationship over abstractions causes the battles to be fought over entirely different terrain. Examples of this framing or reframing of fields of discourse are countless. Tonight I heard the capitalist media (how different does the terrain of discourse look if we call it the "mainstream media," or "the news"?) call U.S. soldiers (how different does this terrain look if we call them "servicemen" or "servicewomen" on one hand, or "mercenaries" or "invaders" on the other?) "America's best." How does that frame all that comes after? Tonight I did not hear in the capitalist media any mention of biodiversity. How does *that* frame further discourse? Look around. Pay attention to the way you shape discourse, sometimes accidentally, sometimes manipulatively, sometimes perforce, most of the time entirely unconsciously, by choosing what will and won't be spoken, what terms will and won't be used. And pay attention to the way your discourse is shaped for you.

This is a central reason we have to tell lies to each other, and especially to ourselves. If Bill Clinton and the timber industry can frame the debate over deforestation as "jobs versus owls," the deforesters have already won before we start. If they can frame the debate such that people believe forests need to be cut down so they won't be killed by beetles, they've already won. If George W. Bush and the timber industry can frame the debate over deforestation such that people believe forests need to be cut down to keep them from burning, they've already won. If abusers can keep you talking about anything and everything except their abuse and how you're going to stop them, they've already won.

If those in power can frame the "debate" over the murder of the planet into the question of how to implement "sustainable development" (look how they've already framed it by calling industrialization "development") they've already won: we are fighting over techniques to salvage civilization, not ways to save the planet. Worse, those in power routinely frame battles of discourse over whether or not damage is even being done. And even as those we love die of cancer, we let them do this. The discourse should be: How do we stop these psychopaths from continuing to kill those we love?

All of this is why I've been hammering so hard on the notion of questioning premises. Not questioning someone's premises (including mine) amounts to ceding the choice of battlefield to whomever chooses these premises. It's like Burnside throwing his troops again and again against Lee's entrenchments. You're going to get slaughtered. Actually it's worse, since at least Burnside *saw* the stone wall, yet so often in our discussions the premises remain partially or entirely hidden.[273]

This analysis applies not just to big armies, and not just to discourse, but to all conflict. It certainly applies to stopping civilization from killing the planet. Right now, what are the battlefields on which we are encouraged—allowed— to fight? We are encouraged to vote.[274] But of course we all know the old Wobbly saying: If voting made a difference it would be illegal. And in any case, our choices of whom/what we can vote for are strictly limited. No matter whether a Republican or Democrat wins, we lose. We are encouraged—allowed—to use the courts, and while of course we may get the occasional win there, we must never forget by and for whom the courts are set up. We must never forget that their authority ultimately comes from the ability of the state to inflict violence. We are encouraged—allowed—to write, so long as we never mention social change and violence in the same paragraph. We are encouraged to recycle, to shop green (so long as we shop!), and so on.

Much more interesting are the fields we are not encouraged—allowed—to choose for our battles. Who chooses for us? What fields are off-limits, and why? Who has declared them off-limits? Why have we ceded this territory?

What do we want? How will we accomplish it? I return to the salmon (you can of course return instead to what you love). As I already mentioned, for salmon to survive, dams, industrial logging, industrial fishing, industrial agriculture must go, the oceans must survive, and global warming must cease. Choose one of these, say, dams. What do we need to do to remove dams? (And notice the difference in implication even between using the verb "would" and "do," as in "What would we need to do to remove dams?" *Would* implies theory, which means we're not really going to do it, while *do* implies reality; the choice has been made, and now we're asking *how*.) What battlefields do we choose?

Try for a moment to think for yourself. I'm not being snide, condescending, or sarcastic. Thinking for yourself is one of the most difficult things to do, especially within a culture where we're inculcated into irresponsibility. How do you know if you're thinking your own thoughts, or if you're thinking the thoughts of the people who produce television programs, or thinking the thoughts of your teachers and preachers in junior high, or thinking the thoughts of some guy who writes books about taking down civilization?[275] But try, really try. If you follow your own thoughts, if you follow your own morals, if you choose to protect those you love most, and to protect your landbase (presuming that you love your landbase, but if you do not then you can choose something else[276]), if you choose your own battlefields, what battles do you choose? What do you do? How do you act? Who are you?

A couple of times at talks just to see what would be the response, I've said, "Picture someone you hate. Not just someone who bugs you, but someone you really hate. It can be personal, as in someone who has sexually or physically abused you, or a pusher who hooked someone you care about on drugs. It can be social, as in a politician or CEO. It can be a historical figure, like Hitler or Christopher Columbus. It doesn't matter. And if you don't hate anyone, that's okay, you can still participate. Now, if you knew for sure that you could get away with it—one hundred percent sure—and you had the opportunity, would you kill the person?" I am not, of course, looking for any answer, nor am I judging any answer. I just am interested in finding what people think and feel. When I ask this question, about half the people in the audience nod *yes*. Many of the others look away, and many frown, disapproving of the question itself. Of course everything I've read about killing suggests that if you put a gun in a person's hand, create the opportunity, and then ask the question, there's a *much* smaller chance the person will, say, pull the trigger. Theory is always easier than practice, and that's especially true in this case.

When I ask this question, the next thing I say is, "It doesn't matter what your answer is. I'm just pointing out that if you say *no* then we have one discussion. If you say *yes* we have an entirely different discussion."

It should be clear by now that I do not care what fields you choose for your battles. I do not know you. I do not know your strengths. I do not know your weaknesses. I do not know your loves, and I do not know your hates. I do not know where or how or over what you should fight. And I would neither dare nor even care to make suggestions as to what you should or should not do when I do not know you or your circumstances.[277]

Here's the point: if you allow your enemy to choose the battlefield, you will probably lose. Choosing the field upon which you will fight is the first step to winning. Choose your battlefields wisely.

❨ ❨ ❨

"Get there first with the most."

Next: *First.*

It's generally easier to defend than attack. Defending, you can hunker down in a protected place and wait for your enemy to come to you. Attacking, you have to expose yourself in terrain known to the defender. The defender can use both static and dynamic weapons while the attacker can use only the latter. If you get their first, you can claim the best defensive position. Then you can entrench that position. You can attempt to force your enemy to either remove you from this entrenched position or quit the battle. You can also do nothing. This latter option is not available to your enemy, unless your enemy is willing to accept the status quo. In other words, getting there first *forces* your enemy to either accept your position or to do something about it. This gives you a powerful advantage. Once again, all of this applies on every level of conflict.

To take the example of Fredericksburg, Lee got there first, and was able to take strong positions and make them stronger. Had Burnside gotten there first, his troops could have sauntered up the hill instead of dying at its base. But because the Confederates were able to get there first they were able to stand and calmly fire from behind a wall while the Federals had to cross open ground.

The same is true of discourse. From watching courtroom dramas, we're all probably familiar with one of the fundamental rules of trying cases: never ask a question in court to which you don't already know the answer. You want to get to the territory first, claim it, know it, and be able to defend it.

Once you've claimed some battlefield—and this is true in all areas of conflict—you can hold it until you abandon it or are dislodged. That is the primary reason I devoted a couple of months to the discussion of pacifism and a few years to this discussion of taking down civilization. Within this culture pacifism has in many circles been able to claim the moral high ground, having presumably found it empty after its previous holders—those who defended themselves and those they loved—had their landbases, cultures, bodies, and souls destroyed by the relentless physical, rhetorical, and spiritual attacks of the civilized. Whether or not pacifism *deserves* the moral high ground, the fact is that within great swaths of this culture it holds it, just as the civilized hold most of the physical ground around the world, and for similar reasons. Pacifism as moral high ground has become the default, the position we are taught to accept, the position we *do* accept, without thinking. It will remain the default until it is made or shown to be untenable. That was one purpose of my analysis: to attempt to shake the strength of pacifism's hold on that ground. It's the same with my larger scale analyses. The positions we accept as defaults don't need to be proven, don't need to be defended, don't even have to make sense so long as no attacks are made on them. Civilization is a high stage of social and cultural development. Industrialization equals development. High technology is good for humans. Civilization is separate from and more important than any landbase. Power is more important than relationship. The world is organized hierarchically, with (rich, white, civilized, male) humans at the top. It is just and moral and *right* (based ultimately on might) for those at the top of this artificial (yet claimed and for the most part recognized) hierarchy to exploit those perceived as below. Nonhumans do not speak. Part of what I've been attempting to do with all of my work is to dislodge these assumptions from their positions. Consider how different our behavior would be if the default positions within this culture—the positions we are taught to accept, and we *do* accept, unless they are shown to be untenable—were that nonhumans do speak, and they have something to tell us; that the world is not organized hierarchically but rather in a complex interweaving, and that to hyperexploit one's landbase is to destroy the tapestry that supports one's own life; that relationship is more important (and more fun) than the acquisition and wielding of power, and that beings are more important than things;[278] that one's culture must spring from and be a part of one's landbase; that high technology (and much not-so-high technology) springs from, manifests, and leads to a form of intelligence alienated from one's landbase, and springs from, manifests, and leads to a preference for machines and machine-based social structures over life and living relationships; that industrialization destroys native cultures and landbases, and has as a primary purpose the hyperexploitation of resources, that is, the conversion of the living to the dead; that civilization originates in conquest abroad and repression at home, and is a social order so psychopathological in its formulation and in its manifestations that it would kill even the planet, indeed, that it is already doing that; that there is no shame, dishonor, or sin in using violence to defend one's life or landbase.

As I listen, then feel, then think, then write, then rewrite, I try to always remain aware of the phrase *getting there first*. I try to come up with every reasonable—and even possible—counterargument and get there first: meet it before it occurs to readers. Some readers, for example, may dismiss my arguments about saving salmon because I say the salmon (and salmon flesh) speak to me. Not a problem. I merely need to anticipate this argument and claim the ground where I will be attacked by pointing out explicitly that for this particular conclusion it doesn't matter whether or not you believe salmon speak: if you want to eat salmon, you have to remove the barriers to their survival. Well, some readers may then argue that they can eat farmed salmon. Not a problem. I just have to anticipate this argument, too, and once again claim the ground where I will be attacked by pointing out that salmon farms will not survive the crash of civilization (and that salmon farms are already incredibly destructive, and that farmed salmon are not the same as wild salmon). So it goes, argument by argument.

Get there first. I want to tell you a story. When I was a high jump coach, before every track meet I made my jumpers get to the stadium before anyone else. They were to be the first to put down their athletic bags near the approach. They were to be the first to tape their mark. For overnight trips sometimes I'd take them to the field the night before, as soon as we got into town. A couple of times we climbed fences to get to where the high jump competition would be held.[279] Even for away meets—*especially* for away meets—I wanted them to claim the high jump pit. It was now theirs, and it was up to other jumpers to take it away from them.

I think we now need to do a similar thing. We need to claim the land where we live. We need to fall into it, to treat it as though it's ours—as in a family we love and protect, not as in something we have the right to trash—and we need to defend it. If someone is going to destroy our landbase, they'll have to come through us to do it, because we were here first. We have the higher claim, and we will defend that claim.

It's no wonder we don't defend the land where we live. We don't live here. We live in television programs and movies and books and with celebrities and in heaven and by rules and laws and abstractions created by people far away and we live anywhere and everywhere except in our particular bodies on this particular land at this particular moment in these particular circumstances. We don't even *know* where we live. Before we can do anything, we have to get here first.

Finding out everything you can about the people whose land you live on and allying yourself with its rightful owners is vital, but there's something even deeper. Whose land is it? Yes, it's Tolowa land, or Apache, or Seneca, or Choctaw, or Seminole land. But even before them, whose land is it? The land belongs to the salmon, to the redwoods, the Del Norte salamanders, the red-legged frogs, the pileated woodpeckers, the marbled murrelets, the spotted owls. The spiders, solitary bees, and huckleberries. They are the land. They define it. They in all physical truth make it what it is. Get to know them. Ally yourself with them. They were here first. They—or their local equivalent—know the land where you live far better than you do. After all, they live there. And when civilization comes down, there will be much you need to learn. There will be much they can teach you, if you are willing to learn, and if they are still alive.

The way things are going, they won't be. For the truth is that right now, no matter how completely we may understand that the land we live on belongs to the indigenous, and no matter how completely we may understand that the land we live on belongs to those nonhumans who have lived here forever, civilization and the civilized have overrun nearly all of the planet, with plans to overrun the rest. Civilization and the civilized hold nearly all of the ground. They have, so far as we are concerned, gotten there first. If we are to recover this ground, we must force them to quit it. I am not speaking metaphorically.

One of the things I've always hated about being an environmental activist is that nearly all of our work is defensive, as we try to stop this or that area from being destroyed. That's necessary work, of course, but it's not enough. We need to begin to beat back the civilized, to reclaim land to let it recover. In addition to the purely defensive work of stopping new roads from being busted into native forests, we need to rip out roads that are already there, whether or not we have the permission of those in power. We need to take out dams. We need to turn croplands back into forest, marsh, and meadow.

The good news is that this is all pretty easy. It takes an extraordinary amount of work and energy to impede succession, and for many places all we need to do is force the civilized off the ground they've stolen and the landscape will do the rest. Bust a dam, and the river will take care of itself. Take out a parking lot, and it won't be long till paradise comes back home.

The bad news is that we live in occupied territory, and those in power will try to maintain that occupation to the very bitter end. This is another sense in which getting there first is critical to bringing down civilization. Since those who are exploiting and killing your landbase will not without a fight relinquish their perceived entitlement to exploit and destroy, any threat to their perceived entitlement is fraught with danger. If they catch you. So do not let them. How do you not let them? By getting there first. Know what you are doing, and know where you are doing it. Practice, like the former Marine told me at the baseball game, until you can perform your tasks in your sleep. Know the terrain. Have escape routes. Plan for contingencies. Plan for more contingencies. Plan for even more.

Get there first. Just as I do when I write, prepare for every possible response to your actions. The state *will* respond. You need to get to each response first and close off that avenue of attack. The state, for example, uses informants. So don't tell anyone what you're going to do, or what you've done. I mean anyone, including your new girlfriend who happens to be the daughter of the deputy sheriff, including your old friend whose new girlfriend happens to be good friends with someone whose mother goes to church with someone whose cousin believes that while dams need to come down some of those environmentalists just go too far and need to be turned over to the police before they hurt someone. I mean anyone.[280] Emails are traceable. Don't send emails, especially traceable ones. Forensics labs can pick up shoeprints. Cover your shoes, and then throw away your shoes and covers. Burn them. Burn all evidence, and then make the ashes disappear. Anticipate every response. No matter what you do, get there first.

☾ ☾ ☾

"Get there first with the most."

Next: *With the most.*

When it comes to winning battles, local superiority means almost everything. It doesn't matter who has the most troops all over the world: the important thing is who has the most troops right here right now. The United States can have more than 1.4 million soldiers in 135 different countries, and it can have about a million cops just in this country, but if there are four of you and none of them standing next to a cell phone tower, you have achieved local superiority. You got there first with the most, and you will probably win this particular battle. If the four of you show up and find you have not achieved local superiority, don't fight right here right now.

If Nathan Bedford Forrest encapsulated most military strategy into six words, baseball Hall of Famer Wee Willy Keeler accidentally distilled most guerrilla strategy into only five. Someone asked him what was the secret of his batting success, and he responded, "Hit 'em where they ain't."

Until those in power find ways to put surveillance microchips into each and every one of us—something they're feverishly working toward, by the way—they will not be able to be everywhere. This means that so long as we do not identify with them, so long as we have driven them from our hearts and minds, so long as we identify with our own human bodies and the land where we live, we will be able to hit 'em where they ain't.

Their security often stinks. We have been so long and so deeply pacified that for the most part we don't strike back, which means for the most part they do not have to defend the lands they've seized, nor even much that could very easily be attacked.

A report a few years ago revealed that security was so lax at the Grand Coulee Dam that local teens used the dam's interior as a skateboard park. This shouldn't be surprising. With 2 million dams and a more or less fully pacified populace, why bother with security?

Hitting 'em where they ain't is not the only way to win. But I don't believe our movement is large enough yet to allow us the luxury of pitched battles, which generally favor the larger army.[281] To return to the American Civil War, Federal General U.S. Grant had far more soldiers at his disposal[282] than his enemy, and so knew he could afford to hammer away with assault after doomed assault. At Cold Harbor, for example, men pinned their names and addresses to their shirts before charging, so that later their remains could more easily be identified. In that summer's campaign the Army of the Potomac suffered more casualties than there were soldiers in the entire Army of Northern Virginia. But Grant knew that even though he did not get to any of these battlefields first, he sure had the most. And he knew he would continue to have the most. And that was enough.

There is a sense in which for the foreseeable future we will never have the most. This is a problem everyone who has ever tried to stop civilization has faced. It was a constant complaint of the Indians. The Sauk Keokuk, who was highly esteemed by the whites for his conciliatory attitudes, argued that to fight back was tantamount to suicide, saying, "Few, indeed, are our people who do not mourn the death of some near and loved one at the hand of the Long Guns [pioneers], who are becoming very numerous. Their cabins are as plenty as the trees in the forest, and their soldiers are springing up like grass on the prairies. They have the talking thunder [cannon], which carries death a long way off, with long guns and short ones, ammunition and provisions in abundance, with powerful warhorses for their soldiers to ride. In a contest where our numbers are so unequal to theirs we must ultimately fail."[283] Keokuk's warlike rival Makataimeshiekiakiak (Black Hawk) spoke of the civilized in similar terms after he was defeated: "Brothers, your houses are as numerous as the leaves on the trees, and your young warriors, like the sands upon the shore of the big lake which rolls before us."[284] Recall the words of the Santee Sioux Taóyatedúta (Little Crow), who also spoke against fighting back: "See!—the white men are like the locusts when they fly so thick that the whole sky is a snow-storm. You may kill one—two—ten; yes, as many as the leaves in the forest yonder, and their brothers will not miss them. Kill one—two—ten, and ten times ten will come to kill you. Count your fingers all day long and white men with guns in their hands will come faster than you can count.... Yes, they fight among themselves, but if you strike at them they will all turn on you and devour you and your women and little children just as the locusts in their time fall on the trees and devour all the leaves in one day."[285] The Wyandot Between The Logs, who also was a friend of the whites (specifically the Americans) dropped the metaphorical language and put it bluntly: "I am directed by my American father to inform you that if you reject the advice given you, he will march here with a large army, and if he should find any of the red people opposing him in his passage through this country, he will trample them under his feet. You cannot stand before him.... Let me tell you, if you should defeat the American army this time, you have not done. Another will come on, and if you defeat that, still another will appear that you cannot withstand; one that will come like the waves of the great water, and overwhelm you, and sweep you from the face of the earth."[286] It's important to note that the Indians who cautioned against fighting still lost their land.

Each of these declarations by each of these Indians is in some ways a restatement of Thomas Jefferson's line, with subject and object inverted: "In war they will kill some of us; we shall destroy all of them."[287] I do not know any environmentalist or other type of activist who has not experienced the despair that comes from facing civilization's juggernaut of destruction. Let's substitute some words: "Few, indeed, are our people who do not mourn the death [clearcutting, damming, extirpation] of some near and loved one [forest, river, species] at the hand of the Long Guns [timber corporations, energy corporations], who are becoming very numerous. Their cabins [fellerbunchers, caterpillars] are as plenty as the trees in the forest, and their soldiers [police] are springing up like grass on the prairies.... In a contest where our numbers are so unequal to theirs we must ultimately fail." And, "See!—the white men [CEOs, clearcutters, developers, police] are like the locusts when they fly so thick that the whole sky is a snow-storm. You may kill one—two—ten; yes, as many as the leaves in the forest yonder, and their brothers will not miss them. Kill one—two—ten, and ten times ten will come to kill you. Count your fingers all day long and white men [CEOs, clearcutters, developers, police] with guns [palm pilots, chainsaws, maps] in their hands will come faster than you can count." And one more time: "Let me tell you, if you should defeat the American army [timber corporation, developer, police unit, or plain old American army] this time, you have not done. Another will come on, and if you defeat that, still another will appear that you cannot withstand; one that will come like the waves of the great water, and overwhelm you, and sweep you from the face of the earth."

Civilization has from the beginning devoted itself almost completely to conquest, to war. It's sometimes hard to say—and I'm not sure I care anyway—whether the civilized hyperexploit resources to fuel the war machine, or need a war machine to seize resources (which are then hyperexploited to fuel the war machine). It's probably a bit like asking whether the dominant culture is so destructive because most of its members are insane, suffering from a form of complex PTSD; or whether the dominant culture is so destructive because its materialistic system of social rewards—overvaluing the acquisition of wealth and power and undervaluing relationship—leads inevitably to hatred and atrocity; or whether the physical resource requirements of cities necessitate widespread violence and destruction. The answer is yes.

As George Draffan and I asked in *Welcome to the Machine*, "Why does our machine culture outcompete and overwhelm real live cultures?" We answered our own question: "Because machines are more efficient than living beings. Why are machines more efficient than living beings? Because machines do not give back. All living beings understand that they must give back to their surroundings as much as they take. If they do not, they will destroy their surroundings. By definition, machines—and people and cultures that have turned themselves into machines—do not give back. They use. And they use up. This gives them short-term advantages in power over the ability to determine outcomes. They outcompete. They overwhelm. They destroy."[288]

The point as it relates to the current discussion is that just as there are functional and systematic reasons we will never be able to outspend civilization, there are functional and systematic reasons we'll never be able to outgun them. *In a pitched battle.* But there are other ways to fight.

Hit 'em where they ain't.

❨   ❨   ❨

I just finished reading an account of Osceola, a Seminole Indian who fought against white theft of Seminole land. Having seen the difficulty of defeating the war machine in open contest, Osceola, according to this account, "had no intention of opposing the white men's armies in force. Instead, he conveyed the women and children to a safe place deep in the swamps and organized his warriors into small parties instructed to buzz about the whites like so many elusive bees, killing where they could and retreating into the safety of the swamps at the slightest evidence of superior force."[289]

It sounds like a pretty good idea.

Hit 'em where they ain't.

❨ ❨ ❨

Hallelujah! Stop the presses! Everything I've written is wrong! Things are going to turn out all right! We don't have to fight back! The President of the United States has had a miraculous revelation and gained ecological understanding. Check out this transcript of George W. Bush's acceptance speech for his second nomination as Republican presidential candidate: "So we have fought the corporations and the earth-destroyers across the Earth—not for pride, not for power, but because the lives of our citizens are at stake. Our strategy is clear. We have tripled funding for landbase protection and trained half a million first responders, because we are determined to protect our landbase. We are transforming our military and reforming and strengthening our intelligence services. We are staying on the offensive—striking corporations and earth-destroyers abroad—so we do not have to face them here at home. And we are working to advance liberty for indigenous peoples, because freedom will bring a future of hope, and the peace we all want."[290]

Isn't that wonderful? I'm so happy.

I'm so . . . deluded.

Okay, I'll admit it. That isn't *quite* what he said. But it was close! We just have to substitute a few words. Instead of *corporations* he said *terrorists*, and presumably meant the poor brown kind, not the kind dressed in U.S. military fatigues, or even more scary the kind dressed in business suits. And he didn't *actually* say the words *earth-destroyer*. And he didn't mention landbase protection but homeland security. In fact he didn't mention landbases at all. And he didn't mention indigenous peoples, either. But wouldn't it have been cool if he had?

❨ ❨ ❨

Okay, okay, I guess we have to fight back. Those in power aren't going to do it for us. And they're not going to do the right thing.

SYMBOLIC AND NON-SYMBOLIC ACTIONS

What do we mean by the defeat of the enemy? Simply the destruction of his forces, whether by death, injury, or any other means—either completely or enough to make him stop fighting.... The complete or partial destruction of the enemy must be regarded as the sole object of all engagements. . . . Direct annihilation of the enemy's forces must always be the dominant consideration.

Carl von Clausewitz [291]

EARLY IN THIS BOOK I MENTIONED THE BLACK BLOC, A GROUP OF anarchists who during the 1999 Seattle WTO protests broke windows of targeted corporations, and I hinted I had a quibble with their tactics. It basically boils down to their violation of Wee Willie Keeler's accidental dictum of guerrilla warfare. If your goal is to break windows at Starbucks or otherwise cause economic damage, why do it in the middle of a huge protest with phalanxes of cops in full riot gear just blocks away? Wouldn't it make more sense to hit and run at four in the morning? You'd probably be able to cause a lot more damage.

But of course my analysis is superficial. The primary purpose of the Black Bloc actions was—and I'm guessing because I was neither involved nor to my knowledge have I spoken about it with anyone who was—never to simply break windows. The purpose was to break the illusion that significant social change can come through means deemed acceptable or moderately unacceptable to those in power and to those, for example police, who serve those in power. The primary aim of the Black Bloc was to send a message. The economic damage caused to Starbucks and its insurers was secondary. The Black Bloc's actions, then, were primarily symbolic.

I have nothing against symbolic actions—I am, after all, a writer, and to write a book is almost as purely a symbolic action as you can get, since it is first and foremost an attempt to send a message—and I have nothing against the Black Bloc except that they've not yet succeeded in bringing down civilization. Of course I have the same problem with the rest of us, including myself.

But I want to take this opportunity to explore a distinction almost entirely ignored among those seeking social change, which is the difference between symbolic and nonsymbolic actions. A symbolic action is one primarily intended to convey a message. A non-symbolic action is one primarily intended to create some tangible change on its own.

A lot of people can't tell the difference. Some have told me, for example, that I should never blow up a dam because then my message would get lost amidst all of the dramatic action. I always reply that if I want to send a message I'll write a book. If I remove a dam it's to liberate a river. The symbolism would be at most secondary.

Not only activists fail to think clearly about distinctions between symbolic and nonsymbolic actions. I don't think many people in general think clearly about them. This shouldn't be surprising, since we're systematically trained to not be able to think at all.

The way the word *terrorist* is thrown around provides an example of the fuzziness of most people's thinking, and also a case study into how whether something is or is not a symbolic act has far less to do with the act itself than with the motivation behind it.

Let's talk about terrorism. We're going to ignore legal definitions of terrorism, because they're designed explicitly by those in power to protect themselves and to demonize their enemy du jour: U.S. soldiers or police (or to be fair, any government's or corporation's soldiers or police) murdering unarmed, unresisting (and often surrendered) civilians is not under their definition called terrorism, whereas when someone who opposes (especially U.S.) governmental or corporate interests kills unarmed unresisting civilians it *is* called terrorism.[292] To make even more clear the absurdity of allowing those in power (and those who serve those in power) to define the term, recall the editorial naming glassy-winged sharpshooters *terrorists* because they suck sap from grapevines. Just to let you know I'm not picking one insane example, the Oregon state legislature has repeatedly considered bills defining terrorism as any act that impedes commerce. As clear as that makes the primacy of premise five of this book, it makes a mockery of reasonable discourse.

Let's define terrorism, much more reasonably I think, as any act motivated by a desire to inspire terror or extreme fear in another in an attempt to change this other's (or a third person's) behavior. An act of terrorism is then an attempt to send a message. It is primarily a symbolic act.

If an Iraqi civilian kills a U.S. soldier (or civilian) to try to frighten others into leaving Iraq, this would be an act of terrorism. If the civilian were to kill the U.S. soldier or civilian in order to reduce by one the number of invaders—one down, a hundred and eighty-some thousand to go—that would *not* be an act of terrorism. Note that the person is just as dead in either case. Note also that the motivations can be mixed, with any particular killing being to varying degrees an attempt to both intimidate and eradicate invaders.

Similarly, if U.S. (or other) troops kill civilians (or soldiers) because they're fighting back (or because they're standing in the way, or for any other tangible reason) that is not an act of terrorism. It is not a symbolic act. If the purpose of the killing is to frighten other would-be resistors into compliance—the phrase *shock and awe* comes to mind—that *would* be an act of terrorism. It would be a symbolic act.

Likewise, if someone were to kill a vivisector in order to stop that particular vivisector from torturing animals, that would not be an act of terrorism. That would not be a symbolic act. If someone were to kill a vivisector in order to dissuade others from pursuing careers in vivisection, that *would* be an act of terrorism. That would be a symbolic act.

I don't want to beat this to death,[293] but I want to make sure this distinction between symbolic and nonsymbolic actions is clear, because to be honest it's not something we often talk about. Rather we more often confuse or conflate the two. So I want to give two more examples. If the state imprisons someone who has committed some crime, and the primary purpose of this imprisonment is to remove this person from society—to prevent this particular person from committing more heinous acts against society at large—that would *not* be a symbolic act, nor an act of terrorism. If on the other hand the state imprisons this person in an attempt to intimidate others out of committing similar acts—sentencing Jeffrey Leuers to more than twenty-two years in prison for burning three SUVs would certainly qualify—that would then be a terrorist, and symbolic, act.

Likewise, if someone were to kill a CEO to remove that person from society, either for retribution for acts of murder, theft, and ecocide the CEO has already committed—and you know that most CEOs have committed these acts, by the very "nature" of corporations—or to prevent that particular CEO from committing any more heinous acts against society at large, that would not be a terrorist act. It would not be a symbolic act. If, on the other hand, someone were to kill a CEO as a warning to other CEOs that they should stop committing murder, theft, and ecocide, this *would* be an act of terrorism, and would be a symbolic act.

I'm not conflating symbolism and terrorism, of course. I'm just using this as an example. Other symbolic acts could include writing letters, creating paintings, throwing pies, holding protests: any act intended primarily to send a message, as opposed to accomplishing something tangible. Me going to get my mail is a nonsymbolic act. I am attempting to make no statement to my neighbors or to anyone else by walking to the mailbox. I'm just picking up my mail.

Note that throughout all of this I am simply trying to be clear. I am not saying that symbolic acts are either better or worse than nonsymbolic acts. Clearly there are times when one is more appropriate than the other and times when the other is more appropriate than the one. It's the same, by the way, for acts of terrorism: I would have no moral problem, to use an obvious example, killing a tyrant like Hitler both to remove him from society and as a possible deterrent to other tyrants.

The problem comes, as it so often does, when we confuse or conflate things that should not be confused or conflated, in this case symbolic and non-symbolic actions. This misperception can go either direction. Sometimes people sending messages forget or ignore the fact that their message carries with it huge costs paid by those who, to them, are nothing more than the medium for their message. Those who run the U.S. military machine may be sending a message to would-be militants when they drop bombs on villages in a display of "shock and awe," but that message is written in the splattered blood of all those blown to bits who were, prior to their extinguishment, beings with lives and purposes all their own. When Harry Truman and the U.S. war machine dropped atomic bombs on Hiroshima and Nagasaki a potent message was delivered to the Soviet Union, but the message was delivered on the charred and sloughing flesh of those dying from radiation poisoning. These messages had their costs, paid by those who had little, if anything, to do with the message or its primary recipients.

This is not to say that one should only send cost-free messages—this book is written on the pulped flesh of trees, delivered to you through the use of oil, with all of its attendant costs, and I hope the message that *is* this book ultimately (and proximately) helps forest communities to survive. It's just to point out the blitheness with which those in power and the servants of those in power (I'm thinking again of the judge who sentenced Jeffrey Leuers) send messages involving costs paid inevitably by those perceived as lower than they on the social hierarchy.

This is yet another variation of premises four and five of this book. Within this culture it is acceptable, often desirable, for those higher on the hierarchy to use the bodies of those lower on the hierarchy to send messages. It follows that the messages of those higher on the hierarchy are, like their property, considered to be worth more than the lives of those below.

This whole situation can get very complicated and messy very quickly. This last week Chechen militants took over a school in the North Ossetian town of Beslan in southern Russia. For fifty-three hours they held more than 1,000 children and adults hostage, and in the end killed about 320 of them. Why did they do this? A former hostage reported that one of the hostage-takers said to her, "Russian soldiers are killing our children in Chechnya, so we are here to kill yours." Chechen commander Shamil Basayev, whom many think planned the takeover of the school, expanded on this: "However many children in that school were held hostage, however many of them will die (and have already died) . . . it is incomparably less than the 42,000 Chechen children of school age who have been killed by Russian invaders." He continued, "Dead children, dead adults—

brutal murder of more than 250,000 Chechen peaceful civilians by the invaders—all of it cries to heaven and demands retribution. And whoever these 'terrorists' in Beslan might be, their actions are the result of [Russian leader] Putin's policies in the Caucasus [and] in response to terrorism and crimes committed by the Kremlin's camarilla, which is still continuing to kill children, flood the Caucasus with blood and poison the world with its deadly bacilli of Russism." The website where this was posted then quotes the Bible: "What measure ye mete, it shall be measured to you." Basayev's got a point, in that this well-publicized atrocity is tiny compared to the routine atrocities committed by Russians in Chechnya that go unmentioned in the world at large (quick: name three massacres in Chechnya by Russians—heck, name three towns in Chechnya). It was noted in many accounts of the massacre that some of the killers were women. What was less often noted are the mass rapes of Chechen women by Russian soldiers. As Dr. Cerwyn Moore, a senior lecturer at England's Nottingham Trent University who has been studying the emergence of female suicide bombers, said, "There has been widespread use of war rape by contract soldiers. The subject is very delicate and hard to get facts on. But when you have Russian contract soldiers looting and raping—and I believe it's the accepted norm— you're going to have things happen later." And it's not just rape. It's murder. Moore noted that about 60 percent of confirmed female suicide bombers had lost husbands, and commented, "When you have a woman who's lost much of her identity because of her husband and family being killed, it's easier for her to be recruited."[294]

I think we can take Basayev and the killers at their word, that this killing was done in retribution for the killing of their own children: *you kill ours, we kill yours, fair enough?* But I believe it's also true that the Chechens were trying to send a message which I believe would run something like this: *stop killing our children.* The next question is: to whom are they trying to send the message? If they're trying to send it to the people of Beslan, I think they're trying to send it to the wrong people. I think it's safe to say that Russia is no more of a democracy than the United States, which means even if the people of Beslan receive the message loud and clear—even if they're terrorized into not supporting Russia's occupation of Chechnya—it probably won't cause the Russian government to withdraw from Chechnya. The people from Beslan almost undoubtedly have no more influence on Russian policy than the people of Crescent City, California have on United States policy.

I'd imagine Basayev and the others are fully aware of this. This makes me suspect that their message was intended not just for the people of Beslan but for

Putin and the others who run the Russian government, those who could actually make the decision to withdraw from Chechnya. But there's a big problem with this logic: it presumes that Putin and others of the Russian elite give a shit about the people of Beslan, an extremely doubtful proposition. Consider the United States: do you think George W. Bush and Dick Cheney care about your life, or the lives of your family? Their rhetoric aside, do you think they honestly care about the lives of American citizens? Do you think they care more for human beings than for corporations, production, personal financial gain, or increasing their personal and political power? If so, how could they possibly promote the use of pesticides? How could they promote the toxification of the total environment, with the consequent deaths of hundreds of thousands of Americans each year? If Bush, Cheney, and company cared about human lives, they would help us to prepare for the end of civilization. But they don't. They don't care about humans in general. They don't care about American citizens. They don't care about this or that small town. If Chechens obliterated the entire town of Crescent City, California, certainly the United States government would use that as an excuse to bump up repression at home and to conquer yet another oil-extracting country, but I can guarantee you George W. Bush and Dick Cheney would feel no pain.

The same holds true for retribution. The point of retribution seems to be: you cause me pain, and I cause you pain so you know how it feels. But I'm guessing Putin feels no pain over the deaths of these children. He undoubtedly feels a bit of a panic as he tries to deal with the public relations nightmare this situation has created. But pain? No.

Putin will almost undoubtedly follow Jefferson's lead in saying, "In war they will kill some of us; we shall destroy all of them."[295] But I realize now that Jefferson was lying all along, and what he really meant was: "In war they shall kill some of those whose lives we don't much care about anyway, and the troops we command shall destroy all of them."

I'm not saying that killing hundreds of children in some small town in southern Russia is a morally acceptable way to send a message to those in power. Nor am I saying it is not understandable that if some group is systematically killing your sons and daughters and husbands and wives and sisters and brothers and mothers and fathers and lovers and friends that you may want to lash out at members of that larger group. I am saying that there are much longer levers they could have used. If they were trying to send a message to Putin or others of the Russian elite, it probably would not have been a bad idea to strike closer to their home.

How would this play out differently if instead of killing children in Beslan, the Chechens killed Putin's children and the children of the others who command Russian soldiers to loot, rape, and kill in Chechnya? What if they skipped the children and went straight after the perpetrators? Would Putin then feel pain? Would that be a more understandable retribution? Would that send a message Putin could understand? Would Putin be so quick to commit more troops to this murderous occupation if he knew that by doing so he was placing his own life and the lives of those nearest to him at risk? Let me put this another way: Do you believe that George W. Bush and Dick Cheney would have been so eager to invade Iraq—oops, to order other people's sons and daughters to invade Iraq—if they themselves would have been in serious danger of being maimed or killed, and if they knew their children would be the first to die?

Not on your life. Not on theirs either.

❨ ❨ ❨

Now, Basayev could probably respond that they were in fact doing a smart thing—although still not a good thing—by hitting 'em where they ain't. But I would say that he's not even hitting 'em—by which I mean in this case his enemy—at all. He could argue that Putin and those close to him would be too well-protected to hit. I would say, *tough.* Proper creativity could find a way to get at them. And certainly Basayev could find a way to get closer than some semi-random school kids in some town far away from the center of Russian power. But to this Basayev could argue, correctly, that North Ossetia hosts one of the region's biggest Russian military installations and plays a key role in Russian efforts to keep the Caucasus under its control, and could argue further that the people of North Ossetia are generally more pro-Russian than those of the other small states between the Caspian and Black Seas. What, he might well ask, are our options?

To which I'd respond that going after this school would have been like anti-Nazi partisans blowing up a German school. What's the use of that? If you can't get to Hitler, why not hit a munitions factory, an oil refinery, or a train switching station? That, I would say, is hitting 'em where they ain't. Hit those places, hit them hard, hit them again, and hit them again, until you've crippled the economy.

Perhaps it's time to add something to Wee Willie Keeler's statement. If you want to win a guerrilla war, you should not only hit 'em where they ain't but you should also hit them where it hurts them the most. Every time. As hard as you can. As often as you can.

❨ ❨ ❨

Environmentalists and other social change activists often make the opposite mistake: we pretend that symbolic victories translate to tangible results. We hold great protests, make great puppets and witty signs to carry, write huge books, hang banners from bridges, get mentioned in newspapers, sing empowering songs, chant empowering prayers, burn sacred incense, hit politicians and CEOs with vegan cream pies, and participate in thousands of other symbolic acts, but the salmon still go extinct, phytoplankton populations still plummet, oceans still get vacuumed, factories still spew toxins, there are still 2 million dams in this country, oil consumption still continues to rise, ice caps still melt, hogs and hens and cows still go insane in factory farms, scientists still torment animals in the name of knowledge and power, indigenous people are still driven off their land, still exterminated.

We especially in the environmental movement are so used to losing that we have come to celebrate and often even live for whatever symbolic victories we can gain, whatever symbolic victories are allowed us by those in power. We lock down on logging roads and stop logging at this one place for one hour, one morning, one day, one week, and then we're removed, and so is the forest that has stood on this ground for thousands of years. But we do sometimes get some good press.

I have no problem with symbolic victories. Sending a message can be important, and is indispensable for recruiting as well as for shaping public discourse. Sometimes sending messages even makes a difference in the real world. I often think about a story a friend told me, about symbolic actions in response to violence. She wrote: "My friend Erica worked at the local Whole Foods Market (I know, the evil empire). There's a small but visible Tibetan immigrant community in the area, and a number of them had jobs at Whole Foods. This one Tibetan man had immigrated a few years back, saved up money and sent for his wife. She arrived and one day soon after he hit her. First thing to note is that she packed her bags and never went back to him. Second thing to note is that everyone else in that community would have taken her in for as long as she needed—she had multiple safe places to go. And third—this is the part I like the most—he was shunned by everyone. Nobody would speak to him. They would turn their faces sideways so as not to look at him and even put their hands up over their eyes to block the sight of him. If he tried to speak to people they would ignore him or walk away. When Erica started working at Whole Foods, this had been going on for two years! So it can be done. We could live in a world where violence against women had severe consequences that wouldn't even necessitate the removal of body parts."[296]

This is an important case study, but I hesitated to include it in this book because I have a concern that too many of us—I'm thinking especially of pacifists, but this concern applies to all of us—will attempt to too generously generalize this example. Too many of us will be tempted to say that because this shunning might have had an effect on this Tibetan man that we should do the same to George W. Bush and Charles Hurwitz. If only we avert our eyes, if only we make these larger-scale abusers feel supremely unwelcome, they will stop destroying the planet. But turning one's face away would only work within a face-to-face community. It only works when the other cares what we think. Hurwitz no more cares what I think about him than he cares about the forests he is destroying. Part of the key then is to *force* these others to care what we think.

For a symbolic act to bring about change in the real world, at least two conditions must be in place. The first is that the recipient of this message must be reachable. We can send all the postcards we want to Bush and Hurwitz beseeching them to stop killing forests, and it won't make a bit of difference. Here's an example. Although Bill Clinton's environmental record was disgraceful, he *did* enact a moratorium on punching roads into the relatively few remaining roadless federal lands.[297] He did this only after a public comment period lasting three years, during which the feds received more than 2.5 million comments, approximately 95 percent in favor of the moratorium. Now Bush has rescinded the moratorium, citing insufficient public comments and support.[298] The response by most environmental organizations has been to ask their members to send letters to Bush respectfully requesting he reinstate the moratorium.[299] Neither Bush nor Hurwitz—and by extension most of those in their positions—is reachable through these means. Neither of them gives a shit about forests, except as dollars on the stump. Sending them a million postcard messages—or a million messages on signposts, or a message from streets filled with a million protesters—will make neither of them do the right thing.

And it's not just those directly in power who are not particularly reachable. As we discussed earlier, I do not think the mass of the civilized will ever rise up to stop the destruction of the world.

But there are those who will.

The second necessary precondition for a symbolic act to bring about change in the real world is that the recipient of the message must be in a position to bring about that change. That is, the person must not only be willing, but able. This doesn't mean the person has to be in power, in fact I would say that for the most part those in power don't fit the first necessary precondition, in that they're not reachable, and if they are reachable (and get reached), as we've discussed earlier, they'll simply lose their power to someone else more well-suited to the psychopathology the system requires of those who make the decisions. It just means that while broadcasting a message can certainly be a good thing, we should also prioritize our efforts to recruit. One willing person with the right skills may help us more than a hundred semi-willing people who can write postcards.

The necessary preconditions for symbolic actions leading to significant social change are not often in place. Most of our actions are frighteningly ineffective. If that weren't the case we would not be witnessing the dismantling of the world. Yet we keep on doing the same old symbolic actions, keep on calling the making of this or that statement a great victory. Now, don't get me wrong, symbolic victories can provide great morale boosts, which can be crucial. But we make a fatal and frankly pathetic error when we presume that our symbolic victories—our recruiting and our morale boosting—by themselves make tangible differences on the ground. And we should never forget that what happens on the ground is the *only* thing that matters.

There comes a time in the lives of many long-term activists when symbolic victories—rare even as these can be sometimes—are no longer enough. There comes a time when many of these activists get burned out, discouraged, demoralized. Many fight despair.

I think fighting against this despair is a mistake. I think this despair is often an unacknowledged embodied understanding that the tactics we've been using aren't accomplishing what we want, and the goals we've been seeking are insufficient to the crises we face. Activists so often get burned out and frustrated because we're trying to achieve sustainability within a system that is inherently unsustainable. We can never win. No wonder we get discouraged.

But instead of really listening to these feelings, we so often take a couple of weeks off, and then dive back into trying to put the same old square pegs into the same old round holes. The result? More burnout. More frustration. More discouragement. And the salmon keep dying.

What would happen if we listened to these feelings of being burned out, discouraged, demoralized, and frustrated? What would those feelings tell us? Is it possible they could tell us that what we're doing isn't working, and so we should try something else? Perhaps they're telling us, to switch metaphors, that we should stop trying to save scraps of soap and try to bust out of the whole concentration camp.

I hate wasting time on makework. It's not that I'm lazy, far from it. I love accomplishing things that I want to accomplish, and love working furiously when I see movement. But I'm extremely sensitive as to whether the work I'm doing is actually accomplishing anything. And the feeling I get when I'm working futilely feels a lot like burnout, discouragement, frustration, and so on. I've felt this sensation often enough to know that it doesn't mean I need to take two weeks off and then come back and do the same damn useless job, nor does it mean I need to work even harder at this damn useless job. Nor does it mean I need to collapse into a sobbing heap of self-pity. None of those do any good. It usually just means I need to change my approach so that I accomplish something in the real physical world.

Useful work and tangible accomplishments make burnout go away quickly.

《 (《

This once again raises the question of what we really want. Do we want to slow the grinding of the machine just a little bit? Do we want to stop it completely? Do we want the Giants to win the World Series and oh, by the way, it would be nice if we still have a world? Do we want to keep our cars and computers and lawns and grocery stores even at the expense of life on the planet? More to the point, do we want to allow others to keep their cars and computers and lawns and grocery stores even at the expense of life on the planet, which of course includes at the expense of poor humans?

Most of the people I know recognize that the choice really is between life and civilization, and if they could snap their fingers and make civilization go away they'd do it in a heartbeat. I know many people who were hoping and praying that Y2K would bring it all down. Now the big hope is peak oil. Some would not be unhappy if a virus took us all out. Anything to stop civilization's grotesque destructiveness.

Look, however, at what these three hopes all hold in common: they're beyond our control. There are many of us who want civilization gone, and who would even conjure it away through magical means if we had them, but in the real physical world don't know how to bring it down, or if we have some useful knowledge, we do not want to take responsibility for actually doing what needs to be done. That's the bad news.

The good news is that there are those who are willing to take on that responsibility.

《 (《

I talk again to Brian. This time we're alone. This time we meet at a Chinese buffet, not too long before closing time.

I love buffets. Partly because I have Crohn's disease—a disease caused by civilization—my metabolism is remarkably inefficient, and I need to consume prodigious amounts just to keep going. I'm usually the skinniest person in line at the buffet, but my plate is fuller even than those who clearly don't need the food. Quantity is always perforce more important to me than quality, and when I'm planning on eating some fine food or when someone takes me to dinner, I usually try to eat beforehand so when the real meal comes I can eat more like a normal person. Only at buffets do I ever get enough to eat, and Chinese buffets are the best.

But I also hate buffets. They're too-perfect mirrors of the way we the civilized perceive the world. There laid out before us is the entire planet ready for us to eat. Plate after plate of pieces of the world, a steady stream from the kitchen brought to us by smiling servants, plates without end, all for us to consume, with no reckoning save an entirely too-small financial payment.

This buffet has piles of disembodied crab legs, frozen shrimp, lobster, pork, beef, broccoli, chicken, noodles, rice.

I do not mind the fact that when I eat I am consuming death. Life does feed off life. Someday someone will consume my life, and my death. I do mind that I live in an entire culture that keeps itself willfully ignorant of the fundamental predator-prey relationship, else it would cease to be the culture that it is. This means that when I eat wild creatures I am too often contributing to their terminal decline. I also mind that eating domesticated creatures—and I include plants in this—means I'm eating misery. And perhaps more even than this I hate the unforgivable wastefulness of it all, the by-catch and poisons and other disregarded side effects of gathering or growing, transporting, storing, and preparing this seemingly never-ending buffet.

That said, I'm glad the restaurant is near closing, because that way I won't feel guilty abut consuming all this food, which would otherwise just go into the dumpster. *Fill the plates, boys!* I'm surprised to see the heapings on Brian's plate: he's as thin as I, and doesn't even have an intestinal disease as an excuse for how much he eats.

He dives right in: "Why are people afraid of hackers? Actually their fear has little to do with us, but instead it's with the realization—a realization they avoid as much as they possibly can—that they have entrusted their lives and the fabric of their communities to a device about which they have no understanding."

"What device is that?"

He looks as though he's disappointed I even had to ask. "Computers," he says. "Despite the media claims, when serious hackers write viruses, worms, and so on, we generally act not out of depravity, boredom, or any other petty thing. It's not mere caprice; the vast majority of us are highly aware of the politics of computing and information. We understand the weight of our actions. That's a prerequisite for any sophisticated hack. When people realize that the macro-based mail viruses that make worldwide headlines affect only Microsoft and particularly Outlook—many free software alternatives are immune—then they might stop getting mad at the hacker nuisance and start looking at the fact that one corporation controls international communications."

"But most people don't care about the politics," I say. "They just want to send emails."

He throws back his head and laughs, a gesture I'm already growing fond of. "And most people don't care about salmon or forests or songbirds or phytoplankton. They just want to watch their televisions after working the jobs they hate. Just because people don't care about the political or environmental underpinnings and consequences of the technologies in their lives doesn't mean these underpinnings and consequences don't exist."

"But messing with people's internet access won't help. It's just going to piss them off."

Brian laughs again. "And taking out dams won't piss people off? How about taking down civilization? You're being silly. Besides, if people's email stopped working, they might have to talk to each other. And if their computers stopped working completely, they might go outside and begin to directly experience the world around them. I'd personally like to see some more people pissed off at the corporations whose inferior products and nefarious business practices have made even simple letter writing so burdensome."

I break open a crab leg.

Brian chews on some broccoli. He swallows, then says, "I've read articles in the mainstream press, what I guess you would call the capitalist press, that say that a lot of hackers are interested in control or power, in that we want to control the way people access the internet. But that's not true. Hacking is about breaking—or at the very least braking—the control large corporations and governments have over that access. Those in power use computers to control us: do you think the current levels of surveillance, for example, would be possible without computers? And if you think today's levels are obscene, wait till tomorrow. No, we don't want power. We want to fuck up those in power. We want to restore freedoms."

LIKE A BUNCH OF MACHINES

This is a kind of crude sketch, but it's easy enough to follow. And, you know what? The rewards of following it don't have to be deferred until the aftermath of a cataclysmic "revolutionary moment" or, worse, the progressive actualization of some far-distant Bernsteinian utopia (which would only turn out to be dystopic, anyway.) No, in the sense that every rule and regulation rejected represents a tangibly liberating experience, the rewards begin immediately and just keep on getting better. You will in effect feel freer right from the get-go.

Ward Churchill[300]

THOSE IN POWER DO NOT SO OFTEN WIN ONLY BECAUSE CIVILIZATION'S social order is organized around converting living landbases into raw materials and raw materials into weapons which are used to conquer further landbases. They do not so often win only because civilization's social order is organized around not giving back. They do not so often win only because within civilization's value structure the acquisition, accumulation, and mass exploitation of resources is more important than morality or community: worse, the acquisition, accumulation, and mass exploitation of resources has within this social structure been converted into a virtue, probably the highest. They do not so often win only because we so often do not fight back. They do not so often win only because they have more soldiers than we do.

No, one reason they win is because they're so very single-minded. Destroying the planet—the current euphemisms for this include "developing natural resources" and "making money"—is the most important thing in the world to these people (by whom I mean those who make the primary decisions for this culture, the mass of the civilized, and civilization taken as a whole). They are psychotically driven, with an energy far beyond the rational. Destroying the world—called, once again, "developing natural resources" or "making money" or "Manifest Destiny" or "making the world safe for democracy" or "fighting terrorism" or "expanding free markets" or any other claim to virtue—is not an avocation nor even a vocation. You could say it's a passion, if you use the dictionary's fourth definition: "intense, driving, or overmastering feeling or conviction."[301] But it's beyond that. It is their obsession, their compulsion, their necessity. It is their conscience and their compass. It is their master and they are its slave. It is their God and their king, their spur and their whip. It is their subjection, their burden, and their source of strength to carry that burden. It is their crisis and their obligation, their desire and their demand. It is the demand made upon them. It is their ultimatum and the ultimatum given to them. It is their charge, their mandate, their command and the command made to them. It is their food and water (indeed, it is obviously more important to them than their food or water). It is their air. It is their life. It is their reason for living. It is their *raison d'être*. It is their essence. It is *who they are*.

But it is deeper even than that.

❰ ❰ ❰

The American Indian activist John Trudell, whose wife, young children, and mother-in-law were burned to death in a house fire almost undoubtedly set by agents or allies of the federal government,[302] said, "We must never underestimate our enemy. Our enemy is committed against us twenty-four hours a day. They use one hundred percent of their effort to maintain their materialistic status quo. One hundred percent of their effort goes into deceiving us and manipulating us against each other. We have to devote our lives, we have to make our commitment, we have to follow a way of life that says that we are going to resist that forever."[303]

❰ ❰ ❰

Have you known many abusers? I mean really known many people who really are abusers. If so, did you notice how quickly and completely and seemingly effortlessly they spin manipulative webs, how they so often say precisely the things that will make you feel the absolute worst, how they so often so perfectly know where to strike at your weakest point? It's uncanny. To argue with an abuser sometimes feels like playing chess with some grand master who plays out every possibility ten moves in advance, and anticipates your every response. And who cheats.

It may sometimes seem like intelligence allows abusers to weave these seamless manipulations, but that's not it. And it may seem they stay up all night scheming of ways to hem you in, but that's not it either.

It's much worse than either of these.

I want to tell you four stories. The first: Years ago I got into an argument with someone in Nevada over a pending wilderness bill for the state. The other person argued that Nevada already had more wilderness than the rest of the United States combined. I'd heard this argument from anti-environmentalists before, and so I'd recently checked out a book from the library listing all of this country's wilderness areas. Not only did Nevada not have more than the rest of the U.S. combined, it had less than any single state west of Nebraska. I told the person this. He said he didn't believe me. I told him again. He said he still didn't believe me. I said I had a book in the car that would prove it. He said he still didn't believe me. I went to the car, brought in the book, showed him the figures. Without hesitating he said, "That's what I've been saying all along: there's too much wilderness in this country already." What impressed and appalled me most about this exchange was not that he was cheating—when facts proved him wrong he pretended he'd been making a different argument all along, just so he could be right—but that he'd changed tack *immediately*. Not quickly. Not after a moment to think. Immediately. It had been essentially a reflexive response.

The second: Also when I lived in Nevada, I often went with my then-brother-in-law to the dump. We'd pile all their trash in the back of his ancient white pickup, and then we'd all pile in the front. Al would drive, I'd ride shotgun, and his two older daughters, aged seven and four, would either sit between us or on my lap. We'd drive the few miles to the dump, and then have a great time throwing the overfull bags from the back and watching them explode on contact with the ground. One day my oldest niece brought a friend. The cab was full, with the four-year-old near the gearshift, the seven-year-old on my left leg, and her friend on my right. Even before Al pulled out of his driveway, I knew something was very wrong. My niece's friend slightly spread her legs and squared her labia through her shorts against my lower thigh. She twined her fingers in mine, then forced each hard against the crotch of each of my fingers. She slid them up and down. I was naïve enough to not understand what was going on, but also experienced enough with little girls to know how they're *supposed* to sit on your leg, and how they're *supposed* to play with your fingers. I was extremely uncomfortable and confused. All became clear a few weeks later when the news broke around this very small town that this little girl's older brother was raping her. I suddenly at least slightly understood her inappropriately sexual behavior. She was not, I don't think, consciously attempting to express herself sexually. Instead her behavior came from two primary sources. The first is that she was acting as she had been trained. The second is that she was unconsciously trying to tell the story of her trauma: she could not speak it verbally but she could speak it with her body.

The same is true for many of us, about many traumas.

The third story is that many years ago I for a time became close friends with a self-described sociopath. I admired Lauren's brilliance and wit. She was an extraordinary sculptor and a tireless activist. Her politics matched mine. But I quickly discovered that any personal information I shared with Lauren she stored and used against me in startlingly creative ways. I introduced her to my friends, and it took me several weeks to see the pattern that soon after each introduction the old friend and I would get into terrible arguments, always in some way about Lauren. Sometimes I'd find myself defending some inappropriate behavior on Lauren's part, sometimes I'd find myself defending myself against something—often some lie—Lauren had said about me to my old friend, and sometimes it would be something trivial between my old friend and me that somehow got blown out of proportion through some strange connection to Lauren. She came on to every one of my male friends, whether or not they were involved with someone, and continued to do this even after I pointed out how inappropriate it was. As with the incident with the little girl, I was very disturbed and confused by all of this until the pieces started coming together. When I saw the common denominator in all the arguments was Lauren, I confronted her. She told me about the abuse—physical and sexual—she had suffered as a child, and tearfully told me she would never do any of this again. But the next time she saw a male friend of mine, she once again made suggestive comments. And the time after. And the time after that. She couldn't help herself. But there were no more arguments with my friends, because seeing the pattern had immunized us against her manipulations. I was no longer particularly upset by her behavior. I may as well have gotten upset at the little girl on my lap. One was chronologically an adult, but both were acting as they had been trained to act. And both were trying to tell their trauma stories, stories they could not speak with their voices but only with their actions.

The fourth story: A friend of mine with a doctorate in psychology worked for a time at a psychiatric outpatient clinic. Although the clinic served a great variety of patients, she primarily worked with low-functioning psychotics. In plain English this means the people she worked with had below average intelligence as well as impaired mental health but were with some assistance still able to live on their own, to marginally function in society. She did not mention if her clients included the forty-third president of the United States.[304] The thing that most struck her about some of her more disturbed clients, she often told me, was their capacity to spontaneously manipulate those in their surroundings, pitting each against the others, now sucking up to one, now cajoling another, now giving seductive vibes to a third, and now giving off an air of innocence to a fourth. These actions seemed almost like autonomic body functions, like breathing, like digesting and defecating, like pulling back from pain. The manipulations would pile one atop the other, with small lies covering big lies, big lies covering small lies, just enough specks of truth sprinkled in to cause confusion. So long as one wasn't sucked into the manipulations, my friend said, watching the clients was like watching jugglers or prestidigitators, as these psychotic people moved without thinking faster and faster to maintain their multiple webs of manipulation. All this from people who didn't have the cognitive ability to read a bus schedule or set an alarm clock.

My friend and I talked often about the implications of all this. Clearly the clients were cognitively incapable of consciously perceiving the weak points of every person in the room, consciously creating plans of attack, consciously creating simple declarative sentences that would most effectively pit three or four people per sentence against each other, consciously anticipating each person's response to the attempted manipulations, consciously planning counterresponses, and consciously keeping track of all this. To do all of this consciously would require a lot of cognitive facility and energy. Just listing it out gives me a headache and makes me tired.

My friend and I soon began to talk about the psychoses somehow having intelligences and energies all their own, almost independent of the individual's native intelligence and energy. Ted Bundy spoke about this. He was asked by police to help them profile the Green River Killer. He said that many serial killers in some ways have a certain clarity and awareness, an ability to read people and situations instantly, "not in an analytical way but in an intuitive way."

I recently called another friend with a Ph.D. in psychology to ask her what's behind all this. She said, "You have to go to the neurological literature, but basically what you've got is someone whose brain has been trained to live and operate in trauma. In order to survive trauma you have to have an extraordinary ability to read and respond to others. Not in the soft way lovers read and respond to each other but in a fear-based way: if you don't read the others accurately you could be beaten, raped, or killed. If those are the conditions under which you're living, those are the rules you must live by. If you're living in a place where you're constantly under threat of attack you have to learn to outstalk or outfox your attacker. It's not about morals. It's about what works. Their behavior looks and *is* perfectly normal and functional when they're in a room full of people who actually could attack at any moment, but when you put them elsewhere, their behavior looks and *is* really odd. You can tell them that they're no longer under threat of attack but their brain is wired for threat. And to actually retrain the brain is very difficult."[305]

I thought about Lauren, and how her behavior was adaptive to the circumstances of her childhood. Her father often beat her mother, her siblings, and her. He often raped her. Her mother often beat her and her siblings. Her brother often raped her and her sister. A neighbor raped her and her sister. And she was raped by others. It's no wonder she came on to every man she met: not only is that *how* one interacts with men, but if they're going to take it anyway, you may as well maintain control by giving it first. It's no wonder as well that she became brilliant at pitting people against each other: *If Father is mad, better his anger is directed at Sister or Brother than me.* Within that context her behavior not only made sense but was inspired. Out of that context, it drove me and her other friends away.

What does this have to do with civilization killing the planet?

Everything.

Let me tell you another story. My old friend the forest activist John Osborn has long said, "The reason we always lose is that the other side knows what it wants, and we have only the faintest clue what we want. They want every last tree, every last stick, and they want it now.[306] We don't know if we want smaller clearcuts, fewer clearcuts, better clearcuts, or what. They are driven to deforest, and are rewarded financially for doing it. Most of us are not driven in the same way. For most of us it's a sideline to our main career, and certainly doesn't pay our bills."

❲ ❲ ❲

A friend of a friend worked on a Democratic senate campaign. He told my friend, "Fighting the Republicans is hopeless. We can usually only come up with two or three dirty tricks per day to play on them, but they come up with five or six before we've even had our first cup of coffee. They're like a bunch of machines."

❲ ❲ ❲

Like a bunch of machines. That's it, isn't it? That's what happens when you remove relationality from your worldview.

❲ ❲ ❲

John Osborn was right, and he was wrong. The truth is that those in the timber industry only think they know what they want. But they don't really want what they want. They want something else. They don't really want trees. They don't really want money. They don't even really want power. Lauren didn't really want to sexualize every relationship. She didn't want to pit me and my friends against each other. She didn't even *see* the men she sexualized, and she didn't see me and my friends. We existed no more than trees do to deforesters. The little girl on my lap didn't really want to sexualize a trip to the dump and the antienvironmental person I talked to in Nevada didn't really want to talk about wilderness areas.

Recall the conversation I had with Luis Rodriguez about gang kids standing on street corners shooting at mirror images of themselves. They don't really want to kill or to die; they want to undergo a transformation, in this case spiritual and metaphorical. They want to grow up.

These kids could kill all the mirror images of themselves they find, and they would still not find what they are seeking. Timber corporation CEOs could deforest all the continents they could find, and they would still not find what they are seeking. My sociopath friend—sociopath former friend—could destroy all the relationships she could find, could come on to all the men she could find, and she would still not find what she is seeking. We can talk to them all we want, and it will not make any difference because we are never talking about what we are talking about.[307] It's as my friend wrote me so long ago: "The point is that I strongly believe that unmetabolized childhood patterns will *always* trump adult-onset intellectualizations."

This energy—this energy to destroy the world—then is literally insatiable. This is precisely the sort of energy and intelligence we have to deal with. This is precisely the sort of relentlessness we must defeat.

❰ ❰ ❰

If they don't really want to deforest, if they don't really want to destroy relationships, if they don't really want power, if they don't really want to kill the planet, what do they really want? They want their fear to go away. Normally fear comes either from outside (e.g., someone pointing a gun at you) or inside (e.g., seeing someone whose looks remind you of someone who long ago shot you). The former sort of fear can be diminished or eliminated by changing one's circumstances. The latter kind of fear requires exploring and coming to understand how you got that fear in the first place. But because abusers and psychopaths both blame others for their actions, they cannot acknowledge that this current fear could originate inside, not out. This means the fear can never be abated, which means in practice that their desire to make their fear go away manifests as a desire to control. They want to control everything around them, so that everything around them does not hurt them. Raised in a culture of trauma, the rules of trauma are those they must live by. But the only way to control everything in this way is to kill everything.

As we see.

I hesitated before writing down what they really want, because at this stage I don't think it so much matters. Remember the monkeys made permanently psychopathological. Remember the rates of recidivism among abusers. Remember the civilized who time and again have met the indigenous and who have not learned from but killed them. The psychopathology is permanent because the void they feel can never be filled, the fear they feel (or sometimes don't even acknowledge they feel, yet nonetheless drives their lives) can never be resolved or even sated. What matters at this point is stopping them from killing all we hold dear.

❧ ❧ ❧

The other day I was out with my machete hacking away at blackberries, when I realized that so long as I approach this task the way I do I will never get rid of the berries. I cut them out only when I feel like it, which isn't all that often. And even when I do, I only take out the blackberries at the edges, where they're encroaching into territory now held by other plants. I never do get all of them, which means they're constantly expanding. And since I never go after the roots even the plants I hack come right back. It seems for every vine I take out, ten more vines pop up to take its place. Further, I'm conflicted about taking out even the vines I do. What right do I have to kill these others who are merely trying to live, even as I am trying to live? I feel bad each time I take on the responsibility of killing one, even though I know that when I don't kill them they kill native plants. So far as I can tell, the blackberries aren't quite so conflicted about crowding out these others.

Standing here sweating, machete in hand, I think of a few more reasons the Indians were unable to stop the whites from stealing their land. The first is that for most of the Indians fighting and war were not a way of life, but rather avocations. Fighting was something you did for fun, in your spare time. Wars in their cultures were the equivalent of sports in this culture, exciting, scary, strenuous, and not all that dangerous. And when you no longer felt like playing, you went home.[308] If one side is psychotically driven to war, and the other fights for fun, guess which side is ultimately going to win? The second is that like me with the blackberries, the Indians did not strike at the root. Their wars were strictly defensive, in that they killed settlers and armies moving onto their landbases. The Shawnee, for example, killed whites moving into what is now Ohio and Kentucky, respectively where the Shawnee lived and hunted. They burned forts on the frontier, but they did not sack Philadelphia. They did not strike at the infrastructure that allowed civilization to expand. If one side is always invading the other's territory, and the other is forced to fight only defensively, guess which side is ultimately going to win? The third is that the Indians did not fight wars of extermination. If one side fights a war of extermination, and the other does not, guess, once again, which side is going to win?

BRINGING DOWN CIVILIZATION, PART II

Alright, let's follow things out a bit further. The more disrupted, disorganized, and destabilized the system becomes,
the less its ability to expand, extend, or even maintain itself. The
greater the degree to which this is so, the greater the likelihood
that Fourth World nations struggling to free ourselves from sys-
temic domination will succeed. And the more frequently we of
the Fourth World succeed, the less the ability of the system to
utilize our resources in the process of dominating you.

Ward Churchill[309]

BRIAN SAYS, "BROADLY DEFINED, HACKING IS A PROCESS OF EXPLORING in and negotiating with any given universe, understanding it and interacting with it on an intimate, personal, and unrestricted level. Hacking, then, is incompatible with civilization."

"Why?" I ask.

"We both know the answer to that," he says.

"I want to hear you say it."

"From the beginning civilization has been based on enclosure. Land held in common for the common good was closed off and became the property of the powerful, with any exploration or experience of that area possible only with their permission. Personal direct experience of the divine was closed off, too, made possible only through the priests of the powerful, and the books they wrote. The Bible comes to mind. The gaining of knowledge through personal direct experience has been closed off as well, as we're supposed to trust scientists—the *other* high priests of the powerful—to tell us what's what. Those in power close off our water. They're closing off our air, or I guess you'd say they're making it so the air is so foul it closes off our throats: our throats actually refuse to take it in. They close off our time: by closing off our access to land they make it so we have to pay rent and pay for food, which means they're stealing our time, stealing our very lives: labor for the common good and labor for our own good got replaced with working for the man. It's all of a piece. For several thousand years now those in power have been hemming us in, closing off our possibilities. And hacking is all about reopening those doors that have been closed on us."

"How?"

"Well, here's a literal example: if you encounter a locked door and you want to get to the other side, you can pick the lock, or you can break down the door. I prefer the former. A hack puts brain over brawn, but I achieve the same end as somebody who forcefully overcomes these artificial limitations on his movement. This concept applies to many such devices."

"Makes sense."

"Hacking is at least as much about ideas as about computers and technology. We use our skills to open doors that should never have been shut. We open these doors not only for our own benefit but for the benefit of others, too."

《 《 《

I have heard two other primary arguments against bringing down civilization. The first is that the collapse of civilization will be a disaster for women, and the second is that the collapse of civilization will be a disaster for the natural world. Both of these arguments have come from people who fully recognize that civilization is already a disaster for both women and the natural world. They just think its collapse will be worse.

The first argument is that as the current social fabric unravels—especially if it unravels cataclysmically—women will, as always within a culture that oppresses women, bear the brunt of it. Civilization's collapse will bring with it a world where women are used, a world foretold in movies like *Mad Max* and *A Boy and His Dog*. A line from the latter gives a flavor of this possible future and also makes clear the movie's gender politics: when the character played by Don Johnson comes across the body of a woman who has been gang-raped and murdered, he says, "What a shame! She could have been used three or four more times." Unfortunately I do not think we have to wait for some science fiction future for this worldview to be made manifest. That these gender politics already manifest a strong cultural desire is revealed not only by the fact that this movie is considered by some a cult classic (or in fact that it was made at all), but even more by the audience response. A typical post from a film fansite runs: "How can you not like this film? Boy gets girl, boy loses girl, boy recovers girl, AND FEEDS HER TO HIS DOG! YES! I recorded this movie off of PBS and I'm glad I did. . . . First and only time I ever saw nudity on PBS." Here's one more, just to show the first one wasn't a fluke: "One of the best (sickest) post-nuclear nightmare flicks I ever seen. What could be better than a horny teenage Don Johnson walking around the deserts of America with a telepathic dog, helping him get laid [*sic*]? I hope my post-nuclear apocolypic [*sic*] life style would be this good."[310]

It's late at night. I'm in the studio of a college radio station. I'm on air, talking about taking down civilization. A woman calls in, raises the concern about mass rapes as this society collapses.

I respond, "Mass rapes already occur. We already live in the midst of a culture of mass rape. How many women do you know who haven't been raped?"

"Not many," says the woman on the line.

"None," says a woman sitting at another microphone in the studio.

I continue, "It's just that most of these rapes are committed by those close to these women. Their fathers, brothers, uncles, cousins, neighbors, lovers, and so on."

I see from her face that the woman in the studio knows what I'm talking about.

I go on, "That's not to say rates of rape might not go up. You raise a great point about the relationship between a breakdown of communities and an increase in rape. I'm guessing there are two main reasons for this. The first is that as communities collapse individual men may take out their frustration on women through battery and sexual abuse. Is that your understanding, too?"

Both women say, "Yes."

"And the second reason is probably that as 'law and order' breaks down police will be less able to protect women."

A positive murmur from the other end of the phone.

"But the first presumes that men don't already do that and in any case is an explicit acknowledgment of the abusive—and terrorist—nature of current gender relationships: if anything causes me as a group frustration, you as a group are going to pay the price. And the second presumes police protect women now. But they obviously didn't protect the women you know who've been raped, and in most cases I'll bet they didn't do anything to help the women afterwards, or to protect other women from the men who raped them."

"That's right," says the woman on the phone.

The woman in the studio leans forward, says, "I know of exactly one man who got arrested for raping a woman. The trial was an absolute nightmare for the woman. Most of her family sided with the assailant, her uncle. The defense and the judge both somehow simultaneously acted like it didn't happen and like she brought it on herself. Never mind that both of those couldn't have been true. She regretted that she even brought charges." The woman pauses, then says, "She told me later that if she wouldn't have been fully aware of what the cops and the courts do to women who kill their rapists, she would have shot him dead. But *that's* a crime the judicial system loves to sink its teeth into."

I agree, then add, "Here's something else to think about. If the destruction of communities leads to higher rates of rape, as it often seems to, then right now the rapes are just being exported. Civilization and the global economy destroy communities all over the world. That's what they *do*. In order for us to maintain our lifestyle, we have to import resources. The importation of those resources requires the destruction of communities in the colonies. The communities are destroyed so the resources can be stolen, or the resources are stolen and in the process the communities are destroyed. Either way this exploitative lifestyle leads to atrocities. I can see how women in the colonies

would be praying for civilization to come down before it destroys their communities, before it leads to their rapes."

I pause a moment before continuing, "If someone is really concerned about rape there is plenty of work to be done right now both here and in the colonies. Plenty. And even if your primary concern is whether there will be mass rapes—or rather more mass rapes—when civilization comes down, there's still plenty you can do. Teach women self-defense. Teach them how to use guns, and because guns will eventually run out of industrially manufactured bullets teach them how to use bows and arrows. Teach them how to use different types of knives. Teach them how to kill assailants with their hands. Not only teach them how to fight back, but even more important, teach them to fight back at all. Once they've determined to fight back everything else is technical. And more important even than this, form protective collectives where both men and women are prepared to defend each other. Those collectives need to be put in place now, because whether or not we bring it about, civilization is going to come down. And in fact, even if civilization doesn't come down for a while—for far too long—then these protective collectives would be important anyway. We need to protect each other from rape and other forms of exploitation in any case: the cops sure as hell aren't doing it."

Both women agree.

"And one more thing about bringing down civilization," I say. "If you care about women, it means you care about human beings. And if you care about human beings you have to care about landbases, because destroying landbases destroys human beings. Civilization is destroying our landbases. Civilization needs to be destroyed."[311]

❨ ❨ ❨

I break open more crab legs. I don't normally eat crab, but these are destined for the dumpster anyway.

Brian says, "Hackers are frequently cast as disobedient; that's a shame. We don't disobey the law, per se. We simply ignore it. Our actions operate independently of the artificial constraints that have been put on us."

I put down my crab legs to take some notes.

He smiles, then states, "You flatter me, writing down what I say. Usually people only do that when they're preparing an indictment."

I start to write that down.

He chuckles.

I look at him hard and say, "Who are you?"

He laughs again, throws back his head. He says, "I'm just like you. I'm a human being who has a fondness for his species, and other species, and life. I'm someone who tries very hard to think clearly."

"That's difficult," I respond. "No one wants clear-thinking slaves, because they might start to think about the whole system that enslaves them."

"It's a very strange game we all end up playing. It's a very strange game we *have* to play if we're to allow this system to continue. We have to think we're clear-thinking as we don't think at all. We have to feel like we're in control as we have almost no control over our lives or our communities. We have to pretend we own things as they in fact own us."

"Just the other day someone said to me that 'we' need to continue deforesting because 'we need' the paper and wood products."

"All of these devices are ultimately superfluous," Brian says. "The needs are artificially created. Who *needs* printed circuit boards? We need air and water and food. We don't need software (especially Microsoft). This culture specializes in giving people diseases, then selling them the cures."

"But they're not real cures . . ."

"Oh, no, or you wouldn't have to keep on coming back. Maybe I should say the culture specializes in giving people addictions, then selling them the smack."

❮ ❮ ❮

The Dutch sociologist and drug addiction counselor Kees Neeteson has written, "Modern Western culture has to contend with a shortage of satisfying existential ideologies. For centuries a reduction has taken place from spiritual toward materialistic thinking, culminating in today's technological consumption society. This society depends on mass production and mass consumption, on ideologies which are superficial [and] therefore easy to manipulate, and on advanced technology and military power. One of the results of this process is that the average individual cannot obtain enough meaningful satisfaction from common social life."[312]

Addictions move in to fill the void of meaning once filled by relationship to community and most especially relationship to landbase. In his book *Rational Madness*, drug and alcohol abuse counselor Ray Hoskins calls addiction "a false path to meaning based on false beliefs, inept coping behaviors, and a basic self-centeredness which treats symptoms instead of coping with reality." Sound familiar? It might not, because this pattern of only treating symptoms (if even they are treated) is so pervasive in this culture as to be almost invisible: water to fish. According to Hoskins, "When this symptom-treating model becomes a major part of a person's life, he is in an addictive process, a process in which he regularly uses addictive behavior to cope with internal and external problems."[313]

Hoskins says this another way: "Addictive Process is a coping style in which a person habitually responds to reality by using fix-oriented behaviors to produce desired feelings rather than by responding directly to the immediate demands of his life."[314]

He gives a silly example, which on reflection is no sillier than most of our behavior: "Imagine yourself at a friend's house and the friend is sitting with you in the living room and complaining that his house plants are dying from a lack of water. All the while he complains, he eats from a large bowl of chocolate-covered raisins. You know there is a sink in the kitchen and you have even seen a pitcher for watering house plants. Yet your friend is just sitting there eating candy and complaining, rather than solving his problem."[315]

Of course this perfectly describes our collective response to a planet being killed. We complain about it as we eat our chocolate-covered, pesticide-laden raisins.

He continues, "Addictive coping follows the above pattern. It always focuses on self-medicating feelings, rather than on solving problems. When you look closely at it, it is always just as crazy as the behavior in the preceding example. Yet, for some reason, it is widespread."[316]

It doesn't really matter whether we're talking about an addiction to heroin, television, consumerism, power, or civilization, the process of addiction emerges when a person enters a closed circle of self-medication that is not directed at solving one's problems but rather at providing means to temporarily forget they exist. This addictive behavior then takes on its own logic for the person who is addicted, a logic that makes no sense to those outside the addiction. This nonsensical logic is based, according to Hoskins, on unrealistic fears, an immature perception of the world, and on faulty yet unchallenged premises. By this point in the trajectory both of this analysis and most especially this culture, I would hope that readers can articulate for themselves the unrealistic fears, the ways the culture inculcates us into immature perceptions, and the faulty and unchallenged premises that are guiding this culture toward its self- and other-destructive end. Hoskins further states that addictions are attempts to fabricate feelings of security when security is otherwise absent, physical sensations for the benumbed, and feelings of control or power over oneself or others for the powerless. And finally, Hoskins makes clear that so long as addictions are present, the primary problems the addictions are meant to mask can never be solved. I'm sure by now readers can fully grasp the implications of this statement.

❰ ❰ ❰

Brian says to me, "There is no ignoring the black blood. This culture lives on black blood. Something needs to cut that artery."

"The sooner the better."

"To impair the oil and electrical systems as much as possible would, I think, be a great start toward taking down civilization. And there are a lot of ways we can do that."

"Like what?"

"I've always been partial to joining hands and singing songs. We can burn candles, too. That will bring down civilization, I'm sure."

"Yeah, but only if we simultaneously send pink bubbles of love floating toward the refineries."

He rubs the bridge of his nose between his forefinger and thumb, then says, "You'd really want a team of people to do it right. One approach won't work in every circumstance, and one set of skills won't do it alone. This culture has a myth of the lone superhero saving the day, but a lone hacker can't bring it all down any more than a lone bomber, a lone writer, a lone tree-sitter, or a lone candle-wielding pacifist."

"We need it all."

He nods. "My interest is computers and related electronics. We've discussed software; there is the hardware side, too. Have you ever heard, for example, of HERF guns?"

"I've heard the name, that's all."

"You've heard of e-bombs, right?"

"Oh, yes."

"They're in a similar class of weapons. E-bombs send a violent pulse in broad patterns. HERF guns, if designed and used carefully, provide a merciful, high-precision method of targeting hardware without involving explosives. You can take out computers but not the operators."

"Are they expensive to build?"

"Well, I usually dumpster dig the materials, so the most I've really spent for parts off the shelf is ten dollars. So there are no major economic barriers to home manufacture and use. You can spend more, according to how you acquire the parts, but I've never handled a model that would have cost more then two hundred, even with all new parts."

"Are they easy to build?"

"Well, that's the problem; if you do build and use them properly, you can take out any susceptible electronics within fifty to one hundred feet. If, on the other hand, you don't build or use them properly, you could end up electrocuted, and very dead."

"That's very bad."

"The radiation poses a threat, also. It must be carefully directed to its target and away from the user."

"So it actually is quite dangerous."

"No. All I'm saying is that poorly assembled, they become human cookers, usually the operator. Well assembled and well used, they're quite safe. I've designed and built several, with delightful results. Enthusiasts just need to be taught how to build and use this particular piece of technology by someone who already knows how to do it. Then if you're reasonably careful it's not dangerous at all. I think everyone who has an interest on a personal level in fucking up computers should explore e-weapons. But, and here's the point, only under experienced direction."

"I see."

"Programs, on the other hand, aren't going to smoke anyone. I definitely don't see any of these sorts of concerns emerging from someone exploring the use of computer viruses or worms."

I take another bite. I really like Chinese buffets.

"But I'm actually far more interested in information. The best I can offer there is a disruption of communications, not only human-to-human, but human-to-device and device-to-device."

"How does that work?"

"Well, so far as the former, much human-to-human communication these days doesn't include voices, with their inflections and so on. It's just messages of dehumanized numbers and text. That means messages can be forged with tremendous ease. This is just one way that, fantasies of the elite aside, a technological society is inherently less secure than a natural society. Barriers to forging communications are no longer sociological, but technological. That's a fundamental issue for us: relationships inhibit fraud.

"Disrupting communications between humans and machines, or machines and machines, is easier still. Suddenly the entire stream of communications is inhuman. These transmissions could be forged by a machine, an invisible third party, a virus, a bot, or even by the device itself."

"And why would someone disrupt these communications? Would someone hack into an oil pipeline or something?"

"That's not the direction I prefer. Eventually the fail-safes kick in. Similarly, the reason Y2K didn't do what many hoped is that the power grid isn't really computerized. Picture big red manual switches, and you've just about got it. The main thing that's computerized is billing."

"Oh, damn."

"As long as they've got these physical fail-safes, computers are relegated to communication instead of control. Computers give the orders and humans hit the switches, if they deem it fitting. Hacking this is good for mischief, but not disabling or impairing the machinery."

"Damn, again."

"Right. We want things to go down so hard they can't come back up. In order to accomplish that we have to recognize their thresholds of operation and we have to understand their various recovery features. We want to target devices or processes that could not be easily restored. So far as power generation, to provide one example, if you take out a part of the facility for which they have no spare parts (for instance, a giant stator that had to be specially manufactured), it can't be easily replaced. Suddenly you've moved from the realm of inconvenience to impairment.

"That much we can do without computers. But here's where hackers could help, specifically using computer systems. Now remember, computers get involved because they carry information. So a hacker could distort the information between one computer and a person, or between a person and a person."

"Meaning . . ."

"Let's say that stator or another vital component of the power plant is supposed to run below a certain temperature. Excessive heat destroys it. How do technicians know how hot it is? They don't touch it with their hands, and they can't keep checking dozens of thermometers. They read information they get from a computer. What happens if you feed information to them that causes them to run the component hot, for example by telling them that the cooling system is operating at a higher capacity than it actually is, and further convinces them that the component in question is running along fine, while in all physical truth it's destroying itself?"

"You could do that?"

"I have, in fact, for the sake of demonstration. I haven't smoked anything; my interest is merely to make certain it's possible. Little has been done to secure these computers and I'm quite confident they would come down hard under a coordinated effort. Hell, just before those cascading blackouts, we had a crash without even trying, and those effects were minor compared to what we could have had on our plates."

"Speaking of which . . ." I get up to fill a plate of my own.

When I get back, Brian is even more excited than normal. He says, "I just remembered, do you want a Linux server? I brought you one in my car. They're great fun. You can run a chatroom, host your own websites, interrupt satellite communications . . ."

"Brian, when we talked before, you said a dozen people could bring it all down. Do you still think that?"

"All of civilization? I think that's optimistic, and I actually don't think it will all come down at once. It will come down in waves. I think that twelve hackers could take down the electrical grid of all of North America, a blackout lasting for months. That blackout itself would take out key components. Of course those in power would immediately start retooling, and because they have more resources than we do they'd eventually be able to come back online. We'd have to hit them again in the meantime."

"What would it take?"

"Guts. And a small, tight, harmonious group of people achieving a degree of intimacy that is foreign to the West, who are ready to live and die together, who are each aware of what the others are doing. When the numbers get bigger, you'll need other cells with other goals. And of course if you're going to hit the power grid you'll have to be the last one standing, which means you have to be off the grid yourself."

"Of course," I say, as though I, too, have thought of that before.

❆ ❆ ❆

I want to be wrong. I want to not believe that 90 percent of the large fish in the oceans are gone, and I want to believe that the climate isn't changing. I want to believe that you can import resources without those resources having to come from somewhere else, and without there being any cost to taking those resources. I want to believe I'm not living on land stolen from the indigenous, and that indigenous peoples aren't still being driven off their land. I want to believe that men don't rape women, and that parents don't beat children. I want to believe wild salmon still run strong up the stream behind my home.

Or maybe I want to believe that the way things are is the way they've always been. Salmon never ran this stream. Passenger pigeons never flew over great forests of American chestnuts in the east. Bison never ran the plains. Rivers have never been safe to drink. Breast milk has always been contaminated with carcinogens. Cancer has always killed our loved ones. Men have always raped women, parents have always beaten children. Social structures have always been dishonest, authoritarian, and repressive.

Or maybe I want to believe this culture and most of its members are not insane. They have not fabricated quadrillions of lethal doses of plutonium-239, dangerous for 250,000 years. I want to believe the authority figures who run this culture are beneficent, or at least not malevolent—driven mad by a social structure that rewards the unbridled acquisition of power—and that I can trust them to do what is best for me and those I love.

I want to believe there are not two million dams in this country, that animals are not tortured in vivisection labs and factory farms. I want to believe we can continue to live the way we do and not (continue) to destroy the planet.

I want to believe the culture is reformable, that if we just make several minor and one or two not-so-minor changes that things will be all right.

I want to believe that none of it matters anyway, that after I die I'll go to heaven or some other place as beautiful and unpolluted as this place once was. I want to believe that the purpose of life is to detach myself from this world I once loved and thought wondrous but I've now been taught is a source of pain.

I want to believe that if I'm just a good enough person, if I can just love enough, be kind enough, that the atrocities will stop on their own. I want to believe that bearing witness to the suffering is enough. I want to believe that writing will take out dams. I want to believe that symbolic action is a substitute for nonsymbolic action.

I want to believe that the natural world will take care of the problems this culture has created, that hurricanes and heat waves will destroy the power grid, that earthquakes will destroy dams. I want to forget that I too am part of nature, and that just as I am asking hurricanes to do their part that they are asking me to do mine.

I want to believe that hackers will solve our problems for us, that *they* will destroy the power grid, with no effort, no responsibility, on our part. I want to believe I have no skills to offer, and I want to believe that the world doesn't need all of our skills, no matter what they are.

I want to believe there is nothing I can do, or better, that there is nothing *to* do. Then in either case I need do nothing, I need take no responsibility. I need be answerable to no one, to no landbase.

I want for everything I've written to be wrong. I so want that.

                              ☾ ☾ ☾

"So," I ask Brian, "will hackers save the day?"

He laughs again, but this time does not throw back his head. "Not by a long shot. Most of us are more eager to escape civilization than dismantle it; that's why we jack into the machines in the first place. Hell, everybody living here can feel that things just ain't workin' on this side of the screen.

"Reclaiming the Earth needs to be a combined arms operation, including people who know their way around the boondoggles that have been erected: telecommunications, programming and logic, hardware, information systems, databasing, pyrotechnics, heavy machinery. We'll need specialists familiar with power generation systems, types of machinery, valves, and so on, energy specialists, people capable around extreme high voltage and current, people who have studied the energy infrastructure, and so on. Even people without those specialized skills are necessary. Each devotee can make a unique contribution, and together we can do quite a lot."

I respond, "It's like I always talk about: if space aliens were doing this to our planet, or if the godless commies had invaded, suddenly lots of us would discover we had skills we hadn't before thought about. And suddenly lots of us would *use* those skills."

"You and I have been talking about the big blows," he says, "and those are very important, but I think it's also important for people to make smaller strikes wherever they can. Anything to impede and impair the functioning of this extractive economy."

"Once again, what steps would we take if we recognized the government was a government of occupation, the economy was an economy of occupation, the culture was a culture of occupation?"

"I think we'd obviously see a lot more sabotage, and we'd also see this sabotage move up the infrastructure."

"That's a critique," I say, "I have of the Earth Liberation Front. It seems they're not really leveraging their efforts. It's one thing to burn an SUV, but what would happen if they began to move their way up the production pipeline? Where would be the most effective places for them to hit?"

"Those are questions," Brian says, "that more people need to ask."

"The good news," I respond, "is that more people are."

☾ ☾ ☾

I'm in St. Petersburg, Florida, and it's hot. It's late November, but it's still hotter than hell. I'm talking with another military man, and I'm marveling at all of the useful knowledge taught to GIs at taxpayer expense. I'm also thinking that this knowledge might be a sparkling example of some of the master's tools coming in handy for dismantling the master's house, or rather his economic system.

We go to the beach. The sand is white, almost blinding. There aren't many people here. That's good. We want to talk.

He says, "We were taught in the Army that when we move into a country one of the most important things we want to do is disrupt the delivery of raw materials. If you can disrupt that flow, you disrupt the entire economy. If you disrupt the economy—and keep on disrupting it—you stand a much better chance of winning. Simple as that."

I think of the American and British bombers pounding the Nazi rail lines, and I think of Russian, Belgian, Dutch, French, Czech and many other partisans doing the same. I think of Federal forces in the American Civil War slowly strangling the Confederacy through a blockade and through cutting rail and river lines for transport of materials. I think of German General Erwin Rommel's complaint, looking back on his loss at El Alamein, the turning point of World War II in North Africa, that "the battle is fought and decided by the quartermasters before the shooting begins. The bravest men can do nothing without guns, the guns nothing without plenty of ammunition, and neither guns nor ammunition are of much use in mobile warfare unless there are vehicles with sufficient petrol to haul them around."[317]

The military man says, "The vast majority of stuff is delivered via three methods: train, truck, and ship. We can ignore air since the amount transported is trivial. Of these, trains and trucks are the easiest to get to. The U.S. rail system, and I would be amazed if Europe's was any different, is wide open. There are tens of thousands of miles of unobserved track that could be taken out with nothing more than a crowbar. When I was a kid we used to pull spikes all the time, just for fun. It takes no time at all. Similarly, millions of miles of roads could be disrupted temporarily by any number of means."

"You learned this in the military?"

"Absolutely. What did you think they taught us in all those classes? What do you think the purpose of the military is?"

He's smiling, so I don't feel chastised for my naïveté.

He continues, "The purpose of the U.S. military is to fuck up the infrastructure of the countries where the United States wants to steal resources or maintain a military presence to use as a staging area to steal somebody *else's* resources. That's what they taught us how to do. Roads (especially junctions), rail lines, ports, and virtually everything else associated with transporting goods has been a military target since day one. So we talked about them quite a lot. But even more than that we were taught to think about the system as a whole and to analyze it looking for the flow of production and goods: both those required locally and those essential elsewhere. We were taught especially to look for choke points, places that some necessary items *must* pass through."

I think about my term for these places: *bottlenecks.*

He continues, "We were also taught to look for transportation segments that are secluded or otherwise isolated, and to look for ways to disrupt the flow of materials even without overt actions."

"What do you mean?"

"If we couldn't blow a bridge, we could still stage a traffic accident. One of those at the right place at the right time could be very useful."

"You were taught that . . ."

"Yes. And you paid for it."

"What else did they teach you?"

"Probably the most important thing is that for this type of activity to be effective it has to be directed and sustained. You must know your area and what resources it requires to function economically, and you must direct your efforts toward interdicting the flow of those resources. You have to know how to get the most bang for your buck."

I smile, thinking I should have known it would be impossible to talk to two military personnel without at least one of them using that phrase, and thinking also that he's talking about what I call *leverage.*

"Of course we were taught about security as well. Never get caught. Never get caught. Never get caught. That was hammered into us."

We sit on the sand in the shade, still hot, and look at the water. It too, is white, and blinding.

He says, "Although they taught us many technical skills, the main thing the classes and the practice did for me was to shift my way of thinking so that now I am constantly evaluating where are the points of greatest stress in any structures and infrastructures that I see. Once you've made that shift in your thinking and perceiving and once you get some experience you begin to understand that all of this work is much easier than it seems."

It's still hot. I don't know how anyone lives here.

He says, "Natural gas."

"What?"

"That's something else they taught us. You know, folks talk all the time about how the military dehumanizes people, destroys their individuality and creativity, and that may be true in some ways, but there are other ways that they taught us to be very creative. We were repeatedly taught to survey our surroundings and find what commonly available resources we could use to achieve our ends. Our exercises always included—no, required—the innovative use of household materials."

"Like what?" Even theoretically, I find all this stuff incredibly fascinating.

"You'd be amazed at what you can do with gasoline and soap flakes. . . ."

"You gonna tell?"

"Napalm. And you can make pipe bombs out of black powder."

"They taught you how to make pipe bombs?"

"It's the military, Derrick. It isn't Boy Scout camp."

"Where does natural gas come in?"

"The delivery of energy is even more fundamental to the system than the transport of raw materials. And not only is the electrical grid wide open but so is the natural gas supply line. Here's an example of just how easy it would be to safely take out a natural gas pipe. Buy a car battery, a piece of glass tubing, and some plastic gloves. First, pour the acid from the battery into a suitable container, then go to one of the millions of gas pipes or relay stations around the world, attach the tube to the pipe with tape molding putty or whatever, pour the acid into the tube, and then walk away and let the acid eat through the pipe. Your onsite time is maybe two minutes."

"They taught you this in the military?"

"Care of Uncle Sam."

Have I mentioned lately that I'm glad I'm a writer?

❈ ❈ ❈

Do you remember the bolt weevils, the farmers who downed power lines in Minnesota? It ends up that not only farmers down power lines. Just today I saw a newspaper headline: "Sabotage blamed for power outage: Bolts removed from 80-foot Wisconsin tower." The article reads, "Someone removed bolts from the base of a high-voltage electrical transmission tower, causing it to fall on a second tower and knock out power to 17,000 customers, police said. The bolts were removed from a plate connecting the legs at the base of one of the 80-foot towers, causing it to knock down the other as it fell Saturday evening near Oak Creek, a Milwaukee suburb, police Chief Thomas Bauer said. 'It does look like it's for the purpose of weakening the structure so it would fall,' Bauer said. The incident caused a four-hour outage Saturday for 17,000 customers, including General Mitchell International Airport in Milwaukee, Bauer said. Screening equipment was shut down and flights were delayed as passengers and luggage were screened by hand, said Pat Rowe, airport spokeswoman. Downed wires from the towers lay across railroad tracks much of Sunday, delaying passenger and freight trains from Amtrak and Canadian Pacific Railroad, Bauer said. Train service resumed Sunday evening after authorities cut the wires, he said."[318]

BREAKING FAITH

Nothing appears more surprising . . . than the easiness with which the many are governed by the few, and the implicit submission, with which men resign their own sentiments and passions to those of their rulers. When we inquire by what means this wonder is effected, we shall find, that, as force is always on the side of the governed, the governors have nothing to support them but opinion. It is therefore, on opinion only that government is founded; and this maxim extends to the most despotic and most military governments as well as the most free and popular.

David Hume[319]

"THE CURRENCY OF THE ECONOMIC SYSTEM IS FAITH," BRIAN SAYS. "The whole system runs on faith. Insurance runs on faith. Banks run on faith. When people stop believing in it, the system crashes, and quickly."

❨ ❨ ❨

There is yet another reason that the civilized have been able to defeat the indigenous. This is perhaps the most important reason of all. Many of the indigenous began to believe that the gods, spirits, mysteries, processes who had protected them forever on their land had abandoned them. This destruction of their faith came about because these protective forces did not save them from the civilized. The civilized burned their villages. The indigenous fought back. Their villages were burned again. They fought back again. And then their villages were burned again. Fewer fought back. Their villages were burned again. Still fewer fought back. Their villages were burned again. They capitulated. No longer did the conquerors need to burn their villages, except once in a while to remind them who's in charge. The conquerors' new subjects had lost faith in their old ways. They had become demoralized.

This is how abusers work. One insult, one threat, one strike, is rarely enough to defeat a woman or break a child's will. The wearing away is repeated, often timed so that just when she begins to recover faith in herself he slaps her down again. Timing is important. Too soon and the demoralization isn't maximized, too late and she might begin to build up reserves of confidence. Even more important than timing, though, is repetition. She must be forcefully taught that there is no escape, that resistance is futile. She must be pacified.

We've all experienced demoralization in ways large and small. Here are a couple fairly small ones I've experienced this summer.

This May I decided I "needed" a new computer. My old one was freezing a couple of times per day. This was a bit of an annoyance, but what really got me was that each time it froze I had to turn it off and then back on, and after it came on it spent about fifteen minutes scanning the hard drive for errors. Fifteen minutes! Of *course* I couldn't just go do something else—like walk in this beautiful forest—for fifteen minutes when I could be on the computer! So I went to the local electronics store—the owner is a really nice guy, and if you don't watch yourself you could easily end up talking to his chatty assistant for an hour—and priced new computers. They were a bit spendy for my pocketbook, so I checked eBay and got seduced by the low prices there. So not only did I support the industrial economy by purchasing a computer, I didn't even support a local business. Before you smack me down for this, let me assure you that the computer gods already did. I was excited when the computer arrived—But $70 for shipping? What a rip off!—and started to set it up. The computer gave me error messages even before I finished installing Windows.[320] No problem. I did what most of us would do: I ignored the error messages and persevered. Then the drivers that came with the computer didn't work. No problem. I just used my old computer to find drivers on the internet and download them. The CD drive still only worked part time (and the sound system not at all), but I thought if I were able to studiously enough ignore these problems they would fix themselves. I installed antivirus software, and tried to get my Windows updated at the Microsoft site. The download was supposed to take several hours, but it aborted after forty-five minutes, told me there were problems with my system.[321] I tried again. This time the antivirus software told me the computer had some viruses (I first wrote that the software told me that *I* had some viruses: talk about identifying with the machine!).[322] I swept the computer, found and deleted the viruses. I tried the download again. This time it aborted after three hours and turned off my computer. I turned it back on. It turned itself back off. (If only we could get all of civilization to turn itself off as simply as that computer! Note to self: figure out how to do that. Have a plan ready by next Thursday. Do it on the following Monday, sometime after lunch.) I turned it back on. It turned itself back off. I called a friend who works for Microsoft. He said I had a virus (I guess we're both identifying me with the machine). Much as I like his politics, I cursed Brian, even though he didn't write this particular virus. I realized I was stupid for entrusting communications to a device and a corporation over which I have no control but I was still mad at hackers because it was two o'clock in the morning and the damn computer didn't work. I took a deep breath and thought, *No problem*. I reformatted the hard drive,[323] started to reinstall Windows. I got error messages, just like the first time. I ignored them, just as we all ignore the larger error messages given to us by the planet. I followed the same procedure, got the same results, and by now the sky was turning gray in the east. I could hear birds. Disgusted, I went to bed. I tried the same thing the next day, and the day after that. By the sixth or seventh time I no longer cursed, but pleaded. By the eighth time I was finally prepared to send back the computer. I was defeated.

Unfortunately I wasn't quite defeated enough, because I still went to the local electronics store and bought a computer. At least I supported a local business (and the computer works great), but you'd have to hit me just a bit harder than this to demoralize me enough to break my addiction to having a computer.

The second demoralization this summer has been more personal. I somehow ended up with a nasty prostate infection. The last thing you need to read is an extended discussion of trouble in my privates, so I'll simply say that for the past several months the pain has come in waves that last a couple of days, then subside for a couple of days. At first I attempted to ignore the pain, and simply hoped it would go away. (Sound familiar?) When that didn't work I tried thinking good thoughts. That didn't work either. The first several times the pain subsided, I was convinced I was on my way back to health. When the pain returned I cursed. When the pain went away I hoped. But this constant raising and dashing of hopes has been wearing, probably at least as wearing as the pain itself. I've become demoralized, or maybe I've moved through the first three of Elizabeth Kubler Ross's five stages of grief—denial, anger, and bargaining—and into the fourth, depression. If it doesn't get better soon I may move into acceptance.

How does this apply to taking down civilization?

We all—even those of us who are wildly anti-civ—buy far too much into the myth of the primacy of the machine. We believe civilization works. We believe civilization is resilient. Whether we want to admit it or not, we believe in the *deus ex machina*, the god in the machine who will save us in the end.[324] This is certainly true of those who believe that science or technology will save us from problems partially created by science and technology. But nearly all of us believe in the machine far more deeply than that.

What do you do when you're thirsty? If you're like me, you go to the sink, and you're utterly certain that when you "turn on" the water it will flow from the tap. It's automatic. It's a complete, and completely invisible, belief backed up in the short term by consistent experience. Likewise when we flip a switch we're absolutely certain that ghost slaves will light up the night. We're certain that when we go to the grocery store we'll find food (which we can purchase from transnational corporations). It may surprise us to learn that for nearly all of our existence humans have had this faith not in technologies but in landbases. They knew for certain they could drink water from streams. They knew for certain the salmon would come, or the passenger pigeons, or the bison, or the char, or whatever creatures they relied on for food. But no more. Our faith has been replaced. Our new faith—deep, abiding, unshakable—is in civilization, that it will one way or another take care of us, that it will continue. This faith is strong enough that nearly all of us no longer perceive it as faith. Indeed, most of us do not even think about it at all. I would imagine the possibility that civilization will not take care of us into the foreseeable future is a thought that never once occurs to nearly all Americans in their entire lives. Not once. Civilization with all it entails is simply the way things are. Not many people consider themselves to have faith in gravity. Gravity just is, and if you trip you fall down. No faith is involved. Civilization is perceived the same way.

Even those of us who oppose civilization also generally have an unshakeable faith that civilization will win, at least in the short run. It has defeated so many who have tried to fight it before, so surely it will defeat us too. This faith, too, is so commonly held as to no longer be considered an article of faith, but rather the way things are.

But civilization is not gravity. It is not an immutable force of nature. It is nothing more nor less than one social organization among many. It is a social organization centered around war and maximizing the exploitation of resources. Civilization is a great mass of people who have been driven individually and collectively insane, driven equally out of their minds and bodies by the exploitative violence that characterizes this mode of social organization. Civilization is nothing more nor less than cities using increasingly sophisticated technologies and increasingly more force to steal increasing amounts of resources from increasingly depleted and ever-increasing parts of the globe. That's it. That's all it is. And it will not last.

Civilization will not win. I know this as surely as I know that rain falls and carries away exposed soils. I know this as surely as I know the sun shines, bringing light and heat, and I know this as surely as I know there is night. I know this as surely as I know that you cannot use something up and expect to use it again. I know this as surely as I know I am an animal.

Civilization cannot continue. I know this as sure as I know I am alive. I know as sure as gravity that we will win.

❨ ❨ ❨

I've written extensively throughout this book of the need to break people's identification with civilization—as those who rely on and identify with the processes and artifacts of civilization, who rely on and identify with machines and with the machine social structure—and to help them to remember they are human animals reliant on their landbase. I have written of the importance of identifying with one's landbase. It's obviously best if this reidentification can take place through discourse, gentle guidance, and direct personal interactions with wild nature. But the fact remains that cities must not be allowed to continue to steal resources from the countryside. Dams must not be allowed to continue to kill salmon, to kill rivers. Deforesters must not be allowed to continue to deforest. Rapists must not be allowed to continue to rape.

There are, I suppose, at least four generic ways you can get someone to stop doing something. The first is that you can kill the other person. The second is that you can make it physically impossible for this other person to continue. This could happen through incarceration, for example, or also through removing the means the person is using to commit the act. An example of the latter is that it would be almost impossible at this point for those who are killing the oceans to continue at their present rate if they did not have oil to power their ships. Thus, denying them oil would go a long way toward stopping their actions. The third is that you can convince the other person to stop. This can be accomplished through providing rewards for changed behavior. It can be accomplished through teaching better ways. It can be accomplished through threats, backed up by the means and a willingness to enforce them. The fourth is that you can demoralize the other person.

One argument for pacifism, if you recall, is that we should never use violence or sabotage against those in power because they and their servants will hit back hard. As true as this may be, it also only looks at that first act. If you hit them, and hit them again, and keep hitting them, eventually they will grow discouraged. This is precisely the tactic they use to break all of us. It works both ways.

If my electricity goes out I get annoyed. If it happened very often I would get very annoyed. I might even get annoyed enough to start spending some time outside. If it happened even more than that I might begin to no longer be able to feel that I can rely on electricity. And that, my friends, is a very good start.

A MATTER OF TIME

The guerrilla has the initiative; it is he who begins the war, and he who decides when and where to strike. His military opponent must wait, and while waiting, he must be on guard everywhere.

Robert Taber[325]

IN TERMS OF FIGHTING AGAINST THE CIVILIZED, THE INDIANS—AND by extension the indigenous before them running all the way back to those who first opposed this awful way of life—had many advantages over us today. They could move in the open in their own territory, and they could recruit through speeches without worrying about snitches and infiltrators, much less bugs and spy cams. They often had entire communities united against their enemy. Huge chunks of territory were out of the reach of the civilized, where those who opposed it could go to rest or to live. Civilization was not so powerful, with tanks and airplanes and automatic rifles and tasers capable of electroshocking entire crowds, indeed, with bombs capable of killing the planet. Television did not blare into the homes of the civilized, broadcasting propaganda directly into people's brains twenty-four hours per day. The skies were full of birds, the rivers full of fish, the forests full of animals: people could easily feed themselves, and had not been forced into a position of dependency on the very social structures they were fighting. The list of advantages held by these earlier fighters over us is very long.

But we are not without advantages of our own. The first is that diversity leads to resilience, and never has any culture so relentlessly destroyed all forms of diversity. Further, never has any culture so totally relied on one resource. Egyptian nationalist Gamal Abdel Nassar rightly recognized that oil is the "vital nerve of civilization." Without it, he wrote, all of industrial civilization's machines and tools are "mere pieces of iron, rusty, motionless, and lifeless."[326] He's right. As Matt Savinar puts it in *The Oil Age is Over: What to Expect as the World Runs Out of Cheap Oil, 2005-2050*, "In the U.S., approximately 10 calories of fossil fuels are required to produce 1 calorie of food. If packaging and shipping are factored into the equation, that ratio is raised considerably. This disparity is made possible by an abundance of cheap oil. Most pesticides are petroleum- (oil) based, and all commercial fertilizers are ammonia-based. Ammonia is produced from natural gas, a fossil fuel subject to a depletion profile similar to that of oil. Oil has allowed for farming implements such as tractors, food storage systems such as refrigerators, and food transport systems such as trucks." He also states, "Oil is also needed to deliver almost all of our fresh water. Oil is used to construct and maintain aqueducts, dams, sewers, wells, as well as to pump the water that comes out of our faucets. As with food, the cost of fresh water will soar as the cost of oil soars. . . . Oil is also largely responsible for the advances in medicine that have been made in the last 150 years. Oil allowed for the mass production of pharmaceutical drugs, surgical equipment and the development of health care infrastructure such as hospitals, ambulances, roads, etc. . . . Oil is also required for nearly every consumer item, sewage disposal, garbage disposal, street/park maintenance, police, fire services, and national defense. Thus, the aftermath of Peak Oil will extend far beyond how much you will pay for gas. Simply stated, you can expect: economic collapse, war, widespread starvation, and a mass die-off of the world's [human] population."[327]

This is probably a good time to remind readers that the longer civilization continues, the more severe will be the human die-off; and to remind them further that the crash of civilization *will* come, and that right soon, because no matter how strong a culture's denial may be, thermodynamics is infinitely stronger; which means if your primary concern is for the health and safety of those humans who live during and after the crash, then you need to turn off your fucking television, get off your butt, and start forming community gardens, start learning local edible plants so you can teach them to others, start figuring out what local people can do for potable water during the crash, and so on. Instead of being scared of or fighting against those who want to bring down civilization while something still remains of the natural world—or to put this a way that even the most anthropocentric of the civilized can understand, while there are still wild plants and animals for you to eat and still wild water for you to drink—you need to work to soften the inevitable crash in whatever way you can. If you're not doing that, I don't want to hear any complaints from you about those who want to bring down civilization. And if you are doing it, you probably understand why people want to bring it down, and you support them in their efforts.

So, one advantage we have over those who came before is that civilization is far more reliant on one resource than ever before. Bottlenecks—chokepoints—imply vulnerability. You can figure out what to do from there.

Another advantage we have is that this culture has clearly reached its limits of growth. From the beginning civilization has required, for reasons long since discussed, constant expansion. This is especially true of its most metastatic manifestation, modern capitalism. At all times in history, civilization still had new territories to exploit. But there are no more ridges to climb to see the vast expanse of uncut forests on the other side. There are no new masses of people to enslave. At this point all that is left of the vast natural capital—as the capitalists would perceive it—are the scraps. This means the age of grotesque exuberance is over.

Sometimes people tell me it would do no good to take out a big dam because those in power would simply rebuild it. I have three responses. The first is that this would be okay, because any money spent rebuilding that dam would not be available to construct some other undoubtedly destructive artifact. One of the central objectives of partisan warfare is to bleed your enemy however you can. Second, if they build it again, we take it out again, just as they did with Indian villages. If we take it out often enough, they, too, will become familiar with Elizabeth Kubler Ross's five stages. They're already in denial. They'll get angry. If we push them past that anger they'll try bargaining. But I'm not interested in bargaining with them. The time for bargaining has long passed, and besides, for them, bargaining is just a euphemism for stalling, lying, and stealing: as Red Cloud said, "They made us many promises, more than I can remember. But they never kept but one. They promised to take our land and they took it."[328] Next, depression: I don't care if they get depressed. After what they've done to the world and to humans, they should feel not only depressed but deeply ashamed. They won't. But they sure as gravity will end up accepting that the dams are coming down. My third response is that the age of exuberance is over. The big dams won't be rebuilt anyway. Neither the money nor the will is there anymore. The industrialized countries no longer even have enough money to maintain and repair their infrastructure now (roads, water pipes, sewers, dams, bridges, and so on), and nobody is attacking it except wind, water, sand, plants, and concrete-eating bacteria.

Another advantage we have is that this culture has exhausted its soil. I mean this both in the physical sense of destroying topsoil—as goes the soil, so goes the culture—but also in the sense of the Oswald Spengler line I quoted early in this book, that it has exhausted its spiritual and emotional soil. The culture is dead, but just doesn't know it yet. Part of our job is to help make that clear. If I may be allowed to switch literary references, in his massive *Decline and Fall of the Roman Empire*, Edward Gibbon described dynasties' "unceasing round of valor, greatness, discord, degeneracy, and decay."[329] It should be clear that civilization is in the final two stages: even if nothing else convinces us, television should (*Fear Factor* anyone?). It should be equally clear that dynasties in those latter stages are easier targets.

All of this leads to the notion of *Kairos*, a Greek term describing the moment at which something new can break through. Rollo May and others called this a "destined time." The theologian Paul Tillich called *Kairos* "the moment at which history, in terms of a concrete situation, had matured to the point of being able to receive the breakthrough."[330] In his case he's talking about the coming of Jesus, but his definition works on many levels. It's a pretty obvious concept, known to anyone who has ever tried to figure out the right time to ask for two weeks off work or to enter or leave a relationship, or to take down civilization. Timing is everything. And the time is right, or very soon approaching. Can you feel it? It is in the air, the wind, the rain, the streams and rivers. It is rumbling in the ground.

We have an advantage far bigger than any of these. A major reason the Shawnee did not sack Philadelphia or Washington, D.C., that they did not strike at civilization's central infrastructure, is that just as the whites had a hard time infiltrating Indian communities, the Indians had a hard time infiltrating white cities. Had Tecumseh tried to enter New York City he would have been spotted and killed or captured immediately. But *we* have no need of hiding as we enter New York City. We don't even need to enter New York City at all. We're already there. We already walk among them. We are just people who, like Brian, have a fondness for our species, and other species, and life. We are people who try very hard to think clearly. We are people who are saying no to the machine culture and yes to life on this planet. We are people who are tired of living hollow lives guided by abstract moralities expressly created to serve those in power, moralities divorced from physical realities, including the land we love, including the land we rely on.

We are people who do not resign ourselves to the fate we are so often told is inevitable. We are people who refuse to continue as slaves. We are people who are remembering how to be human beings. We are people who are ready to take back our own lives, and to defend our lives and the lives of those we love, including the land. We are people who are at long last ready and willing to fight back. We are people who know in our bones the truth of Robert E. Lee's statement, "We must decide between the risk of action versus the positive loss of inaction." We are people who are ready to take the offensive, or support those people who do.

We are survivors. We have survived domestic violence. We have survived racism, and we have survived sexism. We have survived industrial schooling, and we have survived the industrial economy. We have survived television. We have survived the toxification of our total environment. And we are ready to fight back.

We are lovers, lovers of the land, lovers of each other, lovers of our own bodies, including our emotions. We love. We hate. We feel joy, despair, sorrow, outrage, happiness, anger. And we are ready to fight back.

We are the oppressed. We are prisoners, family farmers, animal liberators, women, children, American Indians, blacks, Mexicans, poor whites, Asians, people of the Third World, the indigenous. We are lesbians, homosexuals, transgendered. We are parents, and we are childless. We are those who hate our jobs, we are the unemployed, and we are those who want no jobs. And we are ready to fight back.

We are those who have listened long with love and sorrow, and who now with their permission speak for salmon, redwoods, rivers, voles, spotted owls. We speak for the bison, the sturgeon, the manatee, the shark. We speak for the soil, for the wind, for the snow, for the ice caps. We speak for the phytoplankton, and we speak for the insects. We speak with voices that are no more and no less than the wind moving in and out of our bodies, over our vocal cords. We speak for our homes, and for our neighbors, and we will be heard. They will be heard. And we are ready to fight back.

We are activists. We are teachers. We are students. We are workers in strawberry fields. We are visual artists. We are small business owners. We are construction workers. We are genetic engineers. We are librarians. We are bioweapons specialists. We are ex-navy SEALS. We are demolitions experts. We are hackers. We are clerks at Wal-Mart. We are prisoners. We are single mothers. We are punks. We are fishermen. We are hunters. We are those who oppose hunting. We are writers. We are killers. We are former loggers. We are saboteurs. We are nurses. We are farmers. We are great-grandmothers. We are attorneys. We are ex-cons. And we are ready to fight back.

We are in Los Angeles, Detroit, Boston, New York. We are in St. Louis and we are in Asheville, North Carolina. We are in St. Petersburg and we are in Seattle. We are in tiny towns in Montana and we are in southern Mexico. We are in Canada and we are in Korea. We are in China, India, Australia. We are in the Congo and in Tanzania. We are in Macedonia, Austria, Denmark, Finland. And we are ready to fight back.

We are people who have realized that unless it is stopped, civilization will kill everything on the planet. We are people who have realized that civilization is guided by an urge to destroy, and we are people who have realized that civilization is not reformable. We are people who have learned the lessons of those who have tried to make treaties with those who are killing the planet, and we are people who, with all the world at stake, are finally ready to fight back.

We are people who no longer hope that civilization will stop killing the planet, but we are instead people who will do whatever it takes to stop it. We are people who no longer hope salmon survive, but are people who will do whatever it takes to stop their extinction. We are people who say the same for bison, prairie dogs, desert tortoises, whales, dolphins, lions, great apes, rhinos. We do not hope. We act. And we are ready to fight back.

We are people who understand deep within our bodies that fear is the belief that we have something left to lose, and with all the world at stake we are ready to fight back.

We are people who are putting those who would kill the world we love on notice. You must stop. Now. You will stop. Hear this as you have never heard anything before. You will stop. We are ready to fight back.

And we are going to win.

(((((

We are those who will never forget that the Jews who participated in the Warsaw Ghetto Uprising had a higher rate of survival than those who went along.

(((((

We are those who are on the side of the living, and we are going to win.

BRINGING DOWN CIVILIZATION, PART III

At this point, we've arrived at an understanding of a conflu-
ence of interest that utterly transcends the old "three worlds"
paradigm, harkening an entirely different praxical symbiosis,
one which is not so much revolutionary as it is devolution-
ary. We don't want China out of Tibet so much as we want
China out of China. We don't just want the U.S. out of Southeast Asia or Southern Africa or Central America, we want it out
of North America, off the planet, out of existence altogether.
This is to say that we want the U.S. out of our own lives and
thereby everyone else's. The pieces dovetail rather well, don't
they? Indeed, they can't really be separated and only a false
analysis might ever have concluded that they could.

Ward Churchill[331]

I'M STILL WITH THE EX-MILITARY MAN. I'M STILL IN FLORIDA. WE'RE STILL on the beach. It's still hot. I ask him the same question I asked Brian: "How many people do you think it would take?"

He says, "If we could count on people having reasonable common sense and if we could count on them doing their research and if we could count on their commitment . . ."

"In other words, if we could find people we could count on . . ."

"Yes, which can be a problem right there. These are people we'd have to trust with our lives."

"Yes."

"I would say twenty groups of three to five people could do it in a short time. More generally I'd say we'd need groups of three to five people per tristate area. Then beyond those direct action teams we'd need a lot of support, an underground railroad of people who will give us food and supplies, who will let us crash for the night or for the week. While it might be true that those in power can't stop all of us, some of us will have to go underground, and you can't survive well underground without that web of support. This is a place where those not suited for direct action could still make all the difference. And the key will be for us to make those connections beforehand or to have a core group doing it with some reliable way of communicating with us."

"Yes." I think again of the Indians, and their advantage here, with communities already in place where they could be safe at least for a while. But I think of our advantage, too, that we are already at the center of civilization, where we can do the most good.

"But here's the rub," he says. "Let's say we succeed and create short- to mid-term disruptions in the grinding of this horrid economy. The more successful we are the more that the vulnerable sites get secured. What do we do then? Do we keep pushing to completely destroy it, or do we just do enough damage to cause a major depression, then lie low and wait?"

"Have you ever seen any horror movies?"

"Yeah, why?"

"What always happens at the end?"

"Credits, I don't know."

"Before that. The psychopath is killing everyone in the house, right? And finally the plucky babysitter/girlfriend/best friend fights back, right? She stabs the psychopath with a hanger, or shoots him with a gun, or something. The psychopath falls to the ground. And what does she do next?"

"I see where you're going. She sits down on the couch and starts to cry."

"Yeah, and suddenly the psychopath leaps up and she's got to do it all over again."

"So your answer is . . ."

"Of course."

"I agree. My only concern is that we can't win a battle of attrition."

"But we will win a battle of attrition. That's the point. They can't fight the planet and win. We just have to bleed them and keep bleeding them till this awful system collapses on its own. And it will collapse. And every day sooner we can make that happen is that many species not driven extinct, is that much more landbase for people to survive on, is that many more living rivers and living forests. It's that much sooner that the planet and the people can begin to recover."

❨ ❨ ❨

"I have another concern," I said to the man sitting on the sand next to me. "I have no interest in just weakening the U.S. so some other government can invade or take over its imperial role. It won't help anyone for the Japanese, Koreans, Chinese, French, British, or Russians to exploit the same way the Americans do now."

"Do you really think this is only a domestic effort?" he said. "People in the Third World don't just hate the United States, they hate all imperial forces. They hate everyone who exploits them. What I'm talking about needs to happen all over the world. And it's already happening in the colonies: why do you think I was sent overseas? It was to subdue those who don't want their resources stolen. We need to learn from the resistance of those who are already fighting back. We need to follow their lead."

❨ ❨ ❨

There was one final series of questions I asked both Brian and the man from Florida. It was this: "If twelve or fifty or two hundred people really could bring it down, why hasn't it already happened? I mean, it took millions of Vietnamese to fight off the Americans. And certainly around the world more than two hundred people hate civilization. What makes you think it would only take two hundred people, and not two hundred thousand?"

Here's what Brian told me, that night at the Chinese Buffet: "That's a question I've pondered myself a lot. As far as Vietnam was concerned, our objective seems to be different than theirs. The locals were doing their best to survive and repel platoon after platoon of gunmen, while somehow escaping the napalm and Agent Orange raining down from above. They were also trying to seize and hold ground. Were these folks really out to dismantle civilization, even American civilization? As for the Americans who *are* actively organizing to that end, the best explanations I could offer as to why it hasn't happened would be that the people are: a) geographically isolated; b) ideologically divided; c) lacking the necessary skills; d) under-equipped; e) lacking focus and discipline; f) restrained by relationships; and/or g) afraid. I'm sure that plenty of individual exceptions exist, but unless an entire cadre has dealt with these issues, it's difficult to imagine them accomplishing more than one or two major catastrophes. That said, I don't think I have to remind you of the times that some very small groups have orchestrated some very big catastrophes. What is required of any group that seeks a sweeping change in culture is that they shake its fundamental pillars. They don't have to break everything in sight. All they have to do is give the first in each line of dominoes (generators, internet nodes, corporations, dams, and so on) a hearty enough heave. Once the reaction has achieved a critical threshold, if you don't mind me changing metaphors, a fire will feed itself and grow uncontrollably. Part of the key is winning the minds of the people who would otherwise plug all the machinery right back in again. Once they realize they can actually walk away, without repercussions, they'll be able to exercise their human freedoms in prodigious ways. Just because the formation of a Blackout Brigade hasn't happened yet, please don't think it isn't possible. And please don't think it isn't going to happen. There are more people working on this than you or I know about."

The man in Florida was just as helpful. He said, "My estimate of fifty people is definitely at the lower end. Having more certainly wouldn't cause any problems, and would in fact make everything happen more quickly. But I still think fifty people could do it.

"Here's the logic I learned in the military. First, you have to have small groups where everyone knows and trusts each other; everyone is able to set up simple, low-tech communication systems; and each group as a whole is capable of dispersing and reforming easily. So if you had more people I would suggest more and not larger groups. Second, if these groups are dedicated enough—and when I say fifty people can bring it down I'm presuming they're not messing around—they should be able to pull off an action every ten days. That's three a month. That's sixty per month nationwide, with equivalent numbers of actions by members of the resistance worldwide. Thirty pipelines, twenty major rail lines, and ten power lines. Every month. That's a lot. It's what was expected of me in the military. It's what I *did* in the military. Month in and month out. And it works. Trust me on this one. Third, I'm assuming that all targets are pre-selected and scouted at least a month in advance. Fourth, I'm assuming that the groups stay uninfiltrated and unpredictable. The larger the group the greater chance of infiltration, and also the more organized it must be, and therefore predictable.

"It's always hard to get people to believe that so much is possible with so little. But that's part of the way we've been conditioned to think. We always have to remember: who does it serve for us to think that way? We've been trained to think in terms of pitting ourselves against them in a showdown against their main force. But if you look at some of the more successful guerilla wars, time and again you see smaller, lower-tech forces prevailing through smarter tactics and leveraged actions.

"The key we must always remember is that it takes large numbers of troops to take and hold ground. If you don't want to take and hold ground, you don't need large numbers at all. You just hit and run, and then hit and run again, and you keep doing that. There is no fighting force in the world that can survive the death by a thousand cuts of a dedicated partisan movement."

☾ ☾ ☾

I'm sitting now by myself. I'm looking at a dam. It's big, it's ugly, it's killing salmon, and it will not be here forever.

I'm thinking back again on my conversations with Brian. At one point, late in the evening at the Chinese buffet, he said, "In many ways I am entropy's agent. I'm just following my nature. I don't think about how to do it properly. I do what stimulates me, just like any other animal."

I remember looking at him and wondering, *Where did you come from?*

He continued, "There is fertile ground underneath us all. We just have to get to it."

Silence.

He continued, "Every day, I come in contact with disasters that are only a keystroke away. Acting alone I can take out the power to a city or two. The question is how to trigger these things with a big enough impact. And that only requires a series of miracles."

I laughed, bitterly, then said, "That's all?"

"That's not so much. Life is nothing but a series of miracles. Evolution is a series of miracles. You just have to wait for a series of amino acids to . . ."

"Yes."

"The chain reaction has to be right."

"Yes."

"Once you realize that fate favors you it's only a matter of time."

I'm still looking at the dam, but now I'm smiling. Yes. It's only a matter of time.

❨ ❨ ❨

Just today Radio Canada reported on the lax security at the dams on the massive (and genocidal and ecocidal) James Bay project. The journalists were able to walk entirely unquestioned and certainly unstopped into the main control rooms of the dams they visited. The journalists were surprised at how easily someone could have sabotaged the entire enterprise. The response by Hydro-Quebec, the operators of the (genocidal and ecocidal) dams, was to attempt to place a gag order on Radio Canada to stop the story from being reported.

Yes. It's only a matter of time.

BRINGING DOWN CIVILIZATION, PART IV

Hence, we must seek nothing less than the dismemberment
and dissolution of every statist/corporate entity in the world.
All of them. No exceptions.

Ward Churchill[332]

LAST YEAR I GAVE A TALK AT A GATHERING CALLED BIONEERS. YOU CAN guess my topic. Afterwards one of the questions was, "When is it time to use any means necessary to bring down civilization?"

Nobody had ever asked me that before. My answer was, as my first answers too often are, flip. I said, "When the last passenger pigeon gets killed." But that's not what I wish I would have said. I was attempting, feebly, to say that the time has long come. But my threshold was still too late. I should have said, "The first time a man rapes a woman. The first time a parent beats a child. The first time a city tries to steal resources from someone else. The first time a culture enters into a nonreciprocal and thus nonsustainable relationship with a landbase. The first time a culture fails to follow the fundamental predator/prey relationship.[333] The first time a culture drives a species extinct. But we can't blame those who came before for not bringing down civilization. They're dead and gone, and we're here and now. The time to use any means necessary is now. The time to use any means necessary is when the time is right to use any means necessary."

☾ ☾ ☾

Over the years a few people have said to me that we must not bring down civilization because to do so would hurt the natural world. Civilization has fucked up so much of the planet that it cannot now survive without us.

Of all the arguments against bringing it all down, this is probably the most foolish. It is of course the same argument against taking out dams, only this time writ large.

The argument might make some sense if each day more messes were cleaned than made, if each day more forests were restored than cleared, if each day more bombs were destroyed (not by use) than manufactured, if each day the population voluntarily decreased instead of exponentially increased.

But none of those are happening.

I asked an engineer what he thought of this question.

His response was strong, and what I expected.

"All analyses must begin with the salient fact that the system as it is going now is *guaranteed* to destroy the biosphere. Anything which has *less* than a 100 percent chance of destroying it is better than something which will destroy it for sure. That much is obvious.

"That said, let's talk about a couple of examples that are always thrown out as reasons we can't bring it all down. One is oil wells, especially the off-shore variety. We're told that they will spew out oil and continue to contaminate their surroundings more or less forever.

"Well, first, this ignores the fact that right now oil spills are already contaminating their surroundings, and also that burning oil is changing the climate. Better to have a few smaller spots of contamination than a world destroyed by global warming. But it ignores something even more important, which is that the wells wouldn't contaminate for long. Oil wells only pump out oil under their own pressure for a very short time. The 'gushers' you see in movies are a phenomenon of the first few hours or days of a well's operation. After that, the gush slows to a dribble, and after a few weeks or at most a couple of months comes to a stop altogether. From then on, the only way any oil leaves the subterranean cavities is if something else displaces it. Commercial oil wells pump water down into the oil fields to force the crude out of the ground. So what would happen if an offshore rig were abandoned? Effectively, nothing. Certainly nothing worse than what happens while it's in operation. Add to that the fact that the rig will rust and fall away after a few years if nobody is there to keep painting it and replacing rusted metal, and you have a complete nonissue.

"Another question I get all the time is: What will happen to all the nuclear reactors?

"My first response is always to ask people which of the following they would prefer: 1) a world where people are operating nuclear plants and generating nuclear wastes, cutting corners and risking major accidents, and unscrupulously reprocessing fuel with no concern for the environment; or 2) a world where the generation of new waste is zero, and existing materials, contained in the reactors, effectively remain there for a long time.

"Reactors don't melt down unless they're in operation. There are, it is true, several reactors in operation in the world which are not of a 'safe' design, that is, that will experience bad things if their support systems (cooling, control, and so on) fail. Are those reactors more dangerous after they're abandoned than while they're kept fully fuelled and operational? At least if they're shut down they will from that moment be less likely to melt down as time goes on and the fuel degrades.

"Now, I'm not saying it's a rosy picture. I'm not saying I have a perfect answer to everything. However, I do point out that even given the risk, that scenario has less certainty of destruction than leaving industry to continue marching on building *more* reactors and refining *more* fuel and dumping *more* waste. Certainly it is better to stop it now before such things proliferate any more than they already have.

"In addition, outside of the former Soviet Bloc, almost all reactors are designed in such a manner that they shut down in the case of a loss of power. I'm not talking about control systems shutting them down, I'm talking about physics. I could give you an extensive physics lesson on this, but I'm sure it would bore you silly. The bottom line is that aside from the really evil Soviet-style (like the Chernobyl design) graphite core breeder reactors, and some already obsolete but still operating high-enrichment reactors, if the auxiliary systems fail, the thing shuts down. It remains contained in the pressure vessel (a pretty competent vessel given that it is designed to withstand pressures in excess of 100 psi) and cools down to an ambient (still warm, given that it houses a couple tons of radioactive materials, but not meltdown temperatures) level. With time, the environment in the core will slowly burn out much of the remaining fissionable fuel. Eventually, the vessel will rust, crack, or break. At that time, yes, it would be a terrible disaster. I don't have any good answer as to what to do about it. When you let the nuclear genie out of the bottle . . . you get yourself into some deep shit, no two ways about it. The best I can suggest is that at the same time as people work to bring down the whole machine, they also work to mothball, decommission, and phase out the use of nuclear power. Germany has already announced that it's doing a 100 percent phase-out of nuclear energy, and has already begun a schedule for the shutting down and un-installation of all its nuclear capacity. Nuclear power was never popular in the U.S., and while the weapons industry remains a spoilt child here, they have enough materials that they don't need a large nuclear industry to produce weapons anymore. Likewise for anyone else.

"Now here is the real point: Aside from the nuclear issue, the rest of the wastes, poisons, and pollution that are being generated are not going to be any worse without supervision. To the contrary, the supply of new pollution will come to a halt. That stuff is entering the environment as we speak. If it stopped, there would be no more."[334]

❮ ❮ ❮

When we bring it all down, much of the world will recover much faster even than we could dream. Last summer a massive blackout—not caused by hackers, I might add—in the northeastern United States and Canada shut off electricity for some fifty million people. It also shut off electricity to all of the fossil-fueled turbogenerators throughout the Ohio Valley. The response was miraculous. After only twenty-four hours, levels of sulfur dioxide in the air had gone down by 90 percent and ozone was down by 50 percent. Visibility increased twenty miles.[335] That's after only one day.

The world desperately wants to heal. Yes, this culture has broken things that can never be fixed, but if we just stop this culture from killing the planet, there is much that will heal. All we have to do is stop those who are causing this destruction.

❲ ❲ ❲

To bring down civilization will involve, so far as I can tell, six different broad categories of work. The first is the personal. We need to change ourselves. As Gandhi and countless other pacifists have said,[336] we need to become the change we wish to see. Not only must we reject the reward system of capitalism, but we must attempt to eradicate oppression wherever we find it.

Environmentalists are often not saints. Many are assholes. Some are rapists. Just last night I received yet another email from yet another female activist describing yet another sexual assault by yet another male activist. This woman finally came out and said explicitly what many others have hinted at: "The forest defense community is a rape culture."

This should not surprise us: the culture as a whole is a rape culture, so subsets of it will likely carry those same characteristics, unless and until they do the necessary work of cleansing, healing, and changing themselves.

The response by many males within the forest defense community has been what we would expect. At first there is outright denial, and when that doesn't work they begin long meetings where they change the focus from their behavior to questions like, "What is oppression? What is abuse? Who has the right to dictate and define what we call abuse? Why does she get to call it rape if I call it consensual?"

Sound familiar? This is the same sophistic line of leading questions the civilized have been using from the beginning: What is oppression? What is exploitation of a landbase? Who has the right to dictate and define what we call exploitation of a landbase? Why do you get to call it despoliation if I call it developing natural resources?"

And something inside the woman, or the land, dies.

Something inside all of us dies.

I wrote her back: "The answer as to who gets to define is an easy one. Defensive rights always take precedence over offensive rights. One person's defensive right to bodily integrity always trumps another's perceived right to sexual access. More broadly the recipient of any behavior has the absolute right to dictate and define what sort of behavior is acceptable. If you decide you never want to hear someone say the word 'breathe' near you, then you have that right. If I don't like the conditions you place on our friendship I don't have to hang out with you. This is an actual example, by the way. A friend of a friend of a friend was raped by someone who during the rape kept telling her to breathe, and so that word became a trigger for her. Her friends all just agreed that they would never use that word around her. Not a big deal."

So the first personal change we must make is to eradicate the desire to dominate, exploit, use others. To challenge our perceived entitlements. For someone like myself, a well-educated white male, that's a lot of perceived entitlements to attempt to sort through.

At the same time, I think we need to not get bogged down in this self-examination such that we don't accomplish anything in the real physical world. Over the years, people have written me to complain about almost every word I've written or said. One woman complained when at a talk I said that one of the most revolutionary things any of us can do is raise a healthy child.[337] Because her child had a congenital physical disability, she took offense. Telling her that I meant emotionally healthy did not assuage her: "What if," she asked, "the child has an organic brain condition?" Another woman said that because I made a joke about the tattoos on Angelina Jolie's genitals I am obviously sexist. A man said that because I called one of my CDs *The Other Side of Darkness* I am obviously racist. A black man said that the only reason my work is published is because I am white, and a white woman said that the only reason my work is published is because I am a man. I wasn't clear as to what either of them wanted me to do about it. A vegan wrote that because I eat meat my books (which he hadn't read) are useless and so am I. Then there's the nationally syndicated radio interview I did, where after I talked about how oil will not last forever, which means this culture will not last forever, a man from Mississippi called in, barely able to contain his rage, and shouted, "You got *no call* to go and be talking like that. *No call* whatsoever. No call." You'd have thought I insulted his mother. In a sense I did: his mother culture. A survivor of trauma wrote that because I described child abuse that I was as abusive as his own father had been: my book restimulated him. At last count I had received 388 emails saying these sorts of things. That does not include the people who say them to me personally.

It's enough to drive a person to distraction. Which of course is precisely the point: I spent the last three hours talking about why we need to take down civilization, and all you heard was my joke about the tattoo on Angelina Jolie's genitals?[338] If I spent time seriously attempting to achieve a level of politically correct purity with every word I write or say—if I attempted to never offend anyone—I would never get anything done.

But wait! What's the difference between me and the rapists in the forest? Because my words entered these peoples' ears or eyes, don't they get to define what is or is not offensive? The man who wrote that I was as bad as his father accidentally provides the answer: the difference between my book and the beatings by his father is that he could put the book down. His reading of it was voluntary. The same is true for my talks. If you don't like what I say, don't come.

As is true for so many other questions, the answer to this question of how much self-examination we should do is both circumstantial and commonsensical.

When those countless pacifists talk about becoming the change they wish to see, they're usually talking about eradicating domination, at least in their own hearts, if not in the external world. But there is another aspect to becoming the change we wish to see, one that is infrequently examined and even more infrequently put in place. At least as important as getting rid of the desire to oppress is the task of squeezing out every drop of slaves' blood in our bodies, eradicating our deference to power.

This means learning to think and feel for ourselves. It means learning how to distinguish between legitimate, earned authority and illegitimate unearned authority. It means asking how one can earn authority. It means learning how to be accountable to ourselves and the communities we care about, the communities on which we rely. Just as important it means learning how *not* to be accountable to unearned illegitimate authority, and how *not* to be answerable to those individuals and communities who are exploiting us or others. It means learning to fight back when appropriate using whatever means are appropriate. It means learning how to determine when are appropriate times and what are appropriate means. It means learning how to say *no*, and how, then, also to say *yes*. It means not putting up with abuse of ourselves or others.

It's a mistake to think that doing this personal work will bring down civilization. I have no problem with people becoming the change they wish to see. But I have a big problem with them thinking that this alone in any meaningful way slows civilization's destructiveness. Here is an actual list I saw posted labeled "How to Bring Down Civilization": "1. Don't have a television. Read a book. 2. Don't be a conspicuous consumer. 3. Ride a bike; don't own a car. 4. Don't eat out; cook (or raise) food at home. 5. Don't travel away on vacation; enjoy your home area (if you need to escape from your home place, why the hell are you living there?) 6. Throw parties at your own home instead of going out to bars. Tell people to bring some wine or beer so you don't buy much. Fair is fair. 7. Doing steps 1 - 6 will enable you to work less."

If this list were titled something like *How to Live Without a Wage Job*," I would have thought nothing of it. But the list—and I've seen many like it (*50 Simple Things You Can Do to Save the Earth*, anyone?)—trivializes the task we have before us. None of the items on the list seriously threaten civilization. None seriously help landbases. None help salmon in the least. Nor do they in any way threaten the perceived entitlement of those who are destroying the planet. Further, this list presumes that having parties—especially parties where alcohol is the only mentioned item of necessity—can by any stretch be construed as working on oneself in the first place.

I hate to break the news to all of us, but no matter how narcissistic this culture has trained us to be, bringing down civilization involves far more than working on our extremely important and precious selves. There is a real world out there that needs our help.

The next category of work we must do consists of relieving pain. Civilization and the civilized continue to create a world of wounds, and we all desperately need people to apply salve, to put on healing hands. Sometimes I think there is almost no work more important. We need people to work rape crisis hotlines and to run shelters for battered women. We need those rare teachers who cherish their students into becoming who they are. We need people who find homes for stray cats. We need people who sing to the salmon, who hug trees and say, "I'm so very sorry." People who do these things should be very proud.

But it's a mistake to think that alleviating this pain will alone bring down civilization. It requires far more, including the very real work of defending against the attacks of the civilized. These defensive actions take many forms, including but in no way limited to filing lawsuits attempting to force those in power to follow their own laws, publicly exposing the actions of despoilers, putting our bodies between the attackers and those we love, purchasing land to protect it (or as a friend puts it, "using money to deny others their self-perceived right to destroy") and so on. These actions and many more are crucial. Without them there would be no world left to protect.

It is a mistake, however, to think that defensive actions alone will bring down civilization. We must also do the work of restoring the damage caused by this way of life. Great damage has been done to men, women, and children. It has been done to cultures and communities. It has been done to languages.[339] It has been done to rivers, forests, mountains, oceans, ice caps, caverns, plants, animals, fungi, and so many others.

The first step toward healing someone who has been injured is to provide safety from further attacks. This applies equally to psyches and to forests. But there is more that can be done. Those doing restorative work can help process the toxins—emotional in some cases and physical in others—left over from the traumas. Conditions can be created to help these others heal. For humans there is much therapeutic work to be done, whether by those classified as therapists or by good friends. And there is much therapeutic work for rivers and forests and oceans. We can clean up trash. We can when appropriate remove exotic species. We can replant native. We can provide habitat for beleaguered species. We can most of all ask what each place needs, and we can provide it.

But that's still not enough. We cannot bring down civilization simply by restoring particular people or places to how they were before civilization damaged them. We must also do preparative work, that is, we must also prepare for future attacks, we must prepare to fight back, and we must prepare ourselves and others for civilization's collapse.

I'm a big believer in flexibility. A fair number of college students have asked me if I think they should drop out of school: if it's all going to come down anyway, why bother? I've always responded that because we don't know the future, whenever possible we should choose options we'll be happy with no matter what the future brings. I'm no William Miller, arguing that the end of the world will come in 1843 or 1844 (or 2012, for that matter) and I don't want flocks of Millerites selling their earthly belongings and standing in fields waiting for the end, only to be disappointed when the sun comes up on yet another normal day.[340] Just last night after a talk someone asked me if I thought he should move from the city to the country in preparation for the crash. I gave him the same answer.

This understanding of the need to remain flexible was in fact one of the first ways I came to recognize the stupidity of civilization. Its systematic elimination of other options—destroying a people's landbase, for example, eliminates the option of those people feeding themselves—remains perhaps its most unforgivable sin. Flexibility must, I believe, underlie much of our work in preparing for whatever happens.

But no matter how flexible we remain, preparatory work alone will never bring down civilization. We must also take the offensive. We must not let those in power continue to hack with impunity at whatever fragments of ecological or communal integrity remain. We must strike at the root of their capacity to do this great damage. We must bring down not only the physical infrastructure that allows them to commit these great atrocities, but also the destructive mind-set that has created this infrastructure. We must attack on all fronts, wherever and whenever we find the opportunity.

But offensive work alone will never bring down civilization. We must also change ourselves. We must become the changes we wish to see . . .

THE CRASH

The whole human enterprise is a machine without brakes,[341] for there are no indications that the world's political leaders will deal with the realities until catastrophes occur. The rich countries are using resources with an extravagant disregard for the next generation; and the poor countries appear to be incapable of acting to curb the population increases that are erasing their hope for a better future. In such a world, declarations and manifestos which ignore the imperatives of the limits of growth are empty exercises. All the available evidence says we have already passed a point of no return, and tragic human convulsions are at hand.

Stewart Udall, Charles Conconi, and David Osterhout[342]

The sociologist, no matter how gloomy his predictions, is inclined to end his discourse with recommendations for avoiding catastrophe. There are times, however, when his task becomes that of describing the situation as it appears without the consolation of a desirable alternative. There is no requirement in social science that the prognosis must always be favorable; there may be social ills for which there is no cure.

Lewis M. Killian[343]

WHAT WILL THE CRASH LOOK LIKE? THAT DEPENDS, AS ALWAYS, ON YOUR perspective. From the perspective of bluefin tuna, marlins, sharks, or other large fish, it will undoubtedly look like the cessation of a long and horrible war, and a chance to try to recover.

For migratory songbirds it may look like someone has finally turned out the lights on the skyscrapers that kill billions of your brothers and sisters each year. It may feel like someone has stopped poisoning you with pesticides. Over time it may feel like your habitat is returning as forests and grasslands begin to regenerate.

If you are anthropophagic bacteria you might be glum to find your feedlots starting to empty.

If you are a river, you will undoubtedly be glad to receive no new poisons and to begin to flush yourself clean. You will be relieved to shake off your concrete shackles.

If you are a mountain you may breathe a deep rumbling sigh that you will not be decapitated to remove the coal or gold or silver in your guts.

If you are a lowland gorilla, grizzly bear, or tiger, you may go into hiding, try to hold on through the temporary chaos until the threat of the civilized and the immediately post-civilized humans has passed, when you can begin to recover, to live like your ancestors did forever.

If you are a traditional indigenous person living traditional ways you will no longer have to fend off oil corporations, mining corporations, logging corporations, ski resort corporations and the governments which serve them. You will still have to fend off individuals encroaching on your land. But when these encroachers no longer have the police and military to back them up, you know that you will be able to fight off this threat as you, too, hold on until the danger of civilized and immediately post-civilized humans has subsided, and you, too, can begin to recover, to live as your ancestors did forever.

But those aren't the perspectives you thought I was going to talk about, are they? We'll move on, to talk about the only things that most of the civilized care about anyway: the civilized and their machines.

I have never claimed that from the perspective of the civilized and the machines they serve that civilization's crash will be pleasant. Of course it won't.

Civilization is based on a self-perceived entitlement to exploit everyone and everything. Civilization's crash will by definition mean an end to this perceived entitlement, at the very least through a removal of the means of mass exploitation (commonly called technology or machines). No abuser likes to have his entitlement removed, or even threatened. In this case it will probably be particularly nasty for most (civilized humans) concerned, with almost all of this nastiness coming as a direct result of the civilized using any means necessary (and many unnecessary and unnecessarily vicious means) to maintain the lifestyle to which they have become accustomed. I cannot emphasize too strongly that at every step of the way the civilized have the option of turning the crash into a soft landing. But I also cannot emphasize too strongly that at every step of the way the civilized will not choose this option, but will instead kill everything and everyone who stands in the way of their perceived entitlement, who stands in the way of their increased centralization of power, who stands in the way of production, of the conversion of the living to the dead. At every step of the way the civilized will have the option of converting their weaponry to livingry, as we spoke of so very long ago. And the civilized will not choose it. The civilized will choose murder and ecocide, the latter of which ultimately means suicide, over relinquishing the quest for control. Of course. That's what the civilized have done all along. And the civilized will blame everyone and everything but themselves for the violence they create. That, too, is what they've done all along. All of this will be true no matter the cause or course of civilization's crash.

I'll say it again: the *only* reason the crash will be as nasty as it will, will be because the civilized attempt to maintain their lifestyle. We hear all the time, for example, that "we" are running out of water. And it's true that rivers are dying. Lakes are dying. Seas are dying. Or to be more accurate in each case they are being murdered. The "fact" that "we" are running out of water is one reason, we are told, for the damming of every river. And we are told that, within a few years, two-thirds of all humans will be without adequate access to water. Nonhumans obviously have it even worse. We know as well that governments are busy "privatizing" water, which means they are declaring that regular humans do not have access to water while corporations do. We know also the truth of what one Canadian water company, Global Water Corporation, puts on its website: "Water has moved from being an endless commodity that may be taken for granted to a rationed necessity that may be taken by force." And we know who will use that force.[344] But through all of this talk, we are not so often told that more than 90 percent of all water used by humans is not in fact used by humans at all. It is used by agriculture and industry. The Aral Sea has been murdered for cotton fields in Turkmenistan and Uzbekistan. The Colorado River has been murdered for golf courses in Palm Springs and fountains in Las Vegas. If earthquakes—or humans—take out the dams on the Colorado River—or actually I should say *when* earthquakes or humans take out the dams on the Colorado River—the capitalist press (if it still exists) will of course bray as one that this was a disaster for thirsty people in Nevada, Arizona, and southern California. Now, I don't right now want to get into a discussion of the carrying capacity of that region, except to say that someday the region's human population will—voluntarily or not—be much lower, with water as one limiting factor. For now I merely wish to state that I would not want to hear one word of complaint about any water shortage so long as there existed even a single golf course or lawn between Las Vegas and the Pacific Ocean, or so long as one swimming pool remained filled, or so long as one alfalfa or cotton field in Arizona or citrus orchard in California remained unparched. The same is true for manufacturing facilities. *We* are not "running out of water." Agriculture and industry are using it all. Dams could go and people could still have water: water to drink, cook with, and bathe in. Not to keep their fucking lawns green. Make no mistake: these waters are not being murdered to serve humans, and the humans who do not have access to water—and many millions of these people die each year—are not being murdered because there is not enough water. They are being murdered so the civilized can build computers, so they can play golf, so they can grow cotton in the desert, so they can have lawns. And you and I both know that the civilized will not give up their golf and their lawns and their swimming pools and their cheap cotton, no matter the cost to humans, no matter the cost to the planet.

We can say the same for many other "vital resources." We all know by now that the U.S. military consumes more than 50 percent of the oil used in the United States. Imagine how different the crash might play out if that oil were used to soften our landing, or if it were not burned at all. The military's use is not essential for keeping people alive. It is, however, essential for the ongoing theft of resources we call civilization.

Fifteen hundred people died in Haiti yesterday in a hurricane. The *San Francisco Chronicle* carried an article about the disaster, and mentioned how strange it is that so many powerful hurricanes have struck the area in such rapid succession. Of course this capitalist newspaper could not, would not, and did not mention that a dramatic increase in both the quantity and severity of storms is an entirely predictable (and long predicted) effect of global warming. Thus, according the *Chron*, nature and not the oil economy killed these people. The paper failed to mention another crucial cause of these peoples' deaths: many of them were killed in mudslides caused by logging. Thus, according to the *Chron*, nature and not the transnational timber trade killed these people. We could easily stop further deaths by stopping global warming and by stopping the theft of Haiti's forests.

It won't happen.

Or rather, it won't happen through the choice of those who financially benefit from the oil economy or the international timber trade. Not with human lives at stake. Not with the world at stake.

I'll say it yet again, because there is little that is more crucial to say at this point. The crash will be nasty for the civilized. As always the powerful will attempt to force the less powerful to bear the majority of the burden. The powerful will, by their choices, make it nastier on the poor. As we saw in Haiti. As we see everywhere. But the choice is theirs. They could make it a soft landing if they choose. But they will not choose to do this. If they were sane enough to choose life over green lawns and golf courses, they would be sane enough not to destroy the world in the first place.

I got an email from a friend. It read, "As you've said many times, it's all about choices. The problem is that people make choices based on both experience and an understanding of the given situation, and we are not allowed to really experience anything without being told what we're supposed to feel, and we're not given very much of a foundation to really understand anything either. The result is that we've more or less become biodegradable drones crafted around an increasingly starved spirit. Isn't that what capitalism requires?"

The note continues: "If people do something unconsciously, then by definition they will never voluntarily change their behavior. And that's how a lot of people in this culture live: unconsciously. And then there are those who make conscious decisions, but who can't fathom another way of life, who would rather take this life down with them than leave something for those who come after to find another way. It goes back to unmetabolized childhood processes created in the first place within a toxic environment, compounded by a belief that the current standard of living is the only standard worth having, regardless of the cost. I know you know all of this. We all know all of this. But that doesn't alter the fact that ultimately it's all about making choices. Choices made by those who want civilization to continue, who can't fathom existing without their luxuries, and choices made by those who want to bring it all down, who can't fathom living anymore within this culture. The point is all these choices are made by adults, not children. We are all adults: the degree to which each of us is any good at being an adult is really beside the point here. Adults are by definition accountable for the consequences of the choices they make, conscious or unconscious.

"The understanding that this way of life is based on choices, and that the direction the crash will take is the result of choices, makes me even angrier at the human and nonhuman consequences of the choices made by members of this culture. I feel even more anger and more sorrow at the unnecessary deaths of rivers and oceans and forests and the poor. I feel even more anger and contempt toward those whose luxuries—whose choices—are the cause of these murders. These people will be held accountable for their choices, either by me, by others like me, or by the planet itself."

There's one more topic to cover before we talk about the crash from the perspective of the civilized, which is how civilization's endgame will play out even without a crash. I described this at length in my book *The Culture of Make Believe*. I'm replicating the relevant parts here.[345]

CIVILIZATION: ONGOING HOLOCAUSTS

The lesson of the Holocaust is the facility with which most people, put into a situation that does not contain a good choice, or renders such a good choice very costly, argue themselves away from the issue of moral duty (or fail to argue themselves towards it), adopting instead the precepts of rational interest and self-preservation. *In a system where rationality and ethics point in opposite directions, humanity is the main loser.* Evil can do its dirty work, hoping that most people most of the time will refrain from doing rash, reckless things—and resisting evil is rash and reckless. Evil needs neither enthusiastic followers nor an applauding audience—the instinct of self-preservation will do, encouraged by the comforting thought that it is not my turn yet, thank God: by lying low, I can still escape.

Zygmunt Bauman[346]

MANY OF US GO WILLINGLY TO OUR OWN DEATHS. IN THE NAZI HOLOCAUST, in front of the gas chambers and crematoria were well-kept lawns and flower gardens. Often, as those who were about to die arrived they would hear light music, played by an orchestra of "young and pretty girls all dressed in white blouses and navy-blue skirts."[347] The men, women, and children were told to undress, so they could be given showers. They were told, most often pleasantly, to move into the room where they would soon die. As Zygmunt Bauman observes, "rational people will go quietly, meekly, joyously into a gas chamber, if only they are allowed to believe it is a bathroom."[348]

Once the doors were locked behind them, a sergeant would give the order to drop the crystals: "All right, give 'em something to chew on."[349] Soon, but too late, the people would realize that they had signed their final false contract, and at last they would fight for their lives, stampeding toward the doors that were sealed behind them, where "they piled up in one blue clammy blood-splattered pyramid, clawing and mauling each other even in death."[350]

☾ ☾ ☾

The endpoint of civilization is assembly-line mass murder. The assembly-line mass murder of the Nazi Holocaust is production stripped of the veneer of economics. It is the very essence of production. It took the living and converted them to the dead. That's what this culture does. It was efficient, it was calculable, it was predictable, and it was controlled through nonhuman technologies. And it was also, as well as being grossly immoral, incredibly stupid. Even from the perspective of pure acquisitiveness and land-hunger, it was self-defeating. As German troops froze and starved on the Eastern Front, valuable railroad cars were used instead to move cargos that fed crematoria. The Nazis performed economic analyses showing that feeding slaves just a bit more increased their productivity more than enough to offset the extra cost of feed. Yet they were starved. Similarly, slaughtering Russians was foolish. Many Ukrainians and Russians greeted the Wehrmacht with kisses, open arms, and flowers, happy to be out from under the tyranny of Stalinism. The Germans quickly began murdering noncombatants to make room for the Germans who would move in after the war, or because they were told to, or because the Russians were inferior, or for any of the reasons given for these slaughters since the beginning of civilization's wars of extermination. And so Russian noncombatants fought back. They blew up trains, they killed German officers, they picked off individual soldiers. They hurt the Germans. For all their vaunted rationality, the Germans weren't so very rational, were they?

Of course we're different now. We have rational reasons for the killings. There's no silly talk of master races and *lebensraum*. Instead, the economy is run along strictly rationalist lines. If something makes money, we do it, and if it doesn't, we don't (ignore for a moment that to divorce economics from morals and humanity is as evil as it is to do the same for science). But the United States economy costs at least five times as much as it's worth. Total annual U.S. corporate profits are about $500 billion, while the *direct* costs of the activities from which these profits derive are more than $2.5 trillion.[351] These include $51 billion in direct subsidies and $53 billion in tax breaks, $274.7 billion lost because of deaths from workplace cancer, $225.9 billion lost because of the health costs of stationary source air pollution, and so on.[352] This is to speak only of calculable costs, since other values—such as a living planet—do not, because they're not calculable, exist. The fact remains, however, that it is manifestly stupid to destroy your landbase, regardless of the abstract financial reward or esteem you may gain. Yet this culture spends more to build and maintain commercial fishing vessels than the fiscal value of the fish caught. The same is true for the destruction of forests. In the United States the Forest Service loses in a not atypical year $400 million dollars on its timber sale program, or about seven hundred and seventy-nine dollars per acre deforested.

❈ ❈ ❈

Hitler was ahead of his time. Social conditions weren't yet ripe for a government to fully realize the elimination of diversity toward which he aimed. Simply put, his—or any—corporate-governmental state had not yet achieved the sort of power necessary to emplace and maintain that purity of control. This is true for power relative to other corporate states, it's true for power relative to human beings, and it's true for power relative to the natural world.

So far as the former goes, we need to remember that Hitler wasn't defeated by Jews or members of resistance organizations; he was defeated by other imperial powers, the Soviet Union, Britain, and the United States. Had the Wehrmacht not foundered on the Russian winter and been repulsed by Russian troops, our stories of the time after Dachau would now read much differently. And certainly these other powers didn't stop the Nazis because of that nation's mistreatment of Jews, Romani, and so on. Indeed, each has its own august tradition of similarly unabated ruthlessness. At base, these nations stopped the German government because they didn't want it to control resources they themselves controlled or coveted.

So far as the second, control over people, imagine if Hitler had been able to broadcast his message twenty-four hours per day into peoples' homes, and if people had willingly tuned in to these broadcasts hour after hour, day after day. Imagine the effectiveness of his propaganda if teleplays could have insinuated his form of casting and fate into the lives of his subjects from infancy to senescence. I think we do ourselves a disservice if we look at old clips of Hitler strutting, yelling, and gesticulating, and wonder how the hell anyone came under his spell. First, consider who chose those clips: the victors, who as always have an interest in making their enemies look ludicrous. But more importantly, that wasn't even the Nazis' main form of propaganda. Joseph Goebbels, party propaganda chief for the National Socialists, was clear that rather than having the media inculcate people with heavy-handed political messages, it was much better to give them lots of light entertainment. Goebbels also believed that propaganda worked best when it put forth the illusion of diversity, but had a numbing sameness—a purity—to the underlying ideological message.[353]

For those whom light entertainment failed to convince, technology was also not sufficiently advanced to allow such strict governmental control of individuals as Hitler would perhaps have liked. Sure, his state police force was reasonably efficient for its day, but not only did the Nazis have no satellite surveillance systems, they didn't even have satellites. And the forensic sciences were in their early stages. It would not have been possible to track, identify, or apprehend antisocial individuals through computer-matching of fingerprints or facial scans. I'm sure by now you've heard that every person who attended the 2001 Super Bowl had her or his face surreptitiously scanned; these images were cross-checked with computer images, to identify lawbreakers. And now Sacramento's airport has begun scanning the face of every passenger. Hitler had no worldwide network of computers (named *Echelon* or not) capable of intercepting three *billion* phone or email (*What's email?* I can hear Adolf ask) messages per day, sifting through approximately 90 percent of all transmissions. Hitler not only did not have what we would consider computers, but he also did not have the capacity to capture computer signals such as keystrokes or images from monitors through walls or from other buildings. He did not have the capacity to point special types of cameras at people and perform strip searches, or even body cavity searches. Amateur that he was, Hitler did not even have a national system of social security numbers, which, in the words of United States Secretary of State Colin Powell, "allows us to monitor, track down and capture an American citizen."[354] None of this he had. Scientists used such unreliable means as phrenology to identify potential miscreants, having no knowledge of the human genome. And Hitler would not even have recognized the word *genotype*, much less been able to create genetically altered diseases to target specific races. He would have had no idea what an RFID chip is.

So far as the third, Hitler did not have the capacity to irradiate the planet, nor to poison it (organochloride pesticides and herbicides came into common usage after World War II (and in fact were in many ways by-products of the gas warfare programs of World War I; prior to that *every* farmer was organic). He didn't have the capacity to change the planet's climate. He did not have at his disposal a standing army designed to fight two major wars in disparate parts of the globe at the same time. The Wehrmacht couldn't even handle two *fronts*. The economy had not become so integrated, so *rationalized*—in other words, it had not lost so much of its diversity—as to be under the control of so few people who could kill millions of human beings—hell, who could kill the whole planet— by the merest extension of economic pressure.

In his analysis of the social effects of information technologies, Joseph Weizenbaum wrote, "Germany implemented the 'final solution' of its 'Jewish Problem' as a textbook exercise in instrumental reasoning. Humanity briefly shuddered when it could no longer avert its gaze from what had happened, when the photographs taken by the killers themselves began to circulate, and when the pitiful survivors re-emerged into the light. But in the end it made no difference. The same logic, the same cold and ruthless application of calculating reason, slaughtered at least as many people during the next twenty years as had fallen victim to the technicians of the thousand-year Reich. We have learned nothing."[355]

❰ ❰ ❰

Unless it is stopped, the dominant culture will kill everything on the planet, or at least everything it can.

Each holocaust is unique. The destruction of the European Jewry did not look like the destruction of the American Indians. It could not, because the technologies involved were not the same, the targets were not the same, and the perpetrators were not the same. They shared motivations and certain aspects of their socialization, to be sure, but they were not the same. Similarly, the slaughter of Armenians (and Kurds) by Turks did not (and does not) look like the slaughter of Vietnamese by Americans. And just as similarly, the holocausts of the twenty-first century will not and do not already look like the great holocausts of the twentieth. They cannot, because this society has progressed.

And every holocaust looks different depending on the class to which the observer belongs. The Holocaust looked far different to high ranking Nazi officials and to executives of large corporations—both of whose primary social concerns would have been how to maximize production and control, that is, how to most effectively exploit human and nonhuman resources—than it did to good Germans, whose primary concerns were as varied as the people themselves but probably included doing their own jobs—immoral as those jobs may have been from an outside perspective—as well as possible; may have included feelings of relief that those in power were finally doing something about the "Jewish Problem"; and certainly included doing whatever they could to not notice the greasy smoke from the crematoria (constructed with the best materials and faultless workmanship). The Holocaust then also looked different to good Germans than it did to those who resisted, whose main concerns may have been how to bring down the system. And it looked different to those who resisted than it did to those who were considered *untermenschen*, whose main concerns may have been staying alive, or failing that, dying with humanity.

Manifest Destiny looked different to Indians than it did to JP Morgan. American slavery looked different to slaves than it did to those whose comforts and elegancies were based on slavery, and than it did to those for whom free black labor drove down their wages.

What will the great holocausts of the twenty-first century will look like? It depends on where you stand. Look around.

If you're in group one, one of those in power, your postmodern holocausts will be at most barely visible, and at least a price you're willing to pay, as Madame Albright said about killing Iraqi children. The holocausts will probably share similarities with other holocausts, as you attempt to maximize production—to "grow the economy," as you might say—and as, when necessary, you attempt to eradicate dissent. This means the holocaust will look like a booming economy beset by shifting problems that somehow always keep you from ever reaching the Promised Land, whatever that might be. The holocaust will look like numbers on ledgers. It will look like technical problems to be solved, whether those problems are increasing your access to necessary

resources, dealing with global warming, calming unrest on the streets, or figuring out what to do about too many unproductive people on land you know
you could put to better use. The holocaust will look like houses with gates,
limousines with bullet-proof glass, and a military budget that can never stop
increasing.

The holocaust will feel like economics. It will feel like progress. It will feel
like technological innovation. It will feel like civilization. It will feel like the way
things are.

If you're in the second group, the good Germans, you will continue to be co-
opted into supporting the system that does not serve you well. Perhaps the holocaust will look like a new car. Perhaps it will look like lending your talents to a
major corporation—or more broadly toward economic production—so you
can make a better life for your children. Perhaps it will look like working as an
engineer for Shell or on an assembly line for General Motors. Maybe it will look
like basing a person's value on her or his employability or productivity. Perhaps
it will look like anger at Mexicans or Pakistanis or Algerians or Hmong who
compete with you for jobs. Perhaps it will look like outrage at environmentalists who want to save some damn suckerfish, even (or especially) if it impinges
on your property rights, or if it takes water you need to irrigate, to make the desert
bloom, to make the desert productive. Maybe it will feel like continuing to do
a job that you hate—and that requires so little of your humanity—because no
matter how you try, you never can seem to catch up. Maybe it will feel like being
tired at the end of the day, and just wanting to sit and watch some television.

❨ ❨ ❨

An article appeared in today's *Ottawa Citizen* under the headline: "Science turns
monkeys into drones—Humans are next, genetic experts say." The article read,
"Scientists have discovered a way of manipulating a gene that turns animals
into drones that [*sic*] do not become bored with repetitive tasks. The experiments,
conducted on monkeys, are the first to demonstrate that animal behaviour can
be permanently changed, turning the subjects from aggressive to 'compliant'
creatures.

"The genes are identical in humans and although the discovery could help
to treat depression and other types of mental illness, it will raise images of the
Epsilon caste from Aldous Huxley's futuristic novel *Brave New World*.

"The experiments—detailed in the journal *Nature Neuroscience* this
month—involved blocking the effect of a gene called D2 in a particular part of

the brain. This cut off the link between the rhesus monkeys' motivation and reward.

"Instead of speeding up with the approach of a deadline or the prospect of a 'treat,' the monkeys in the experiment could be made to work just as enthusiastically for long periods. The scientists say the identical technique would apply to humans.

"'Most people are motivated to work hard and well only by the expectation of reward, whether it's a paycheque or a word of praise,' said Barry Richmond, a government neurobiologist at the U.S. National Institute of Mental Health, who led the project. 'We found we could remove that link and create a situation where repetitive, hard work would continue without any reward.'

"The experiments involved getting rhesus monkeys to operate levers in response to colour changes on screens in front of them. Normally they work hardest and fastest with the fewest mistakes if they think a reward for the 'work' is imminent.

"However, Mr. Richmond's team found that they could make the monkeys work their hardest and fastest all the time, without any complaint or sign of slacking, just by manipulating D2 so that they forgot about the expectation of reward.

"The original purpose [sic] of the research was to find ways of treating mental illness, but the technicalities of permanently altering human behaviour by gene manipulation are currently too complex, he said. However, he and other scientists acknowledge that methods of manipulating human physical and psychological traits are just around the corner, and the technology will emerge first as a lucrative add-on available from in vitro fertilization clinics.

"'There's no doubt we will be able to influence behaviour,' said Julian Savulescu, a professor of ethics at Oxford University. 'Genetically manipulating people to become slaves is not in their interests, but there are other changes that might be. We have to make choices about what makes a good life for an individual.'

"In a presentation at a Royal Society meeting titled *Designing Babies: What the Future Holds*, Yuri Verlinsky, a scientist from the University of Chicago who is at the forefront of embryo manipulation, said: 'As infertility customers are investing so much time, money and effort into having a baby, shouldn't they have a healthy one and what is to stop them picking a baby for its physical and psychological traits?'

"Gregory Stock, author of *Redesigning Humans* and an ethics specialist from the University of California, agrees. 'I don't think these kind of interventions are exactly round the corner, they are a few years away, but I don't think they are going to be stopped by legislation,' he said."[356]

Remind me again, what are we waiting for? Why are we not bringing down civilization now?

❨ ❨ ❨

If you're in the subsection of the third group who might some day resist but don't know where to put your rage, the holocaust might look like armed robbery, auto theft, assault. It might look like joining a gang. It might look like needle tracks down the insides of your arms, and might smell like the bitter, vinegary stench of tar heroin. Or maybe it smells minty strong, like menthol, like the sweet smell of crack brought into your neighborhood at the behest of the CIA. Or maybe not. Maybe it's the unmistakable smell of the inside of a cop car, and a vision through that backseat window of a little girl eating an ice cream cone, with the knowledge that never in your life will you see this sight again. Maybe it looks like Pelican Bay, or Marion, or San Quentin, or Leavenworth. Or maybe it feels like a bullet in the back of the head, and leaves you lying on the streets of New York City, Cincinnati, Seattle, Oakland, Los Angeles, Atlanta, Baltimore, Washington.

If you're a member of the subsection of group three already working against the centralization of power, against the system, then maybe from your perspective the holocaust looks like rows of black-clad armored policemen, and it smells like teargas. Maybe it looks like lobbying a congress you know has never served you. Maybe it looks like the destruction of place after wild place, and feels like an impotence sharp as a broken leg. Maybe it looks like staring down the barrel of an American-made gun in the hands of a Colombian man wearing American-made camo fatigues, and knowing that your life is over.

For those of the fourth class, the simply extra, maybe it looks like the view from just outside the chainlink fence surrounding a chemical refinery, and maybe it smells like Cancer Alley. Maybe it looks like children with leukemia, children with cancer of the spine, children with birth defects. Maybe it feels like the grinding ache of hunger that has been your closest companion since you were born. Maybe it looks like the death of your daughter from starvation, and the death of your son from diphtheria, measles, or chicken pox. Maybe it feels like death from dehydration, when a tablet costing less than a penny could have saved your life. Or maybe it feels like nothing. Maybe it sounds like nothing, looks like nothing: what does it feel like to be struck by a missile in the middle of the night, a missile traveling faster than the speed of sound, a missile launched a thousand miles away?

Maybe it feels like salmon battering themselves against dams, monkeys locked in steel cages, polar bears starving on a dwindling ice cap, hogs confined in crates so small they cannot stand, trees falling to the chainsaw, rivers poisoned, whales deafened by sonic blasts from Navy experiments. Maybe it feels like the crack of tibia under the unforgiving jaws of a leghold trap.

Maybe it looks like the destruction of the planet's life support systems. Maybe it looks like the final conversion of the living to the dead.

As much as I cannot help but see the similarities between prisons and concentration camps, it seems to me a grave error to count on Zyklon-B-dispensing showers to mark the new holocaust. Perhaps the new holocaust is dioxin in polar bear fat, metam sodium in the Smith River. Perhaps it comes in the form of decreasing numbers of corporations controlling increasing portions of our food supply, until, as now, three huge corporations control more than 80 percent of the beef market, and seven corporations control more than 90 percent of the grain market. Perhaps it comes in the form of these corporations, and the governments which provide the muscle for them, deciding who eats and who does not. Perhaps it comes in the form of so much starvation that we cannot count the dead. Perhaps it comes in the form of all of these, and in many others I could not name even if I were able to predict.

But this I know. The pattern has been of increasing efficiency in the destruction, and increasing abstraction. Andrew Jackson himself took the "sculps" of the Indians he murdered. Heinrich Himmler nearly fainted when a hundred Jews were shot in front of him, which was surely one reason for the increased use of gas. Now, of course, it can all be done by economics.

And this I know, too. No matter what form it takes, most of us will not notice it. Those who notice will pay too little attention. It does not matter how great the cost to others nor even to ourselves, we will soldier on. We will, ourselves, walk quietly, meekly, into whatever form the gas chambers take, if only we are allowed to believe they are bathrooms.

ENDGAME

The assassination [of Hitler] must take place, cost what it will. Even if it does not succeed, the Berlin action must go forward. The point now is not whether the coup has any practical purpose, but to prove to the world and before history that German resistance is ready to stake its all. Compared to this, everything else is a side issue.

Henning von Treskow[357]

The second lesson [of the Holocaust] tells us that putting self-preservation above moral duty is in no way predetermined, inevitable and inescapable. One can be pressed to do it, but one cannot be forced to do it, and thus one cannot really shift the responsibility for doing it on to those who exerted the pressure. *It does not matter how many people chose moral duty over the rationality of self-preservation—what does matter is that some did.* Evil is not all-powerful. It can be resisted. The testimony of the few who did resist shatters the authority of the logic of self-preservation. It shows it for what it is in the end—*a choice.* One wonders how many people must defy that logic for evil to be incapacitated. Is there a magic threshold of defiance beyond which the technology of evil grinds to a halt?

Zygmunt Bauman[358]

HOW WILL THE CRASH PLAY OUT? PREDICTING THE FUTURE IS ALWAYS a sketchy endeavor, and I believe this is especially true for the crash of civilization. There are too many variables, and there will be too many bifurcation points. Will a plague of antibiotic-resistant bacteria hit humans so hard the human population plummets? Maybe the crash will come through a genetically modified virus, whether released by a *Twelve Monkeys* protégé, someone who hates the U.S. government, the U.S. government itself (remember the line from *Rebuilding America's Defenses,* that "advanced forms of biological warfare that can 'target' specific genotypes may transform biological warfare from the realm of terror to a politically useful tool."[359]), or perhaps most likely of all, Bayer or Monsanto. Maybe peak oil will bring it all down. Maybe global warming. Maybe hackers and ex-military people. Maybe soil loss. Maybe water loss. Maybe nukes: I have absolutely no doubt that when those who run the United States feel their power slipping, whether through oil shortages, external invasion, internal revolt, or ecological collapse, they will have no moral qualms about nuking anywhere they feel necessary, including places in the United States (hell, they've bombed Nevada for decades now). Indeed, I have great fears that when they feel their power slipping—and slip it will no matter what anyone does—they may blow up the entire planet before they give up their losing game. I asked Dean and Brian if they thought hackers could prevent this. They said, "No. I'm sure we could hack our way into a dozen or so missile sites and prevent them from being fired but there's no way we could get in to stop them all. There are thousands and thousands. There's just too many."

All that said, I'm going to lay out some possibilities, which may or may not come to pass. Honestly, you might do just as well to skip the next several pages and take a nice long walk. But only on one condition: that you spend the next few days developing your own series of scenarios for what happens next, holding yourself as honest as you can, and making yourself answerable in this honesty to those humans and nonhumans who come after. That's not a rhetorical suggestion. Close the book, put it down, and take a couple of days to think about it.[360]

Now, on to one version of the crash, this one caused by oil.[361]

Oil, as Brian mentioned, is the black blood of industrial civilization. As demand for this cheap energy continues to outstrip supply, the United States and other industrialized nations will continue to invade regions containing oil. Environmental regulations will be systematically gutted or ignored. Those who effectively oppose oil extraction will be bought off, silenced, or killed.

But no matter how many regions the industrialized nations invade, no matter how many landbases they destroy, supply for oil will never again exceed demand. Oil prices will continue to rise, leading to the eventual strangulation of the entire economy.

We can say the same for natural gas, except that its decline may be even more precipitous than that of oil.

Oil and natural gas are used not only as energy sources. Natural gas is the feedstock from which nearly all chemical fertilizers and pesticides are derived. No natural gas, no industrial agriculture. And plastics are petroleum products. No oil, no plastics. But it goes much further than this. Oil is used in the fabrication of at least 500,000 different types of products, including, to provide a very tiny sample: "saccharine (artificial sweetener), roofing paper, aspirin, hair coloring, heart valves, crayons, parachutes, telephones, bras, transparent tape, antiseptics, purses, deodorant, panty hose, air conditioners, shower curtains, shoes, volleyballs, electrician's tape, floor wax, lipstick, sweaters, running shoes, bubble gum, car bodies, tires, house paint, hair dryers, guitar strings, pens, ammonia, eyeglasses, contacts, life jackets, insect repellent, fertilizers, hair coloring, movie film, ice chests, loudspeakers, basketballs, footballs, combs/brushes, linoleum, fishing rods, rubber boots, water pipes, vitamin capsules, motorcycle helmets, fishing lures, petroleum jelly, lip balm, antihistamines, golf balls, dice, insulation, glycerin, typewriter/computer ribbons, trash bags, rubber cement, cold cream, umbrellas, ink of all types, wax paper, paintbrushes, hearing aids, compact discs, mops, bandages, artificial turf, cameras, glue, shoe polish, caulking, tape recorders, stereos, plywood adhesives, TV cabinets, toilet seats, car batteries, candles, refrigerator seals, carpet, cortisone, vaporizers, solvents, nail polish, denture adhesives, balloons, boats, dresses, shirts (non-cotton), perfumes, toothpaste, roller-skate wheels, plastic forks, tennis rackets, hair curlers, plastic cups, electric blankets, oil filters, floor wax, Ping-Pong paddles, cassette tapes, dishwashing liquid, water skis, upholstery, chewing gum, thermos bottles, plastic chairs, transparencies, plastic wrap, rubber bands, computers, gasoline, diesel fuel, kerosene, heating oil, asphalt, motor oil, jet fuel, marine diesel, and butane."[362]

Rising energy costs will undoubtedly hasten the consolidation of the already mammoth conglomerates that control the economy. These state-backed monopolies will act as state-backed monopolies are wont to do, and they will drive prices up and wages down. Unemployment will continue to rise. The gap between rich and poor will continue to widen. Spending on the military, police, and prisons will continue to climb. Starvation will increase, as the poor continue to be denied access to land and water used instead for the production of consumables for the upper classes.

As energy becomes more and more expensive, and as an ever-greater percentage of governmental spending is aimed toward security™—which has always meant providing security *for* those who steal resources and security *from* those whose lives and landbases are ruined—less money will be available to provide basic maintenance for the infrastructure. This is already happening. The infrastructure—or at least that part of the infrastructure which serves the poor—will continue to degrade.

The more the infrastructure degrades, the more that stockpiles of food, oil and gas, chemicals, and pharmaceuticals will be controlled by the military, police, and other warlords.[363] We see this already in U.S.-occupied Afghanistan, and elsewhere.[364]

The already (and always) faint line between corporations and governments will fade not only from reality but from memory as well. The already fundamentally false distinction between "public" armies and private "security forces" will disappear entirely. Mussolini's definition of fascism—the merger of state and corporate power—will be complete. We see this already in U.S.-occupied United States.

I think the fictional reference I'm looking for is *1984*.

The writer and activist Aric McBay has described what may happen next.[365] In order to make up for the "energy gap" in agriculture and manufacturing, slavery and forced labor will become ubiquitous. Production will still falter, since huge numbers of people are required to put out even a fraction of the output of a large machine. Many of these people will be worked to death. These deaths will matter no more than these deaths have ever mattered to those who value production over life.

Portions of cities may be ghettoized—sealed off—to prevent those inside from escaping or getting food. Those inside may be forced to work in the factories for food, water, and so on.[366] Many of these people will be worked to death as well. In the country, people will be worked to death in agricultural labor camps.

But without cheap energy, none of this will be sufficient to supply cities with necessary resources. In time even the rich may begin to go hungry. Urban populations will crash because of disease, starvation, and emigration. The electrical and other infrastructures will fail completely because of deliberate attacks, lack of maintenance, and lack of energy. Most cities will become effectively uninhabitable, and industrial civilization as we know it will be over.[367]

Simultaneously, the breakdown of the infrastructure will reduce the effectiveness of the military and police, who rely partly on high-tech communications and energy-intensive mobility to kill or capture their targets. This reduced effectiveness will lead to a gradual return of power to the local level as it becomes impossible to maintain distant control without massive inputs of cheap energy. This transfer of power back to the regional level will not, however, herald the beginning of an enlightened era of ecovillages where people live sustainably in peace and harmony, in part because there are still too many people for the landbases to permanently support, in part because the landbases have been too degraded to support as many people as they did prior to the arrival of civilization, and in part because most of the people are still insane, that is, civilized, and have no idea how to enter into a relationship with other humans, much less a landbase. So there will be those who attempt to seize power. There will be fights over resources, over the protection of resources, over the depletion of resources.

I think the fictional reference I'm looking for is *Mad Max*.

But there are two pieces of good news. The first is that gasoline degrades quickly, so this *Mad Max* era will not be able to last very long. Soon the cars will sputter to a stop. Soon the chainsaws will cease to roar. Soon there will be only the sounds of living beings. No machines. This leads to the second piece of good news. Local battles are eminently winnable. Years ago, as I mentioned earlier, when I asked a member of the rebel group MRTA what he wanted for the people of Peru, he said, "We need to be able to grow and distribute our own food. We already know how to do that. We merely need to be allowed to do so." Without interference from those distant others who wish to steal their resources, people could grow and distribute their own food. And they could form protective organizations with no fear of being overwhelmed by state power. And they could kill those who try to stop them, those who try to seize power, those who try to steal their food or their land.

Hundreds of pages ago I wrote, about the collapse of civilization, "The urban poor are in a much worse position than the rural poor. They obviously do not have access to land. In the long run, they would of course be far better off without civilization. The problem—and this is a very big one—is that in the short run many of them would be dead: their food is funneled through the very system that immiserates them." I was right, and I was wrong. The most unforgivable word I used is *obviously*. They do not have access to land, true, but why is that? It is not because the rich own the land, since land ownership in this sense is nothing but a shared delusion, backed by the full power of the state. And that last phrase is the key to everything. Without the full power of the state, the rich are no longer rich. They are simply people in big houses with big swimming pools and big piles of paper claiming they own big plots of land. Big deal. Cut these people off from their support by the colonizers, and the poor, including the urban poor, will be able to take back the land that is currently used to produce nonfood crops like cut flowers, dog food, coffee, opium, and cocaine for the rich.

What, at its essence, does military technology do? It allows one person to kill many. That has *always* been the point. With the technology in place you have thousands of starving people being held back by mobile police forces with guns. But without an industrial infrastructure, soon enough a gun is nothing but a metal pipe attached to a piece of wood. Take away these technologies—take away the full power of the state—and you have thousands of starving people with machetes up against a rich guy and the people he used to pay holding guns that will soon run out of bullets. I'll put my money with the starving people. I'll stake my life with them.

❨ ❨ ❨

Let's be clear. The richest fifth of the world consume 45 percent of all meat and fish, while the poorest fifth consume 5 percent; they consume 58 percent of total energy, the poorest fifth less than 4 percent; they have 74 percent of all telephone lines, the poorest fifth 1.5 percent; they consume 84 percent of all paper, the poorest fifth 1.1 percent; they own 87 percent of all vehicles, the poorest fifth less than 1 percent. Taking out the electrical infrastructure will not harm the poor. It will harm only those who are killing the poor, and killing the world.[368]

❨ ❨ ❨

I've read several accounts of the crash that suggest that deforestation will increase. This strikes me as nonsensical, for several reasons, not the least of which is that the international market for pulp and paper will be gone, as will cheap oil, gasoline, and metal. Only a madman would cut down a redwood if he couldn't sell it.[369]

People will of course still need wood to use for cooking, and in cold climates warmth. But it strikes me that trees will not be the first to burn. Trees are hard to fell, and there will be an awful lot of more easily accessible wood. It's called "someone else's furniture."

❨ ❨ ❨

If you're a member of the first group, those in power, the crash of civilization will for a time probably seem like business as usual. You will continue to attempt to increase your access to resources. You will continue to attempt to increase your power. The holocausts you cause—although of course you would never call them holocausts, and it would never occur to you that you are the cause—will become increasingly visible, but time after time you will be willing to pay these prices, so long as it is always someone else who really pays. Faltering production will concern you, but your faith in the system will remain unshaken. Dissent will disturb you slightly, but you will recognize that there have always been those who are jealous of your freedoms, envious of your wealth. And so you will hire more police, install more cameras. And you will keep the charade going as long as you can.

If you're in the second group, the good Americans, you will continue through the crash to be co-opted into supporting the system that does not serve you well. No longer, however, will the holocaust look like a new car. Your expectations will be diminished. And then they will be diminished again. And then again. No longer will the ongoing holocaust that is civilization look like merely continuing to do a job you hate, but it will come to more and more openly resemble slavery. To maintain the façade of pride and dignity—real pride and dignity having long since been stripped away—you will resist calling it slavery. When you do finally call it slavery, it will probably only be after the razor-wire-topped and electrified gates have been shut behind you. You may vaguely recognize the phrase, *Work Makes You Free,* but you may not know what it means.

If you're in the subsection of the third group who might some day resist but don't know where to put your rage, the collapse of civilization might look like opportunities for increasing your personal power. You might attempt to seize and hold territory through force and terror. You might for a time succeed. But as the system collapses in on itself you, too, like those in group one, with whom you have so much in common, may find your power base too reliant on resources brought in from outside. Also like those in group one, unless you have a change of heart, you will to the extremely bitter end try to maintain your power. You will keep the charade going as long as you can. Or maybe you won't. Maybe your hardships have honed your rage, made it sharp against those who caused the hardships in the first place, those who exploited and degraded you and those you love, including your community.

If you're a member of the subsection of group three already working against the centralization of power, against the system, then you may find the system no longer so invulnerable as once it seemed. Its weaknesses will every day be more obvious, more inviting. And you will strike. And you will strike again. You will taste something you have never before tasted: victory. And then victory after victory. Some of you will certainly die, as those who were once fully in power cannot even to the end admit they were wrong and just give it up. But some of you will not die. And day after day your success will become more clear.

If you are of the fourth class, those who are now considered simply extra, you, too, will taste something that you, too, have never before tasted: hope. Hope in lasting change. Subsistence farmers will no longer be threatened with dispossession. The same with indigenous peoples. The landless will take back land. And over time this faint taste of hope will become stronger, and stronger still, until it begins to taste like something altogether different: agency. You will come to know that this lasting change is not merely something to be hoped for but is something you can achieve through working in solidarity with others in your community. And you will come to do it.

❨ ❨ ❨

For as long as civilization continues, many of us will walk quietly, meekly, into whatever form the gas chambers take, if only we are allowed to believe they are bathrooms. But more and more of us will no longer make this mistake. We will begin to allow ourselves to know what we have always known. And once we know that they are not bathrooms—once we see civilization for what it is—then it will be time for us to dismantle the gas chambers—and gas refineries, oil wells, factory farms, pharmaceutical laboratories, vivisection labs, and all of the other cathedrals of civilization—and to make certain that they will never be erected again.

❨ ❨ ❨

Early in this book I described how environmental change is often not gradual, but catastrophic. I quoted one scientist as saying, "Ecosystems may go on for years exposed to pollution or climate changes without showing any change at all and then suddenly they may flip into an entirely different condition, with little warning or none at all." Another wrote that "only in recent years has enough evidence accumulated to tell us that resilience of many important ecosystems has become undermined to the point that even the slightest disturbance can make them collapse."

The natural world is far more resilient than civilization. This culture may go on for years without showing any change at all and then suddenly flip into an entirely different condition, with little warning or none at all. This will happen when the system has been undermined to the point that even the slightest disturbance can make it collapse.

What are you waiting for?

THE RETURN OF THE SALMON

The difference between the White man and us is this: You believe in the redeeming powers of suffering, if this suffering was done by somebody else far away, two thousand years go. We believe that it is up to every one of us to help each other, even through the pain of our bodies. . . . We do not lay this burden onto our god, nor do we want to miss being face to face with the spirit power. It is when we are fasting on the hilltop, or tearing our flesh at the sundance, that we experience the sudden insight, come closest to the mind of the Great Spirit. Insight does not come cheaply, and we want no angel or saint to gain it for us and give it to us second hand.

Lame Deer[370]

If the salmon could speak, he would ask us to help him survive. This is something we must tackle together.

Bill Frank, Jr.[371]

Finis Initium (Finish What You Begin)
   inscription on Count Claus von Stauffenberg's ring.[372]

THIS IS A BIG BOOK ABOUT A BIG SUBJECT. BUT THERE ARE MANY AREAS I'VE left unexplored. I've not talked much about the technical aspects of removing dams or other destructive structures, nor even provided any more than the most basic guidelines for choosing what actions to take. I've provided no detailed examination of the workings of the economic system. I've talked about bottlenecks and levers but for the most part have not pointed out what they are, much less described what to do about them. And I've not talked at all about organizing. Should we act in small cells or as parts of a larger, more cohesive organization? How do these cells coordinate? How does this organization make decisions? Who decides what we do? What do we do about infiltrators? What about snitches? What sort of training do we give people, and how do we do it?

I am not a bomb maker. I am not some sort of infrastructural engineer. I am not an organizer.[373] I am not a general. I am a writer. I am a philosopher. I can do what I can do.

The rest is up to you.

❨ ❨ ❨

The extraordinary writer and activist Aric McBay interviewed the equally extraordinary writer and activist Lierre Keith about why so many of us do not resist, and what it will take for us to achieve a critical mass of resistance.

He said, "One of my favourite quotes is something Dietrich Boenhoeffer wrote while in prison in Germany during World War II, as he awaited execution for resisting the Nazis: 'We have spent too much time in thinking, supposing that if we weigh in advance the possibilities of any action, it will happen automatically. We have learnt, rather too late, that action comes, not from thought, but from a readiness for responsibility. For you thought and action will enter on a new relationship; your thinking will be confined to your responsibilities in action.' Radicals often like to construct imaginary models of their hypothetical utopias and sketch out the improvements they want to see in the future. But as we know, if industrial civilization doesn't come down soon—very soon—there *is* no future for us. (And I'm still surprised at how determinedly oblivious even radicals can be to this simple fact. They really just don't want to hear it.) What does it take to move people beyond mere strategizing and philosophy? How do people acquire a real 'readiness for responsibility'?"

Lierre responded, "I think the biggest reason otherwise radical people don't want to face the necessity of ending industrial civilization is privilege. We're the ones reaping the benefits. We've sold out the rest of life on earth for convenience, creature comforts, and cheap consumer goods, and it's appalling.

"But there's another group of people, who don't think their access to ice cream 24/7 is more important than life on earth. That's good. But they're sunk in a rational, realistic despair. What can I do about any of it? It's all going to hell, and nothing I personally do is going to make any real difference. Why bother to take down a cell phone tower when there's thousands more across the country? But it's not useless to take down that cell phone tower if I know that tonight five hundred other people are doing the same thing. Now my action has meaning, impact. But radical environmentalists haven't moved to that level of organization yet.

"I think the readiness to act is born from two sources: rage and love. And we have to have the stamina to keep loving even when what we love is being destroyed, and we have to have the courage to make that love be an action, a verb.

"I wouldn't bother to recruit anyone who has to be coaxed into action. Focus on the people who want to act but don't know what to do. Give them a serious plan and maybe we have a chance."

❨ ❨ ❨

We will have better than a chance. All it will take is a series of miracles. Nothing could be more natural than that.

❨ ❨ ❨

A couple of days ago I witnessed a miracle. I am blessed to witness similar miracles each year at this time.

I look at a stump. I see nothing out of the ordinary. The stump is hollow, the inside rotted. The tree was cut a long time ago. Huckleberries grow inside and around it. The berries not eaten by me, birds, bears, or insects hang on the branches long into the fall. The berries slowly shrivel, and eventually drop to the ground.

It is a bright day. Warm. At first there is nothing. And then it starts, just as it starts every year. I see one ant, and then another, and then another. They are coming from the stump, they are coming from underground. They climb to the top of the stump, where they gather. More and more. It is now a stream of ants flowing out of the stump, out from underground, out from the only home they have known. Now it is a river. They all have wings. They fly. The sky shimmers with light reflecting off their wings. Birds swoop down to eat as many as they can. A spider hangs motionless in its web, resting one long leg along a strand to feel for any change in tension. The ants fly away. They do not come back. Their wings are meant for one flight only. They fall off when the ants find their new homes. Yet still they fly. I always envy their courage.

That is the miracle I witness each fall.

(((

The world gives us so very much. It gives us our life. All of our neighbors—the ants, spiders, salmon, geese, sharks, seals, cottonwoods, chestnuts—are doing the real work of keeping this planet going. Isn't it time we did our share?

(((

Early in this book I cited as one reason that civilization is killing the planet the fact that many of those in power take their lead from God. It's okay to burn oil and natural gas, they say, no matter the cost to the earth, because increased earthly devastation marks the End Time, and the return of the Christ. It's okay to perform any acts that harm the planet because the natural world is inconsequential to God's plan: "Nowhere in the Bible does it say that America will be here one hundred years from now." George W. Bush invaded two countries because, he has stated publicly, "God told me to strike at al Qaida and I struck them, and then he instructed me to strike at Saddam, which I did."

Bush listens to God, I listen to trees. What's the difference?

I see three possibilities. One is that we're both right. God does actually talk to Bush, and trees actually do talk to me. This leads to another question: Who would you rather listen to, a distant sky God—disconnected from and superior to the earth—who by his own admission is angry and vengeful, and who has preached and justified more rape and rapine than any other god we've ever heard of; or trees, who, to the best of my knowledge have never once told anyone to go forth and subdue the earth, and to have dominion over the fish of the sea, and over the fowl of the air, and over every living thing that moveth upon the Earth; trees who have never once justified a single act of rape or rapine, trees who have never said that they are jealous, trees who live right next to us, and who are our closest neighbors?

The next possibility is that one of us is wrong. Maybe Bush is right, and I am wrong. Or maybe I am right and Bush is wrong. If either of these is the case, you need to decide for yourself, and you also need to ask yourself who—God, trees, rivers, ants, yourself, or some combination—you want to follow.

The third possibility is that we're both wrong. We're both either mishearing, or we're both delusional, or some other explanation. In which case you the reader are left at the same place you started: in a position to choose for yourself.

In some ways it doesn't really matter, since no matter where we got our arguments, these arguments should be evaluated on their own merit. If me having conversations with nonhumans bothers you, I have a very simple solution: grab a black magic marker and cross out every sentence in this book where I mention hearing wild nature speak. Do not cross out what I say I hear. Then reread the book. The arguments still stand. Do the same for those who listen to God: can their arguments stand on their own, without counting on the Big Man's authority to back them up?

Finally, the Son of Bush's God is reputed to have said, "By their fruits ye shall know them." And the bottom line of all of this really is what actions we take. Bush's God tells him to invade other countries, and he does it. An oil executive's God tells him to burn oil, and he does it. A U.S. Senator's God tells him to destroy the earth, and he does it. The trees tell me that if I consume the flesh of another, I must take responsibility for the continuation of their community, and I do that. Take your pick.

Or better, listen for yourself.

❨ ❨ ❨

This book is not, of course, just about taking down civilization. It is about something that needs to happen first. Why do you think I laid out the premises explicitly for you, put you in a position of actively choosing to agree or disagree with them? Why do you think I've approached this from so many directions? Why do you think I've expressed my own fears, expressed my own confusion? Why do you think I've made points, undercut or contradicted them, and then made them again?

Because I'm not the point, and what I understand isn't the point. The point is the process I'm trying to model. The point is that you puzzle your own way through, and figure out for yourself what, if anything, you need to do. I said before I don't want a flock of Millerites. Likewise, I don't want a flock of Jensenites. I don't want to replicate the same old model on which civilization has from the beginning been based: God/King/President/Priest/Scientist/Expert/Author reveals the Holy Truth to those who have ears to listen, and if you don't listen, well, then, off to hell you go. I'm not going to let you off that easily. I'm asking you to be responsible for your own thinking, responsible to your own heart, answerable to your own understanding. I'm asking you to think and feel and understand for yourself.

If you start doing that, civilization will begin to crumble before your eyes. Because above all else, civilization cannot survive free men and women who think and feel and act from their own hearts and minds, free men and women who are willing to act in defense of those they love.

☾ ☾ ☾

Do not listen to me. I do not live where you do. I do not know how to live there sustainably. I do not even know how to live here sustainably. I do not know how to live sustainably at all. If you want to know what to do, go to the nearest river, the nearest mountain, the nearest native tree, the nearest native soil, and ask it what it needs. Ask it to teach you. It knows how to live there. It has lived there a very long time. It will teach you. All you need to do is ask, and ask again, and ask again.

☾ ☾ ☾

People often ask me what sort of a culture I would like to see replace civilization, and I always say that I do not want any culture to replace this one. I want 100,000 cultures to replace it, each one emerging from its own landbase, adapted to and adaptive for its own landbase, each one doing what sustainable cultures of all times and all places have done for their landbases: helping the landbase to become stronger, more itself, through their presence.

☾ ☾ ☾

There's a place I go when the sorrow gets to be too much for me, when I feel I just cannot go on. It's only a few miles from my home, and coincidentally only a couple of miles from a couple of different sites where in the nineteenth century the civilized massacred hundreds of Tolowa Indians. In the 1960s a corporation started to put in a housing division there. The corporation laid out paved roads in neat squares. But then because of environmental concerns it was never able to get permission to build any houses. So for the last forty years the housing division has sat.

And the forest has begun to reclaim its own. Trees push through pavement, roots making ridges that run from side to side of the street. Grass comes up in every crack. Wind, water, sand, and bacteria make potholes that grow year by year. Or maybe we should switch perspective and speak of the ground beneath finding its way back to the surface. Trees and bushes reach from each side of the road to intertwine limbs, at first high above the ground, then lower and lower, until sometimes you cannot even see where there used to be a road.

Forty years, and the land is coming back. That makes me happy.

Someday I know that each year more salmon will swim up the stream behind my home than swam here the year before. Each year more migratory songbirds will return than the year before. Each year more trees will creep out that much further from the edges of forests into clearings. Each year more roads will have that many more holes in them, that many more plants growing first along their shoulders, then all across. Each year that many more bridges will fall, each year that many more dams will fail or be removed. Each year that many more electrical wires will come down.

And someday, someday soon, wolves will return, and grizzly bears, and all those others whose home this is and has been.

And someday, someday soon, the rivers will again be full of salmon.

❦ ❦ ❦

Three nights ago I gave a talk just south of Eureka, California. Afterwards a woman approached me to sign her book. Before I could ask her name she began to cry. I did not say anything, but just looked at her softly.

Finally she said, through tears, "I have been here nineteen years, and I have seen the death of the salmon. When I first moved here, there were so many they would rub against my leg as I walked the river. I remember the first time I felt a thirty-five pound salmon rub his weight against me. The rivers were full, even then. Sometimes the steelhead came up so fast from the ocean, they still had sea lice on them."

She stopped and looked me square in the eye, tears on her cheeks.

She said, "And now they're gone. There are almost none there. I have watched them die, and now I am standing here in front of you crying about fish who are gone."

I didn't say anything.

She said, "I need to stop crying."

I didn't say anything.

"They're never coming back."

"How old are you?" I asked.

"Forty-eight."

"You will live to see them come back," I said. "I make you that promise. They might not be back in five years, or even ten. But they will be back soon. Civilization is going to come down, and the fish will find their way home."

The fish will find their way home. And so will the wolves and bears and spotted owls and sharks and tuna and Port Orford cedars. So will the great apes and the tigers. So will the shad and the sea bass. So will the bison and the prairie dogs.

And so will we.

❨ ❨ ❨

Over the years I have been criticized because I do not suggest models by which people should live. "You're only interested in tearing things down," some people say, "not in providing alternatives."

I do not provide alternatives because there is no need. The alternatives already exist, and they have existed—and worked—for thousands and tens of thousands of years.

Over the years I have heard many of the civilized ask how we could possibly live without civilization. It is a question I have never heard any Indians answer publicly. It is a question I have never asked, because I already know the answer. In private many Indians have answered this question I have never asked. They have said, "After civilization is gone from the earth and from your hearts, we will teach you how to live. We will not do it before then because your culture has been trying to kill us, and also because you would try to make money from what we say, or you would try to paste what we tell you onto your unworkable system. So until civilization is gone we will just hold on to our traditions and hold on to our existence. Later, if you come to us, we will help."

What they say is true. And it is true also of the land. Once civilization is gone, once it is only a terrible, terrible memory, the land, too, will teach us how to live.

❨ ❨ ❨

The Christian Cheyenne Chief Lawrence Hart described one tradition of his people, which he called the Cheyenne Peace Tradition. I want to describe another, called the picket pin and stake. Before a battle, a few of the bravest Cheyenne Dog Soldiers would be chosen to wear sashes of tanned skins called "dog ropes." Attached to each dog rope was a picket-pin, normally used to tether horses. During battle, the pin would be driven into the ground as a mark of resolve. Once the pin was driven, the Dog Soldier would remain staked to that piece of ground, even to his death. Retreat was no longer an option. The pin could only be removed when everyone was again safe, or when another Dog Soldier relieved him of his duty.

It is time. I have driven my picket-pin. I am staked out, and willing to give in no more.

Where will you drive your own picket stake? Where will you choose to make your stand? Give me a threshold, a specific point at which you will finally stop running, at which you will finally fight back.[374]

❨ ❨ ❨

Stand with me. Stand and fight. I am one. We would be two. Two more might join and we would be four. When four more join we will be eight. And we will be eight people fighting whom others will join. And then more people, and more.

Stand and fight.

❨ ❨ ❨

The questions before each of us now are: What are your gifts and how can you use them in the service of your landbase? What can you do? What does your landbase most need from you? How can you achieve it? What do you want to do?

And right now, perhaps the most important of all: What are you willing to do?

❨ ❨ ❨

And finally, a note I will write to you in a few years, using a pencil on a piece of scrap paper, delivered by hand, from village to village, until it reaches you. The letter reads, "We did it! We brought down civilization! There were times I didn't think that was possible. I thought the fascists would kill or imprison us all. I thought they would destroy the world before we would be able to stop them. I am thankful to everyone who worked to bring it down. The hackers, the saboteurs, those who put their bodies in front of the destroyers, those who fought back, those with e-bombs. The people who took out the dams. And all those who supported and protected the underground fighters. I love you all so very much.

"Things were tense here for a while, but people are finally beginning to understand that their lives depend on the landbase. There were some pitched battles fought—and I mean battles—over protecting the last of the salmon from people who wanted to kill them. We won.

"It's so different to fight these battles now. The problems are local, and so are on a scale we can comprehend and deal with. There has been some violence, as I mentioned, but having removed the crazy superstructure that was causing people to kill the planet, we are finally just barely starting to become a community. That process will of course take a thousand years. I mean, people are still as fucked up as they were before. There is just as much child abuse, just as much rape, just as many people who were raised in a culture based on power and control. That we'll have to work on over time. But we'll be able to do it. I'm sure you recall our great slogan, 'Dismantle globally, renew locally.'

"It has been a long dark night. Six thousand years of deforestation, despoliation. But it's over. It can't rise up again. We made sure of that, didn't we? And of course the easily accessible reserves of oil are gone, so there will never be another oil age. And the same with iron, so no iron age. We're all back living the way we were meant to, in a world being renewed.

"It's been a long and hard struggle, and there were many times I was very scared. But we did it. We really did it. I'm so very happy.

"If you recall, I used to close my letters with 'love and rage.' I cannot tell you how good it feels to no longer close that way, but to be able to close this way, and to mean it.

"Love and Peace,

"Derrick"

Acknowledgments

AS ALWAYS, MY FIRST AND MOST IMPORTANT DEBT OF GRATITUDE is to the land where I live, which sustains and supports me. Equal to this is my debt to the muse, who gives me these words, and without whom I cannot imagine my life. Thanks also to the source of my dreams.

My thanks to the redwoods, red and Port Orford cedars, alders, and cascara; the Del Norte, Olympic, slender, and Pacific giant salamanders; the Pacific tree and northern red-legged frogs; the rough-skinned newts; the spotted and barred owls, the phoebes, pileated woodpeckers, hummingbirds, herons, mergansers, and so many others; the coho and chinook salmon, the steelhead; the banana slugs, the flying ants, and the solitary bees. Thank you to the reeds, rushes, sedges, grasses, and ferns, the huckleberries, thimbleberries, salal, and salmonberries. The chanterelles, turkey tails, amanitas, and so many others. My thanks to the gray foxes, the black bears, Douglas squirrels, the moles, the shrews, the bats, the woodrats, the mice, the porcupines, and the shy aplodontia. My thanks to all the others who graciously allow me to share their home, and who teach me how to be human.

There are others who have helped with this book. They include, among others: Melanie Adcock, Roianne Ahn, Anthony Arnove, Tammis Bennett, Gabrielle Benton, Werner Brandt, Karen Breslin, Julie Burke, Leha Carpenter, George Draffan, Bill and Mary Gresham, Felicia Gustin, Alex Guillotte, Nita Halstead, Tad Hargraves, Phoebe Hwang, Mary Jensen, Lierre Keith, Casey Maddox, Marna Marteeny, Mayana, Aric McBay, Dale Morris, Theresa Noll, John Osborn, Sam Patton, Peter Piltingsrud, Karen Rath, Remedy, Tiiu Ruben, Terry Shistar and Karl Birns, Dan Simon, Julianne SkaiArbor, Shahma Smithson, Jeff and Milaka Strand, Becky Tarbotton, Luke Warner, Bob Welsh, Belinda, Rob, Brian, Dean, my military friends, John D., Narcissus, Amaru, Yeti, Persephone, Shiva, Emmett.

All these acknowledgments are in a sense premature. It is customary after finishing a book for authors to acknowledge in print all those who helped bring the book to fruition. But this book isn't yet finished. If it is to be more than mere words, this book will only be complete when this culture of death no longer imperils life on earth. At that point the acknowledgments and gratitude will flow like rivers rushing through canyons once blocked by now-crushed dams.

Notes

We Shall Destroy All of Them

1. Quoted in the *The Sun*, March 2004, 48.
2. I first wrote this sentence using the words *would* and *if* instead of *will* and *when*. I can't tell you how good it feels—and how important it is—to use the latter words instead of the former.
3. Schmitt, 7. It all reminds me of a website I heard about the other day: Masturbate for Peace: Using Self-Love to End Conflict (http://www.masturbateforpeace.com). At first I thought it was a brilliant satire of so much of our so-called resistance, with rhetoric like, "Joining this movement is simple. Just masturbate in your own way, focusing your thoughts and energy towards love and peace. Encourage others to do the same." It also has slogans like "Peace is spiffy, stroke your stiffy," or "War is silly, whack your willy." But the site also has a bunch of fairly disgusting stuff, like a link to a photo of a man who pulled the skin off his penis by trying to masturbate using a running vacuum cleaner.
4. See, for example, the rape of women by men.
5. Last year I read about someone who died after being bitten by a shark. She was wearing a wetsuit and swimming with seals. Seals are a major prey of sharks. Wetsuits can make you look like a seal. I remember thinking when I read about this death that with so many risks one can take in one's life, I do not think that dressing up like shark food and swimming through some sharks' kitchen is one I want to take.
6. To be fair, some of those killed by police are legitimate killings, where people shoot at the cops, or are men who are holding their wives at gunpoint, and so on. But, and this is really the point, a great many are not. And there is no effective accountability for police who kill. Right now there is a big controversy in San Francisco over whether the district attorney will seek the death penalty for someone who shot a cop, with nearly everyone clamoring for the person's life. I actually have no problem with that, or would have no problem with it, if those police who inappropriately killed people were subject to the same penalty. There is no acknowledgment of irony in the fact that the same newspaper carrying opinions about executing this person who killed a cop also carrying an article saying that the government is refusing to prosecute guards at a California Youth Authority prison who were *caught on videotape* beating the hell out of some of the kids under their authority. That is premise four in action.
7. And so far as the war on women, there are plenty of men willing to use their penises to similar ends.
8. The same is true, of course, on an individual level. You cannot make peace with an abuser. You will lose the peace as surely as you will lose the arguments. There is only one way not to lose to an abuser, which is to get that person out of your life. That the same is true on the larger cultural level should by now be obvious. Something else that should be obvious is that on the larger cultural level, we can no longer just leave. If we cannot leave, how will we get the abuser that is this culture out of our lives? The answer seems pretty clear.
9. I much prefer the response I once heard from an activist: "When those in power talk about trying to create a win-win situation, I reach for my gun."
10. And don't give me the same old tired line about how, if we individually give these up, the culture will stop killing the planet. We've already discussed this, but it just doesn't seem to matter. You or I changing our lifestyles will not stop the culture from killing the planet. The system needs to be broken down.
11. I am defining *winning* as I did above: I want to live in a world with more wild salmon every year than the year before, a world with more migratory songbirds every year than the year before, a world with

more ancient forests every year than the year before, a world with less dioxin in each mother's breast milk every year than the year before, a world with wild tigers and grizzly bears and great apes and marlins and swordfish. I want to live on a livable planet.

12. In *A Language Older Than Words* I said I grew up in Montana. I did this only because my former publisher insisted, because he was afraid my father would sue for libel. It is typical for abusers to use such lawsuits—or even their threat—to silence their victims. The publishers at *The Sun* even insisted that I use a pseudonym for an interview I did of Judith Herman about the aftermath of violence. If my father was able to impose his will on two separate sets of publishers he has never met and force me, his adult son, to use a fictitious name and tell lies about where I grew up in order to protect him, imagine the degree of captivity routinely experienced by a significant portion of children and partners every day and every night.

13. The redwood trees respond: "I'm sorry, but we do not mew like kittens or sing like whales: *they* copy *us*."

14. I have never claimed that I am never stupid.

15. Both physically and spiritually.

16. Can you say global warming? Cancer? Premature puberty?

Winning

17. *The Sun*, October 2003, 48.

18. *One* of the starkest? I don't know if I want to hear any that are starker than this.

19. Lean.

20. Kopytoff.

21. Even if we're smart, they're still going to kill or imprison many of us. It's unforgivably naïve to think there will not be casualties on all sides, whether we win, lose, or make no attempt.

22. I don't think that "using the force" would be a tactic, since it's a spiritual attitude. A tactic would be *how* Luke flies and what approach he chooses to take to the tube itself. "Using the force" is a way of being. A spirituality or a way of being is *not* a tactic.

23. I need to be clear that I am not cynical enough to believe all relationships are this way, nor is that my experience. I was speaking specifically of those I've known or heard about whose specific goal of marriage was more important than either integrity or the quality of the relationship. I am also thinking about the extremely popular (and extremely morally troubling) book from a few years ago called *The Rules™: Time-Tested Secrets for Capturing the Heart of Mr. Right.*

24. I guess this would be a strategy. There isn't really a word for plans to achieve operational goals the way that the words *strategy* and *tactics* exist, and in any case the terms are a bit fuzzier than I'm making them seem, as from a general's standpoint the movement of a brigade might be tactical, but from the perspective of a lieutenant commanding a squad in that particular brigade, such movement would be strategic: it's all about perspective.

25. I told you that you might not believe it.

26. You could also end up on an aircraft carrier or in Washington, D.C., or South Carolina.

27. And don't give me any nonsense about how effective we are. If we were effective in the least, the world would not be getting killed.

28. I first wrote "more or less ignoring morality on the larger scale," but that's insane: we are not stopping those in power from killing the planet; this is not "more or less ignoring" larger-scale morality, but ignoring it to an outrageous, unbelievable, unspeakably despicable, and most important of all, unforgivable degree.

29. Men, too, are of course trained to hate women, but we probably shouldn't talk about that, should we?

30. Similarly, women aren't supposed to hate, lest they be called ballbusters.

Importance

31. I have two problems with this, of course. The first is that Wilson equates men with humans, and the second is that he equates humans with civilized humans. If civilized humans disappeared, the world would be a much richer place.

32. Why anyone would move to a beautiful redwood forest and cut it down to put in a lawn escapes me.

33. I'm editing this section while sitting on a beach monitoring ORV use. Every vehicle I've seen so far has been in violation of lax (and unenforced) regulations. For example, just now there are a couple of assholes (note to copyeditor: yes, I am aware that calling an ORV driver an asshole is redundant) chasing down birds and two more trying to pick them off with paintball guns.

34. Along with her extraordinary talent for writing, of course.

35. Larry T. B. Sunderland, "California Indian Pre-Historic Demographics," Four Directions Institute, http://www.fourdir.com/california_indian_prehistoric_demographics.htm. See also the map "Native American Cultures Populations per Square Mile at Time of European Contact," also part of the Four Directions Institute website, http://www.fourdir.com/aboriginal_population_ per_sqmi.htm (accessed June 4, 2004).

36. Eckert, 709, n. 190.

37. Interestingly enough, the day I wrote these words I received an email attacking me for writing in another book that "trees don't belong to us anymore than do water or air. They belong to themselves." The letter writer told me those two sentences pretty much sum up everything that is wrong with my work and my worldview. He also commented, "I suppose a corn plant belongs to itself in your world as well." Well, yes, it does. Women belong to themselves as well, as do children, rocks, rivers, and all of us. Within this culture the notion that everything belongs to those at the top is as common as it is destructive.

38. Mann. Even if he were correct about the amount of change brought about by Indians, which he is not, he is still, unforgivably, conflating change with destruction. I ask myself the same question about him that I did about Stossell. His transparent illogic about passenger pigeons makes me think he's a fool, but his comment that anything goes makes it clear that he is evil.

39. Sigh. Can any of these people who support civilization ever say anything about *anything* without commenting that humans are "thoroughly superior"?

40. Hunn.

Identification

41. Personal communication, October 30, 2001.

42. Or rather your twisted projection of how a predator actually acts.

43. Duh.

44. Also, asbestos in the soil might kill the ORV users. The author apparently does not consider the possibility that this is the land defending itself.

45. Gaura.

46. I mean specifically the wearing of fur by other than those traditional indigenous peoples and other than those who kill and skin animals for their own use, in an ongoing and reciprocal relationship with the communities of fur-bearing animals.

47. I am not suggesting, by the way, that vivisection, factory farming, or factory fishing (or logging, mining, or oil extraction, for that matter) are utilitarian in the broadest sense of helping us to survive, since all are manifestly destructive and cruel. I am talking about perceived utility to the culture.

48. As well as, if I may get all cute and literary, destructively pointless.

49. BLM: three more letters that let me know we're fucked. FWS would be three more, and USFS would be four more.

50. Consider the consistent refusal by governments to halt the spread of pervasive carcinogenic chemicals by the corporations that manufacture them and the CEOs who run these corporations.

51. Bancroft, 21.

52. Of course the Department arrested precisely zero Pacific Lumber employees for their illegal activities.

53. Hawley.

54. Janet Larsen, "Dead Zones Increasing in World's Coastal Waters," Earth Policy Institute, June 16, 2004, http://www.earth-policy.org/Updates/Update41.htm (accessed June 20, 2004).

55. "Deepsea Fishing."

56. Bruno.

57. *Disinfopedia*, s.v. "BP," http://www.disinfopedia.org/wiki.phtml?title=BP (accessed June 22, 2004).

58. Nancy Kennedy.

59. Ibid.

60. Burton.

61. BP, Steph, http://www.bp.com/genericarticle.do?categoryId=2010104&contentId=2001092 (accessed June 21, 2004).

62. BP; Frank, Ellen, and Griffin; http://www.bp.com/genericarticle.do?categoryId=2010104&content Id=2001196 (accessed June 21, 2004).

63. "Study Says Five Percent."

Abusers

64. Caputi, *Gossips, Gorgons, & Crones*, 53.

65. Bancroft, 33.

66. Ibid., 34, his italics.

67. Ibid.

68. I have known many women whose husbands beat them only on their bodies, never on their faces, because that would show.

69. Bancroft, 34.

70. Ibid., 34–35, italics and bold in original.

71. Ibid., 35.

72. Ibid., 54.

73. Ibid., 151.

74. Ibid., 157.

75. Ibid., 152.

76. Faust, 81.

77. Bancroft, 197, his italics.

78. Ibid., 288, his italics.

79. Edwards, *Compassionate Revolution*, 81.

80. Bancroft, 43.

81. Laing, 186.

82. With self and other.

83. Bancroft, 63, his italics.

84. Note that I also disagree with his implication that guilt or empathy are specifically human emotions, and to imply that the abuser distances himself from her humanity suggests that were she not human, there would already be the distance that could enable his abuse. I'm not attacking Bancroft here, who I feel does extraordinary work, but merely trying to point out how easy it is to succumb to this culture's rhetoric of superiority.

85. Bancroft, 196.

86. Ibid., 311.

87. Ibid., 361.

88. Ibid., his italics.

89. And I would say most often not even then.

90. Bancroft, 360.

91. And I would say most often not even then.

A Thousand Years

92. Densmore, 172.

93. DeRooy, 12.

94. "Doing nothing" in this case can include writing letters, holding signs, and other forms of protest if those doing them know their actions are symbolic, in other words, while they may get out a message, they won't stop the destruction. It can also include writing books.

95. You didn't know that Himalayan blackberries can speak a form of Latin, did you?

96. Pampas grass is another invasive exotic that gets shaded out when trees come up.

97. Hurdle.

Dams, Part I

98. Gide.

99. *Dams and Development*, xxx.

100. "Rivers Reborn: Removing Dams and Restoring Rivers in California," Friends of the River, http://www.friendsoftheriver.org/Publications/RiversReborn/main3.html (accessed July 11, 2004).

101. Glen Martin.

102. Ibid.

103. How often have you heard a man say he doesn't know answers?

104. Evidently during the Pliocene a species of salmon called the sabertooth salmon grew up to ten feet and weighed as much as 350 pounds.

105. Crane.

106. Montgomery, 181.

107. Ibid.

108. Crane.

109. Ibid.

110. The patching was ad hoc, as engineers crammed timbers, concrete, rubbish, and anything else they could find into the hole. Interesting, isn't it, how members of this culture act willy-nilly when it comes to destroying things, but have to study everything literally to death before they will act to protect something? It's insane.

111. Crane.

112. Ibid.

113. Ibid.

114. George Draffan, Endgame Research Services: A Project of the Public Information Network, http://www.endgame.org (accessed July 10, 2004).

115. Through its American subsidiary Daishowa America.

116. Gantenbein.

117. I owe the term *Selective Law Enforcement Officers* to Remedy, "Mattole Activists Assaulted, Arrested after Serving Subpoena for Pepper Spray Trial," Treesit Blog, August 27, 2004, http://www.contrast.org/treesit/ (accessed August 27, 2004).

118. "Reviving the World's Rivers: Dam Removal," part 4, Technical Challenges, International Rivers Network, http://www.irn.org/revival/decom/brochure/rrpt5.html (accessed July 11, 2004).

119. Leap years.

120. "What's the Dam Problem?" part 3, Dunking the Dinky Dams.

121. *Dams and Development*, preface.

122. Paulson.

123. "What's the Dam Problem?" part 3, Dunking the Dinky Dams.

124. Ibid.

125. Bromley and Kelberer.

126. *Dams and Development*, xxxi. Note that even these fine people still ignore the natural world, except as it affects things like "downstream livelihoods."

127. When those at the top tell you that something is going to be profitable and that it will help you, it means they're going to rob you and assault you if you resist. You have options at that point. One is to surrender. One is to fight back. I am sure there are others. The point is that we often forget that we have options.

128. *Dams and Development*, xxxi

129. "What's the Dam Problem?" part 1, Out, Damn Dam!

130. Bromley and Kelberer.

Pretend You Are a River

131. Personal communication, October 30, 2001.

132. Lame Deer and Erdoes, 146.

133. Kathleen Moore and Jonathan Moore.

134. Ibid.

135. Ibid.

136. Montgomery, 29. The rest of the paragraph is from Montgomery; Kathleen Moore and Jonathan Moore; and "Herring and Salmon," Raincoast Research Society, http://www.raincoastresearch.org/herring-salmon.htm (accessed July 16, 2004).

137. Kathleen Moore and Jonathan Moore.

138. Ibid.

139. Montgomery, 39.

140. Kathleen Moore and Jonathan Moore.

Dams, Part II

141. *The Sun*, October 2003, 48.

142. *Dam Removal.*

143. Most of this is from Pitt.

144. "About FEMA," FEMA, http://www.fema.gov/about/ (accessed July 21, 2004).

145. Harry V. Martin.

146. Steve Wingate, "The OMEGA File—Concentration Camps: Federal Emergency Management Agency," http://www.posse-comitatus.org/govt/FEMA-Camp.html (accessed July 21, 2004).

147. Harry V. Martin.

148. Ibid.

149. Wingate.

150. Hicks.

151. Ibid.

152. Ibid.

153. Farrell.

154. Hicks.

155. "Learn about EPRI," EPRI, http://www.epri.com/about/default.asp (accessed July 22, 2004).

156. "About CMD," Center for Media and Democracy, Publishers of *PR Watch*, http://www.prwatch.org/cmd/ (accessed July 22, 2004).

157. "Flack Attack."

158. Patton Boggs, "Profile," http://www.pattonboggs.com/AboutUs/index.html (accessed July 22, 2004).

159. The article cites one other Interior Department employee who also insisted on anonymity for fear of retaliation.

160. Harden.

161. Barringer.

Dams, Part III

162. Notes from Nowhere, 148.
163. Dams are one reason, by the way, that rich people's ocean-view houses in southern California are falling off cliffs. The dams stop sediment from reaching the ocean and recharging beaches.
164. McConnell.
165. Bancroft, 311.
166. Ibid., 361.
167. Ibid., 360.

Dams, Part IV

168. Lyons.
169. I'm writing this, for example, while I sit in the waiting room of a doctor's office, and if you can believe it, I'm even having a good time here. I'm writing! Yay! (Although I must admit that after sitting in this plastic chair for two hours listening to cartoons on the TV interspersed with the loud hacking laughter of smokers, I'm ready to write somewhere else.)
170. In case that's an important question.
171. There's a small concrete structure not far from my home that was recently (and legally) refitted to allow fish passage. All that was required was for a notch to be cut into the foot-tall dam so fish could swim through the opening.

Too Much to Lose (Short Term Loss, Long Term Gain)

172. Marcos, 420.
173. Then vacuum packed in plastic. Oh well.
174. By which I mean wild salmon, but it's true for all salmon as well, since factory-farmed salmon won't survive the fall of civilization.
175. Thompson.
176. McCarthy, "Disaster."
177. Bauman, 203.
178. I am indebted to Nita Halstead for this section.

Psychopathology

179. Jensen, *Walking*.
180. Although inside the psychological and psychiatric industry the term *sociopath* has come into more prevalent use than *psychopath* to mean much the same thing, I am choosing the word *psychopath* because it still seems to hold sway in common usage.
181. *New Columbia Encyclopedia*, 4th ed., s.v. "psychopath."
182. Ramsland.
183. Ibid.
184. The bit about growth in marine fish stocks is utter nonsense, of course.
185. Revkin.
186. Atcheson.
187. McCarthy, "Greenhouse Gas."
188. Duh!
189. Blakeslee.
190. "Antisocial Personality Disorder," Mental Health Matters, http://www.mental-health-matters.com/disorders/dis_details.php?disID=8 (accessed August 6, 2004).
191. If you consume the flesh of another, you now must take responsibility for the continuation of the other's community.

192. Whose name has been, of course, stolen by the U.S. military for use in a helicopter used now to kill those who oppose the interests of those who run the United States. Have I mentioned that I hate this culture?

193. Blaisdell, 84–85.

194. I got this simile from Ward Churchill.

Pacifism, Part I

195. Burroughs.

196. Gandhi-ites throw this one at me all the time, but it's possible that Gandhi never actually said it. The quote is all over the internet, and it was in the movie *Gandhi*, but David Lean was not known for the historical accuracy of his films.

197. Lorde, 112.

198. Ibid., italics in the original.

199. I am grateful to Lierre Keith for this paragraph.

200. I am grateful to Mary Jensen for these two paragraphs.

201. I am grateful to Tiiu Ruben for this analysis.

202. What the *fuck* does Gandhi believe that children who are being beaten and raped have on *their* lips, besides immeasurable courage and compassion (and probably their father's cum)? Yet this does not stop their fathers. How dare Gandhi say this!

203. Fischer, 380.

204. Ibid., 348.

205. Gandhi, 32.

206. Bancroft, 288.

207. Ibid., his italics.

208. Once again, I've yet to see confirmation in print of this line. It's all over the internet, and it's in the movie *Gandhi*. The point is that it's thrown out *ad nauseum* and is not out of line with the other quotes we *know* are his. And I need to emphasize that even through this argument, the point is not Gandhi but rather pacifism, and this is one of pacifism's rallying cries.

209. "Too Hot for Uncle John's Bathroom Reader," Trivial Hall of Fame, http://www.trivialhalloffame .com/Gandhi.htm (accessed August 8, 2004).

210. I'm not alone, by the way, in my distaste for Gandhi. My introduction to this disgust came by way of a student from India in one of my classes at Eastern Washington University. She began my education into the real Gandhi. I've since encountered many Indians who do not deify Gandhi the way white activists so often do. In fact, many can't stand him.

Responsibility

211. *Merriam-Webster's Collegiate Dictionary*, electronic ed., vers.1.1, s.v. "responsible."

212. *Online Etymology Dictionary*, http://www.etymonline.com (accessed August 14, 2004), s.v. "responsible" and "respond."

Pacifism, Part II

213. Douglass, 204.

214. Insofar as we can make a meaningful distinction.

215. I'm embarrassed to admit I made this same assertion in my book *A Language Older Than Words*. I don't know what to say, except that I hadn't thought it through. I was wrong.

216. I am grateful to Lierre Keith for this final story.

217. Bettelheim was a terrible person, far worse than Gandhi. He was accused, most probably accurately, of physical and sexual assault on children. His attitudes on autism were despicable: he blames mothers for it. His attitudes on anti-Semitism were essentially as bad: he once shouted at an audience of Jews, "Anti-Semitism, whose fault is it? Yours! . . . Because you don't assimilate, it is your fault. If you

assimilated, there would be no anti-Semitism. Why don't you assimilate?" But the reason I include a few of the despicable Bettelheim's sins in a footnote and a few of Gandhi's in the body of the text is that Bettelheim's position is not one that people claim carries moral high ground (Bettelheim's analysis *here* concerns tactical responses to violence), yet that is, so far as I can tell, the main thing Gandhi really has going for him: his adherents claim (as often and as loudly as they can) that Gandhi's position carries the day because it carries moral weight. This makes an examination of his own morality and the morality of his positions eminently relevant.

218. Bettelheim, vi.
219. Ibid., xiv.
220. Churchill, *Pacifism,* 107, n. 19.
221. Bettelheim, xii.
222. Ibid., vii.
223. Ibid., viii.
224. Churchill, *Pacifism,* 36. As a sign of disrespect, Churchill never capitalizes the word *nazi.*
225. Bettelheim, xii–xiii.

What It Means to Be Human

226. Mason, 14.
227. My thanks to Gabrielle Benton for this paragraph.
228. I tell this story in *A Language Older Than Words.*
229. "Third National Incidence Study."
230. Figures for childhood sexual abuse vary widely. I've chosen representative rates. For much higher figures (53 percent for women, 31 percent for men), see "Child Sexual Abuse." For thorough examinations, see Diana E. H. Russell (both books in the bibliography) and Jim Hopper, "Child Abuse: Statistics, Research, and Resources," last revised February 25, 2006, http://www.jimhopper.com/abstats/ (accessed August 19, 2004), among many others.
231. *Webster's New Twentieth Century Dictionary of the English Language,* 2nd ed., s.v. "civilization."
232. *Oxford English Dictionary,* compact ed., s.v. "civilization."
233. *Effects of Strategic Bombing,* 13.

Pacifism, Part III

234. Churchill, "New Face," 270.
235. Koopman.
236. Hey, stop laughing and bear with me on this one.
237. As part of their activism. There are of course a lot of asshole activists who, like other males within this culture, have raped women.
238. Judith Herman, 90.
239. I shamelessly stole this line from Tom Wheeler, editor of the extraordinary *Alternative Press Review.*
240. Judith Herman, 91.
241. Ibid.
242. Handler.
243. Wyss.
244. And presumably their doughnuts.
245. Remedy, "Mattole Activists Assaulted, Arrested after Serving Subpoena for Pepper Spray Trial," Treesit Blog, August 27, 2004, http://www.contrast.org/treesit/ (accessed August 27, 2004).
246. Brian Martin.
247. My thanks to Lierre Keith.
248. My thanks to Curt Hubatch.
249. Diamond, 1.
250. Cockburn, "London," quoting Jeremy Scahill.
251. Not that it did any good, in this case.

252. My thanks again to Lierre Keith.

253. Once again, I am explicitly excepting those pacifist activists like Gandhi, Berrigan, King, Helen Woodson, and so on.

254. Directly, as opposed to indirectly: I do have Crohn's disease, a disease of industrial civilization.

255. Note, by the way, what concern he places first, and note also his exclamation point.

256. I hate his condescension.

257. Note his identification with the bank and those who work there, as opposed to those who are harmed by the actions of the bank, whom of course he does not mention.

258. Mitford, 272–73.

259. Jensen and Draffan, *Machine*, 220, citing Bauman, 203.

260. Jensen and Draffan, *Machine*, 220.

261. Elliott, 12. The italics were in the original.

Fewer Than Jesus Had Apostles

262. Havoc Mass, 18.

263. The one taught by that pocket dictator Mr. Bush (no relation) who threatened to flunk anyone who so much as chipped one of his precious test tubes.

264. See, for example, Jensen and Draffan, *Machine*, 26–28.

Pacifism, Part IV

265. Huntington, 51. I would of course make two changes in this quote: I would add *temporarily* as the third word, and I would substitute *cultures* for *civilizations*.

266. This sounds familiar. Hmm, where have we heard it? Ah, "In order to maintain our way of living, we must tell lies to each other and especially to ourselves."

267. Churchill, "Appreciate History."

Get There First with the Most

268. *Anderson Valley Advertiser*, July 11, 2003, 11.

269. I know we've mostly heard this as "Get there fustest with the mostest," but that's just not true. As Robert Selph Henry wrote in his definitive *"First with the Most" Forrest*, "Forrest would have been totally incapable of so obvious and self-conscious a piece of literary carpentry. What he said, he said simply and directly—'Get there first with the most men,' although doubtless his pronunciation was 'git thar fust,' that being the idiom of the time and place. Such a phrase, compacting about as much of the art of war as has ever been put into so few words, had no need of the artificial embellishment of the double superlative" (Henry, 19).

270. Note that I'm not saying it's true in all discourse. Not all discourse is antagonistic.

271. Thomas, 100, n. 43.

272. Sorry, wetlands.

273. Which would be like Burnside sending his troops up the hill against an enemy he thought was in the open but who is actually behind an invisible wall. And if the premises are hidden enough, maybe the enemy is actually invisible and so is the hill. Soon enough, it—both the situation and this stupid metaphor—gets out of hand.

274. Or rather, some of us are encouraged to vote; the poor, people of color, felons are all either discouraged or prohibited.

275. That's one reason I could not encourage that sixteen-year-old to burn down a factory. I didn't know him well enough to know if he was thinking his own thoughts. To be an adult one must not only know freedom but responsibility. As I say in *Walking on Water*, freedom without responsibility is immaturity, and responsibility without freedom is slavery. We need people mature enough to think for themselves. This young man may have been, or he may not have been. I didn't know, which is one reason I couldn't advise him.

276. If you do not love your landbase, why have you read this far into this book?

277. Which is my big beef with the pacifists: they seem to have a one-size-fits-all prescription. Well, the truth is that one size only fits one.

278. Or as one writer put it about the differences between the indigenous and the civilized:

"[R]eligion rather than business being the principal business; living to live rather than to get; belonging rather than belongings as a reigning value; apparent rarity of enforced civil or military service and the apparently frivolous nature of much religious service tending to disguise the possibility that it may have been enforced; group ownership of land and wealth, and consequent tendencies toward individual cooperation rather than competition, and apparent rarity of the police and lawsuits necessary to regulate individual possession; dualism and institutionalized factionalism with consequent tendencies toward reciprocating government, toward a world in balance between two opposing forces, whether the world of thought and the spirit or the world of practical politics, rather than the Old World compulsion toward one party rule, insofar as possible, whether in religion or politics" (Brandon, 60).

279. Obviously, I didn't have them put down their equipment or make their marks the night before. I just wanted them to see the place and sit with it, so they could spend the night visualizing their jumps the next day.

280. Did I mention that I mean anyone?

281. Why do you think the rules of war, written by governments—in other words, those who raise large armies—exclude many non-army combatants from their protection?

282. And I mean, at his *disposal.*

283. Blaisdell, 80.

284. Samuel Drake, 662.

285. Gordon, 343–44.

286. Blaisdell, 67.

287. Jefferson, 345.

288. Jensen and Draffan, *Machine*, 74.

289. Tebbel and Jennison, 212–13.

290. The citation is "In His Own Words." I've taken a few *minor* liberties with what he said. Here's what George W. Bush actually said:

"So we have fought the terrorists [*sic*] across the Earth—not for pride, not for power, but because the lives of our citizens are at stake [*sic*]. Our strategy is clear. We have tripled funding for homeland security [*sic*] and trained half a million first responders, because we are determined to protect our homeland [*sic*]. We are transforming our military and reforming and strengthening our intelligence services. We are staying on the offensive—striking terrorists [*sic*] abroad—so we do not have [*sic*] to face them here at home. And we are working to advance liberty [*sic*] in the broader Middle East, because freedom [*sic*] will bring a future [*sic*] of hope [*sic*], and the peace [*sic*] we all [*sic*] want [*sic*]."

Symbolic and Non-symbolic Actions

291. Clausewitz, 226–29.

292. I just read in the capitalist press an account of a U.S. soldier using the butt of his rifle to smash the head of an unarmed, unresisting Iraqi civilian (the civilian was later tortured to death by soldiers and CIA operatives in a shower; this location was presumably chosen to make it easy to clean up the blood), but the capitalist journalist did not use any such indelicate terms as "smash." Instead, the accepted term among capitalist journalists for smashing in someone's head with the butt of a rifle is now "butt stroking." And it's not only capitalist journalists who call it this. They are (of course) parroting the U.S. military, as in this statement from *National Defense Magazine*: "If you can shoot your enemy, then shoot him. If you can't do that, stick him with your bayonet, butt stroke him with your rifle butt, ram him with your rifle barrel" (Harold Kennedy).

293. Or butt stroke it several times, and discover later that it has somehow expired during interrogation (a sandbag having somehow ended over onto its head), which is what I would say if I were a capitalist journalist.
294. Steele.
295. In fact, he is already calling for more conquest abroad and repression at home. No big surprise there.
296. My thanks to Lierre Kieth.
297. The national forests have 380,000 miles of roads in them, more than the interstate highway system, enough roads to circle the globe fifteen times. Note that Clinton's moratorium did not halt logging in roadless areas, which continued at breakneck speeds, only now the murdered trees are removed by helicopter.
298. No, I'm not making this up, and yes, you're right, he has no sense of shame whatsoever.
299. No, I'm not making this up either, and yes, you're right, these organizations have no sense whatsoever.

Like a Bunch of Machines

300. Churchill, "New Face," 270.
301. *Merriam-Webster's Collegiate Dictionary*, electronic ed., vers. 1.1, s.v. "passion."
302. For a quick exploration of the probable causes of the fire, see, for example, "John Trudell: Last National Chairman of AIM," Redhawks Lodge, http://siouxme.com/lodge/trudell.html (accessed September 12, 2004).
303. Trudell.
304. That's a joke, Mr. Gonzalez: put away those electrodes.
305. My thanks to Roianne Ahn for this paragraph.
306. Remember the words of Harry Merlo, former CEO of Louisiana-Pacific: "We don't log to a 10-inch top or an 8-inch top or even a 6-inch top. We log to infinity. Because it's out there and we need it all, now."
307. I am explicitly excluding the little girl from this, who may have been reachable.
308. For a reasonably thorough exploration of this, see my book *The Culture of Make Believe*, 174–85.

Bringing Down Civilization, Part II

309. Churchill, "New Face," 270.
310. "Index of Comments for *A Boy and His Dog*," Badmovies, http://www.badmovies.org/comments/ ?film=185 (accessed September 17, 2004).
311. I thank Tiiu Ruben for this paragraph.
312. Neeteson. Normally when you put things in direct quotes they are direct quotes. But Neeteson's paragraph makes obvious that English is his second language, so I cleaned it up. Here is the original: "Modern Western culture has to contend with a shortage of satisfying existential ideologies. For centuries a reduction took place from spiritual thinking towards materialistic thinking, ending in the technological consumption society. This society depends on mass production and mass consumption, on ideologies which are superficial, therefore easy to manipulate and on advanced technology and military power. One of the results of this process is that the average individual cannot obtain enough meaningful satisfaction from the common social life."
313. Hoskins, 10.
314. Ibid., 11.
315. Ibid., 10.
316. Ibid.
317. Liddell Hart, 328.
318. "Sabotage Blamed."

Breaking Faith

319. Hume.

320. Note that I am projecting animation onto the computer: it "gave" me error messages. We pretend the land has nothing to say and nothing to give, but our computers give us messages.

321. Just like above, I am saying that the download site "told me" something.

322. The computer is still talking to me. Too bad I don't listen quite so well to the trees.

323. I first wrote that I reformatted *my* hard drive. Even after I become aware of the language, I *still* identify with the computer, with the machine.

324. Or at least it will save the machine.

A Matter of Time

325. Taber, 22.

326. Yergin, 487.

327. Matt Savinar, "Life After the Oil Crash: Deal with Reality, or Reality Will Deal with You," http://www.lifeaftertheoilcrash.net (accessed September 21, 2004).

328. Brown, 273, 449.

329. Gibbon, 3:479.

330. Tillich, 3:369.

Bringing Down Civilization, Part III

331. Churchill, "New Face," 270.

Bringing Down Civilization, Part IV

332. Ibid.

333. This one is, I think, the real answer.

334. I somehow lost the name of the person who sent this to me, although I have the full text of the note itself. If you are the author of this and would like me to cite you, send me a note.

335. Marufu et al.

336. And said, and said.

337. This statement presumes, once again, that you're going to have a child at all. Given overconsumption, I think it's far more revolutionary than that to not reproduce at all.

338. And even then you didn't get either of the points I was trying to make with the joke. The other point besides the one I mentioned earlier in this book is that she had a man's name tattooed on her genitals, as though he owns them. Now, that's sexist.

   And here's the most recent complaint I got: because it's just not possible for me to make up new jokes every single night, I obviously recycle them. Someone wrote a very angry blog denouncing me as a phony revolutionary because I reuse jokes. I'm not making this up. And I'm guessing that if she were to go see the Rolling Stones she would complain if they did not make up songs just for her.

339. Both human and nonhuman languages.

340. I don't want a "flock" of anything, except I would love to see wild birds fly overhead in flocks so large they darken the sky.

The Crash

341. I would of course say the whole civilized enterprise.

342. Udall, 271.

343. Killian, xv.

344. "The Free Trade Area of the Americas and the Threat to Water," International Forum on Globalization, http://www.ifg.org/reports/ftaawater.html (accessed September 27, 2004).

345. Jensen, *Culture*, parts of the chapter called "Holocausts."

Civilization: Ongoing Holocausts

346. Bauman, 206.

347. Shirer, 1262–63.

348. Bauman, 203.

349. Shirer, 1263.

350. Reitlinger, 160.

351. George Draffan, Endgame Research Services: A Project of the Public Information Network, http://www.endgame.org (accessed July 10, 2004).

352. Estes, 177–78.

353. Jensen, "Free Press for Sale."

354. *Fox News Sunday*, June 17, 2001.

355. Weizenbaum, 256.

356. Rogers.

Endgame

357. Mason, 147.

358. Bauman, 207.

359. "Rebuilding America's Defenses."

360. Hey, no fair cheating! You were supposed to put the book down. Now go outside. I'll see you in a couple of days.

361. I'm glad you're back. I hope you had a nice couple of days. Now let's compare lists.

362. Thompson.

363. My thanks to writer and activist Aric McBay for this paragraph.

364. Including U.S.-occupied United States.

365. Aric's excellent website is http://www.inthewake.org

366. I guess that describes the current reality as well.

367. My thanks to writer and activist Aric McBay for these paragraphs.

368. My thanks to writer and activist Aric McBay for this as well.

369. Only a madman would cut down a redwood to sell it, as well.

The Return of the Salmon

370. Forbes, 154.

371. Montgomery, 39.

372. Mason, 143.

373. One look at the floor around my workspace would convince you of this.

374. My thanks to Ward Churchill for this paragraph. Also to Richard S. Grimes.

Bibliography

Abel, Annie Heloise. *Chardon's Journal at Fort Clark, 1834–1839*. Lincoln: University of Nebraska Press, 1997.

"About FEMA." FEMA. http://www.fema.gov/about/ (accessed July 21, 2004).

"Accumulated Change Courts Ecosystem Catastrophe." *Science Daily*, October 12, 2001. http://www.sciencedaily.com/releases/2001/10/011011065827.htm (accessed November 29, 2001).

ACME Collective. "N30 Black Bloc Communique." Infoshop, December 4, 1999. http://www.infoshop.org/octo/wto_blackbloc.html (accessed March 16, 2002).

Alcatraz: The Whole Shocking Story. Directed by Paul Krasny. 1980.

American Cynic 2, no. 32 (August 11, 1997). http://www.americancynic.com/08111997.html (accessed June 7, 2003).

"Anarchists and Corporate Media at the Battle of Seattle." *Global Action: May Our Resistance Be as Transnational as Capital*, December 4, 1999. http://flag.blackened.net/global/1299anarchistsmedia.htm (accessed March 16, 2002).

Anderson Valley Advertiser. http://www.theava.com.

Anderson, Zack. "Dark Winter." *Anderson Valley Advertiser*, November 7, 2001.

"Antisocial Personality Disorder." Mental Health Matters. http://www.mental-health-matters.com/disorders/dis_details.php?disID=8 (accessed August 6, 2004).

Atcheson, John. "Ticking Time Bomb." *Baltimore Sun*, December 15, 2004. http://www.commondreams.org/views04/1215-24.htm (accessed February 9, 2005).

Axtell, James. *The Invasion Within: The Contest of Cultures in Colonial North America*. Oxford: Oxford University Press, 1985.

Bacher, Dan. "Bush Administration Water Cuts Result in Massive Fish Kill on Klamath." *Anderson Valley Advertiser*, October 2, 2002, 1.

"Index of Comments for *A Boy and His Dog*." Badmovies. http://www.badmovies.org/comments/?film=185 (accessed September 17, 2004).

Baker, David R. "Living a Fantasy (League)." *San Francisco Chronicle*, September 21, 2004, F1.

Bales, Kevin. *Disposable People: New Slavery in the Global Economy*. Berkeley: University of California Press, 1999.

Bancroft, Lundy. *Why Does He Do That? Inside the Minds of Angry and Controlling Men*. New York: Berkley Books, 2002.

Baran, Paul. *The Political Economy of Growth*. New York: Monthly Review, 1957.

Barringer, Felicity. "U.S. Rules Out Dam Removal to Aid Salmon." *New York Times,* December 1, 2004. http://www.nytimes.com/2004/12/01/politics/01fish.html?ex=1102921137&ei =1&en=1ba893433747ec91 (accessed December 1, 2004).

Barsamian, David. "Expanding the Floor of the Cage, Part II: An Interview with Noam Chomsky." *Z Magazine,* April 1997.

Bauman, Zygmunt. *Modernity and the Holocaust.* Ithaca, NY: Cornell University Press, 1989.

"B.C. Court OKs Logging in Endangered Owl Habitat." *CBC News,* July 9, 2003. http://www.cbc.ca/storyview/CBC/2003/07/08/owl_spotted030708 (accessed July 10, 2003).

"B.C.'s Spotted Owl Faces Extinction Scientists Warn." *CBC News,* October 7, 2002. http://www.cbc.ca/storyview/CBC/2002/10/07/spotted_owls021007 (accessed July 10, 2003).

Beeman, William O. "Colin Powell Should Make an Honorable Exit." *La Prensa San Diego,* March 14, 2003. http://www.laprensa-sandiego.org/archieve/march14-03/comment2.htm (accessed June 20, 2003).

Bettelheim, Bruno. Introduction to *Auschwitz: A Doctor's Eyewitness Account,* by Miklos Nyiszli. New York: Frederick Fell, 1960.

"Biased Process Promotes Forced Exposure to Nuclear Waste; Radioactive Materials Could Be Released into Consumer Goods, Building Supplies." *Public Citizen,* March 26, 2001. http://www.citizen.org/pressroom/release.cfm?ID=600 (accessed January 21, 2002).

Blaisdell, Bob, ed. *Great Speeches by Native Americans.* Mineola, NY: Dover, 2000.

Blakeslee, Sandra. "Minds of Their Own: Birds Gain Respect." *New York Times,* February 1, 2005.

Blyth, Reginald Horace. *Zen and Zen Classics.* Tokyo: The Hokuseido Press, 1960.

"BLU-82B." FAS Military Analysis Network. http://www.fas.org/man/dod-101/sys/dumb/ blu-82.htm (accessed November 19, 2001).

Bonhoeffer, Dietrich. *Dietrich Bonhoeffer: Letters and Papers from Prison: The Enlarged Edition.* Edited by Bethge Eberhard. New York: The MacMillan Company, 1953.

BP, Frank, Ellen, and Griffin. http://www.bp.com/genericarticle.do?categoryId=2010104& contentId=2001196 (accessed June 21, 2004).

BP, Steph. http://www.bp.com/genericarticle.do?categoryId=2010104&contentId=2001092 (accessed June 21, 2004).

Brandon, William. *New Worlds for Old: Reports from the New World and Their Effect on the Development of Social Thought in Europe, 1500–1800.* Athens: Ohio University Press, 1986.

Brice, Wallace A. *History of Fort Wayne: From the Earliest Known Accounts of This Point, to the Present Period.* Fort Wayne, IN: D. W. Jones and Son, 1868.

Bright, Martin, and Sarah Ryle. "United Kingdom Stops Funding Batterers Program." *Guardian,* May 27, 2000.

Bromley, Chris, and Michael Kelberer. *The Alumni Channel: A Newsletter for Alumni and Friends of St. Anthony Falls Laboratory.* February 2004. http://www.safl.umn.edu/newsletter/alumni_channel_2004-12.html (accessed July 13, 2004).

Brown, Dee. *Bury My Heart at Wounded Knee: An Indian History of the American West.* New York: Holt, Rinehart, and Winston, 1970.

Bruno, Kenny. "BP: Beyond Petroleum or Beyond Preposterous?" *CorpWatch: Holding Corporations Accountable,* December 14, 2000. http://www.corpwatch. org/article.php?id=219 (accessed June 22, 2004).

Burroughs, William S., and David Odier. *The Job: Interviews with William S. Burroughs.* New York: Penguin, 1989.

Burton, Bob. "Packaging the Beast: A Public Relations Lesson in Type Casting." *PRWatch* 6, no. 1 (1999): 12. http://www.prwatch.org/prwissues/1999Q1/beast.html (accessed June 21, 2004).

Cancers and Deformities. One part of the extraordinary "The Fire This Time" site, http://www.wakefieldcam.freeserve.co.uk/cancersanddeformities.htm (accessed January 26, 2002).

Caputi, Jane. *The Age of Sex Crime.* London: The Woman's Press, 1987.

———. *Gossips, Gorgons, & Crones: The Fates of the Earth.* Santa Fe, NM: Bear & Company, 1993.

Catton Jr., William R. *Overshoot: The Ecological Basis of Revolutionary Change.* Chicago: University of Illinois Press, 1982.

Center for Defense Information. http://www.cdi.org/ (accessed January 16, 2002). It's very hard to find old budgets on their Web site, but the numbers will be just as startling, if not more so, in more recent budgets.

"Child Sexual Abuse: Information from the National Clearinghouse on Family Violence." The National Clearinghouse on Family Violence (Ottawa, Canada), January 1990, revised February 1997. Available in pdf format at http://www.phac-aspc.gc.ca/ncfv-cnivf/family violence/nfntsabus_e.html (accessed March 13, 2006).

Chomsky, Noam. *Year 501: The Conquest Continues.* Boston: South End Press, 1993.

Churchill, Ward. "Appreciate History in Order to Dismantle the Present Empire." *Alternative Press Review: Your Guide Beyond the Mainstream,* August, 17, 2004. http://www.altpr.org/modules.php?op=modload&name=News&file=article&sid=272&mode=thread&order=0&thold=0 (accessed August 23, 2004).

———. "The New Face of Liberation: Indigenous Rebellion, State Repression, and the Reality of the Fourth World." In *Acts of Rebellion: The Ward Churchill Reader.* New York: Routledge, 2003.

———. *Pacifism as Pathology: Reflections on the Role of Armed Struggle in North America.* Winnipeg, Canada: Arbiter Ring, 1998.

———. *Struggle for the Land: Indigenous Resistance to Genocide, Ecocide, and Expropriation in Contemporary North America.* Monroe, ME: Common Courage, 1993.

Clausewitz, Carl von. *On War*. Translated by Michael Howard and Peter Paret. New Brunswick, NJ: Princeton University Press, 1976.

"CNN Says Focus on Civilian Casualties Would Be 'Perverse.'" *Fairness and Accuracy in Reporting*, November 1, 2001. http://www.fair.org/index.php?page=1670 (accessed March 11, 2006).

Cockburn, Alexander. *Anderson Valley Advertiser*, April 2, 2003, 9.

———. "The Left and the 'Just War.'" *Anderson Valley Advertiser*, October 31, 2001, 1.

———. "London and Miami: Cops in Two Cities." *Anderson Valley Advertiser*, November 26, 2003, 5.

Cokinos, Christopher. *Hope Is the Thing with Feathers: A Personal Chronicle of Vanished Birds*. New York: Jeremy P. Tarcher, 2000.

Combs, Robert. *Vision of the Voyage: Hart Crane and the Psychology of Romanticism*. Memphis: Memphis State University Press, 1978.

"Coming Your Way: Radioactive Garbage." *Rachel's Hazardous Waste News*, no. 183, May 30, 1990. http://www.ejnet.org/rachel/rhwn183.htm (accessed January 21, 2002).

Conot, Robert E. *Justice at Nuremberg*. New York: Carroll & Graf, 1983.

Cook, Kenneth. "Give Us a Fake: The Case Against John Stossel." *TomPaine.com*, August 15, 2000. http://www.tompaine.com/feature.cfm/ID/3481 (accessed March 13, 2004).

Cottin, Heather. "Scripting the Big Lie: Pro-War Propaganda Proliferates." *Workers World Newspaper*, November 29, 2001. http://groups.yahoo.com/group/MainLineNews/message/20262.

Crane, Jeff. "The Elwha Dam: Economic Gain Wins Out Over Saving Salmon Runs." *Columbia Magazine* 17, no. 3 (Fall 2003). http://www.washingtonhistory.org/wshs/columbia/articles/0303-a2.htm (accessed July 8, 2004). The Washington State Historical Society publishes this journal.

Creelman, James. *On the Great Highway: The Wanderings and Adventures of a Special Correspondent*. Boston: Lothrop Publishing Co., 1901.

Crévecoeur, Hector St. John de. *Letters from an American Farmer and Sketches of Eighteenth-Century America*. Edited with an introduction by Albert E. Stone. New York: Penguin, 1981.

Dam Removal: Science and Decision Making. Washington, DC: The H. John Heinz III Center for Science, Economics, and the Environment, 2002.

Dams and Development: A New Framework for Decision-Making. The Report of the World Commission on Dams. London: Earthscan, November 2000.

Davidson, Keay. "Optimistic Researcher Draws Pessimistic Reviews: Critics Attack View That Life Is Improving." *San Francisco Chronicle*, March 4, 2002, A4.

"Deepsea Fishing Nets Devastating the World's Sea Beds, Greenpeace Says." *CBC News*. http://www.cbc.ca/cp/world/040618/w061818.html (accessed June 20, 2004).

DeLong, J. Bradford. "The Corporations as a Command Economy." http://www.j-bradford -delong.net/Econ_Articles/Command_Corporations.html (accessed March 17, 2004).

Densmore, Frances. *Teton Sioux Music.* Bureau of American Ethnology, bulletin 61. Washington, DC: Smithsonian Institution, 1918.

DeRooy, Sylvia. "Before the Wilderness." *Wild Humboldt* 1 (Spring/Summer 2002): 12.

Devereux, George. *A Study of Abortion in Primitive Society.* New York, 1976.

Diamond, Stanley. *In Search of the Primitive: A Critique of Civilization.* Somerset, NJ: Transaction Publishers, 1993.

Dimitre, Tom. "Salamander Extinction?" *Econews: Newsletter of the Northcoast Environmental Center,* March 2002, 10.

Disinfopedia, s.v. "BP." http://www.disinfopedia.org/wiki.phtml?title=BP (accessed June 22, 2004).

Douglass, Frederick. *The Frederick Douglass Papers.* Edited by John Blassingame. Series 1 (Speeches, Debates, and Interviews), vol. 3 (1855–63). New Haven, CT: Yale University Press, 1985.

Dowling, Nick. "Can the Allies Strategic Bombing Campaigns of the Second World War Be Judged a Success or Failure?" *Historic Battles: History Revisited Online.* http://www.historic-battles.com/Articles/can_the_allies_strategic_bombing.htm (accessed March 5, 2004).

Draffan, George. Endgame Research Services: A Project of the Public Information Network. http://www.endgame.org (accessed July 10, 2004).

Drake, Francis S. *The Indian Tribes of the United States: Their History, Antiquities, Customs, Religion, Arts, Language, Traditions, Oral Legends, and Myths.* Vol. 2. Philadelphia: J. B. Lippincott and Co., 1884.

Drake, Samuel G. *Biography and History of the Indians of North America, from Its First Discovery.* 11th ed. Boston: Benjamin B. Mussey & Co., 1841.

Drinnon, Richard. *Facing West: The Metaphysics of Indian-Hating & Empire-Building.* Norman: University of Oklahoma Press, 1997.

Dvorak, Petula. "Cell Phones' Flaws Imperil 911 Response." *Washington Post,* March 31, 2003, B1. http://www.washingtonpost.com/ac2/wp-dyn?pagename=article&node=&contentId=A54802-2003Mar30&notFound=true (accessed June 14, 2003).

Eckert, Allan W. *A Sorrow in Our Heart: The Life of Tecumseh.* New York: Bantam Books, 1992.

Edwards, David. *Burning All Illusions.* Boston: South End Press, 1996.

———. *The Compassionate Revolution: Radical Politics and Buddhism.* Devon, U.K.: Green Books, 1998.

The Effects of Strategic Bombing on the German War Economy. The United States Strategic Bombing Survey, Overall Economic Effects Division, October 31, 1945.

Elliott, Rachel J. "Acts of Faith: Philip Berrigan on the Necessity of Nonviolent Resistance." *The Sun,* no. 331, July 2003, 4–13.

Engels, Frederick. *Herr Eugen Dühring's Revolution in Science.* Moscow: Cooperative Publishing Society of Foreign Workers in the U.S.S.R., 1934.

"Learn about EPRI." EPRI. http://www.epri.com/about/default.asp (accessed July 22, 2004).

Estes, Ralph. *Tyranny of the Bottom Line: Why Corporations Make Good People Do Bad Things.* San Francisco: Berrett-Koehler, 1996.

The Estrogen Effect: Assault on the Male. Written and produced by Deborah Cadbury for the British Broadcasting Corporation, 1993. Televised by the Discovery Channel, 1994.

Extreme Deformities. One part of the extraordinary "The Fire This Time" site, http://www.wakefieldcam.freeserve.co.uk/extremedeformities.htm (accessed January 26, 2002).

"Facing Up to Fluoride: It's in Our Water and in Our Toothpaste. Should We Worry?" *The New Forest Net.* http://www.thenewforestnet.co.uk/alternative/newforest-alt/jan2fluoride.htm (accessed January 21, 2002).

"Fair Trade: Economic Justice in the Marketplace." Global Exchange. http://www.globalexchange.org/stores/fairtrade.html (accessed March 16, 2002).

Farrell, Maureen. "A Brief (but Creepy) History of America's Creeping Fascism." *Buzzflash,* December 5, 2002. http://www.buzzflash.com/contributors/2002/12/05_Fascism.html (accessed July 21, 2004).

"Fast Facts about Wildlife Conservation Funding Needs." http://www.nwf.org/naturefunding/wildlifeconservationneeds.html (accessed January 16, 2002).

Faust, Drew Gilpin. *The Ideology of Slavery.* Baton Rouge: Louisiana State University Press, 1981.

Fischer, Louis. *The Life of Mahatma Gandhi.* New York: Harper, 1983.

Fisk, Robert. "Iraq Through the American Looking Glass: Insurgents Are Civilians. Tanks That Crush Civilians Are Traffic Accidents. And Civilians Should Endure Heavy Doses of Fear and Violence." *Independent,* December 26, 2003. http://fairuse.1accesshost.com/news1/fisk4.html (accessed October 15, 2004).

"Flack Attack." *PR Watch* 6, no. 1 (1999): 1. http://www.prwatch.org/prwissues/1999Q1/ (accessed July 22, 2004).

Flounders, Sara. Introduction to *NATO in the Balkans: Voices of Opposition,* by Ramsey Clark, Sean Gervasi, Sara Flounders, Nadja Tesich, Thomas Deichmann, et al. New York: International Action Center, 1998.

"Fluoride Conspiracy." The Northstar Foundation. http://www.geocities.com/northstarzone/FLUORIDE.html (accessed January 21, 2002).

Forbes, Jack D. *Columbus and Other Cannibals: The Wétiko Disease of Exploitation, Imperialism and Terrorism.* Brooklyn: Autonomedia, 1992.

"Fox: Civilian Casualties Not News." *Fairness and Accuracy in Reporting,* November 8, 2001. http://www.fair.org/index.php?page=1668 (accessed March 11, 2006).

Fox, Maggie. "Largest Arctic Ice Shelf Breaks Up." *ABC News* (Australia), September 23, 2003. http://www.abc.net.au/science/news/stories/s952044.htm (accessed October 28, 2003).

Fox News Sunday, June 17, 2001.

Franklin, Benjamin. *The Papers of Benjamin Franklin.* Vol. 4, *July 1, 1750–June 30, 1753.* Edited by Leonard W. Labaree, Whitfield J. Bell, Helen C. Boatfield, and Helene H. Fineman. New Haven, CT: Yale University Press, 1961.

"The Free Trade Area of the Americas and the Threat to Water." International Forum on Globalization. http://www.ifg.org/reports/ftaawater.html (accessed September 27, 2004).

"Frequently Asked Questions about Anarchists at the 'Battle for Seattle' and N30." *Infoshop.org.* http://www.infoshop.org/octo/a_faq.html (accessed March 16, 2002).

Fromm, Erich. *The Sane Society.* New York: Fawcett, 1967.

Gandhi, Mohandas K. *Gandhi on Non-Violence.* Edited by Thomas Merton. New York: New Directions, 1964.

Gantenbein, Douglas. "Swimming Upstream." *National Parks Conservation Association Magazine,* Summer 2004. http://www.npca.org/magazine/2004/summer/salmon3.asp (accessed July 10, 2004).

Garamone, Jim. "*Joint Vision 2020* Emphasizes Full-Spectrum Dominance." *American Forces Information Service News Articles,* June 2, 2000. http://www.defenselink.mil/news/Jun2000/n06022000_20006025.html (accessed March 8, 2002).

Gaura, Maria Alicia. "Curbing Off-Road Recreation: Asbestos, Rare Plants Threaten Free-wheeling Bikers in the Clear Creek Management Area." *San Francisco Chronicle,* June 13, 2004, B1.

Genesis 1:28. The Bible, silly. What did you think?

"Get the Facts and Clear the Air." Clear the Air, National Campaign Against Dirty Power. http://cta.policy.net/dirtypower/ (accessed September 3, 2004).

"A Ghastly View of Fish Squeezed through the Net by the Tons of Fish Trapped within the Main Body of the Net." NOAA Photo Library. http://www.photolib.noaa.gov/fish/fish0167.htm (accessed July 10, 2003).

Gibbon, Edward. *The Decline and Fall of the Roman Empire: Complete and Unabridged in Three Volumes.* Vol. 3, *The History of the Empire from A.D. 1135 to the Fall of Constantine in 1453.* New York: The Modern Library, n.d.

Gide, André. *André Gide: Journals.* Vol. 4, *1939–1949.* Translated by Justin O'Brien. Champaign: University of Illinois Press, 2000.

Glaspell, Kate Eldridge. "Incidents in the Life of a Pioneer." *North Dakota Historical Quarterly,* 1941, 187–88.

Global Exchange Reality Tours. http://www.globalexchange.org/tours/, and follow links from there for the other information (accessed March 16, 2002).

Goldman, Emma. *Living My Life.* New York: New American Library, 1977.

Goldsmith, Zac. "Chemical-Induced Puberty." *Ecologist,* January 2004, 4.

Goleman, Daniel. *Healing Emotions.* Boston: Shambhala, 1997.

Gordon, H. L. *The Feast of the Virgins and Other Poems.* Chicago: Laird and Lee, 1891.

"Gradual Change Can Push Ecosystems into Collapse." *Environmental News Network*, October 12, 2001. http://www.enn.com/news/enn-stories/2001/10/10122001/s_45241.asp (accessed November 29, 2001).

Grassroots ESA. http://nwi.org/GrassrootsESA.html (accessed January 16, 2002).

Griffin, Susan. *A Chorus of Stones: The Private Life of War.* New York: Doubleday, 1992.

Grimes, Richard S. "Cheyenne Dog Soldiers." Manataka American Indian Council. http://www.manataka.org/page164.html (accessed February 23, 2005).

Gruen, Arno. *The Insanity of Normality: Realism as Sickness: Toward Understanding Human Destructiveness.* Translated by Hildegarde and Hunter Hannum. New York: Grove Weidenfeld, 1992.

Guevara, Ernesto Che. *Che Guevara Reader: Writings on Politics & Revolution.* 2nd ed. Edited by David Deutschmann. Melbourne: Ocean Press, 2003.

Handler, Marisa. "Indigenous Tribe Takes on Big Oil: Ecuadoran Village Refuses Money, Blocks Attempts at Drilling on Ancestral Land." *San Francisco Chronicle*, August 13, 2004. http://www.sfgate.com/cgi-bin/article.cgi?file=/chronicle/archive/2004/08/13/MNGHB86B4V1.DTL (accessed August 19, 2004).

Harden, Blaine. "Bush Would Give Dam Owners Special Access: Proposed Interior Dept. Rule Could Mean Millions for Industry." *San Francisco Chronicle*, October 28, 2004, A1, A4. http://www.sfgate.com/cgi-bin/article.cgi?file=/chronicle/archive/2004/10/28/MNGIE9HQ6U1.DTL (accessed October 31, 2004).

Hart, Lawrence. "Cheyenne Peace Traditions." *Mennonite Life*, June 1981, 4–7.

Hastings, Max. *Bomber Command.* New York: Touchstone, 1979.

Havoc Mass. "Electric Funeral: An In-Depth Examination of the Megamachine's Circuitry." *Green Anarchy*, no. 15, Winter 2004.

Hawley, Chris. "World's Land Turning to Desert at Alarming Speed, United Nations Warns." *SFGate.com*, June 15, 2004. http://sfgate.com/cgi-bin/article.cgi?file=/news/archive/2004/06/15/international1355EDT0606.DTL (accessed June 20, 2004).

Heinen, Tom. "Prophecy Believers Brace for Armageddon: Many Think Apocalyptic Battle between Jesus and the Anti-Christ Could Loom in Not-Too-Distant Future." *Milwaukee Journal-Sentinal Online*, December 31, 1999 (appeared in print January 1, 2000). http://www.jsonline.com/news/metro/dec99/apoc01123199a.asp (accessed May 18, 2003).

Henry, Robert Selph. *"First with the Most" Forrest.* Jackson, TN: McCowat-Mercer Press, 1969.

Herman, Edward S. "Nuggets from a Nuthouse." *Z Magazine*, November 2001, 24.

Herman, Judith Lewis. *Trauma and Recovery: The Aftermath of Violence—from Domestic Abuse to Political Terror.* New York: Basic Books, 1992.

"Herring and Salmon." Raincoast Research Society. http://www.raincoastresearch.org/herring-salmon.htm (accessed July 16, 2004).

Hicks, Sander. "Fearing FEMA." *Guerilla News Network.* http://www.guerrillanews.com/war_on_terrorism/doc1611.html (accessed July 21, 2004).

Hoffmann, Peter. *The History of the German Resistance, 1933–1945.* Translated by Richard Barry. Cambridge, MA: The MIT Press, 1977.

Hooker, Richard. "Samsara." *World Civilizations.* http://www.wsu.edu:8080/~dee/GLOSSARY/SAMSARA.HTM (accessed July 14, 2003).

Hopper, Jim. "Child Abuse: Statistics, Research, and Resources." Last revised February 25, 2006. http://www.jimhopper.com/abstats/ (accessed August 19, 2004).

Hoskins, Ray. *Rational Madness: The Paradox of Addiction.* Blue Ridge Summit: Tab Books, 1989.

Human Resource Exploitation Training Manual—1983. CIA, 1983. http://www.gwu.edu/~nsarchiv/NSAEBB/NSAEBB27/02-02.htm (accessed March 11, 2006).

Hume, David. "On the First Principles of Government." You can find this in any library or all over the internet.

Hunn, Eugene S. "In Defense of the Ecological Indian." Paper presented at the Ninth International Conference on Hunting and Gathering Societies, Edinburgh, Scotland, September 9, 2002. http//:www.abdn.ac.uk.chaggs9/1hunn.htm (accessed May 30, 2004).

Hunter, John D. *Memoirs of a Captivity among the Indians of North America, from Childhood to the Age of Nineteen.* Edited by Richard Drinnon. New York: Schocken Books, 1973.

Huntington, Samuel. *The Clash of Civilizations and the Remaking of World Order.* New York: Simon and Schuster, 1997.

Hurdle, Jon. "Lights-Out Policies in Cities Save Migrating Birds." *Yahoo! News,* June 10, 2004. http://story.news.yahoo.com/news?tmpl=story&cid=572&e=8&u=/nm/life_birds_dc (accessed July 6, 2004).

"In His Own Words: What Bush Told the Convention." *San Francisco Chronicle,* September 9, 2004, A14.

"Information on Depleted Uranium: What Is Depleted Uranium?" Sheffield-Iraq Campaign, 6 Bedford Road, Sheffield S35 0FB, 0114-286-2336. http://www.synergynet.co.uk/sheffield-iraq/articles/du.htm (accessed January 23, 2002).

Jefferson, Thomas. *The Writings of Thomas Jefferson.* Edited by Andrew A. Lipscomb and Albert Ellery Bergh. Vol. 11. Washington, DC: Thomas Jefferson Memorial Association, 1903.

Jensen, Derrick. *The Culture of Make Believe.* White River Junction, VT: Chelsea Green, 2002.

———. "Free Press for Sale: How Corporations Have Bought the First Amendment: An Interview with Robert McChesney." *Sun,* September 2000.

———. *A Language Older Than Words.* White River Junction, VT: Chelsea Green, 2000.

———. *Listening to the Land.* White River Junction, VT: Chelsea Green, 2004.

———. *Walking on Water: Reading, Writing, and Revolution.* White River Junction, VT: Chelsea Green, 2004.

———. "Where the Buffalo Go: How Science Ignores the Living World: An Interview with Vine Deloria." *Sun,* July 2000.

Jensen, Derrick, and George Draffan. *Strangely Like War: The Global Assault on Forests.* White River Junction, VT: Chelsea Green, 2003.

————. *Welcome to the Machine: Science, Surveillance, and the Culture of Control.* White River Junction, VT: Chelsea Green, 2004.

Johansen, Bruce E. *Forgotten Founders: Benjamin Franklin, the Iroquois and the Rationale for the American Revolution.* Ipswich, MA: Gambit Incorporated, 1982. Also available in pdf format at http//:www.ratical.org/many_worlds/6Nations/FF.pdf (accessed June 7, 2003).

"John Trudell: Last National Chairman of AIM." Redhawks Lodge. http://siouxme.com/lodge/trudell.html (accessed September 12, 2004).

Joint Vision 2020. Approval Authority: General Henry H. Shelton, Chairman of the Joint Chiefs of Staff; Office of Primary Responsibility: Director for Strategic Plans and Policy, Strategy Division. Washington, DC: U.S. Government Printing Office, June 2000.

http://www.joric.com/Conspiracy/Center.htm. A great site on the conspiracies to kill Hitler. To find the information on Elser, go to this site, then follow the link to The Lone Assassin, and continue to follow links to the end of the story.

Juhnke, James C., and Valerie Schrag. "The Original Peacemakers." *Fellowship*, May/June 1998, 9–10.

Keegan, John. *The Second World War.* 1st Amer. ed. New York: Viking Penguin, 1990.

Kennedy, Harold. "Marines Sharpen Their Skills in Hand-to-Hand Combat." *National Defense Magazine*, November 2003. http://www.nationaldefensemagazine.org/article.cfm?Id=1263 (accessed September 5, 2004).

Kennedy, Nancy. "Outrage-ous." *Shield* (the international magazine of the BP Amoco Group, U.S. ed.), Summer 1992. http://www.psandman.com/articles/shield.htm (accessed June 21, 2004).

Kershaw, Andy. "A Chamber of Horrors So Close to the 'Garden of Eden: In Foreign Parts in Basra, Southern Iraq.'" *Independent*, December 1, 2001. http://news.independent.co.uk/world/middle_east/story.jsp?story=107715 (accessed January 27, 2002).

Killian, Lewis M. *The Impossible Revolution? Black Power and the American Dream.* New York: Random House, 1968.

Kirby, Alex. "Fish Do Feel Pain, Scientists Say." *BBC News*, April 30, 2003. http://news.bbc.co.uk/2/hi/science/nature/2983045.stm (accessed May 12, 2003).

Koopman, John. "Interpreter's Death Rattles Troops: Iraqi Woman Became Close Friend of U.S. Soldiers." *San Francisco Chronicle*, August 1, 2004, A1. http://www.sfgate.com/cgi-bin/article.cgi?file=/chronicle/archive/2004/08/01/MNGJ57UGB826.DTL (accessed August 16, 2004).

Kopytoff, Verne. "Google Goes Forth into Great Beyond—Who Knows Where?" *San Francisco Chronicle*, May 2, 2004, A1.

Krag, K. *Plants Used as Contraceptives by the North American Indians.* Cambridge, MA: Harvard University Press, 1976.

Laing, R. D. *The Politics of Experience.* New York: Ballantine Books, 1967.

Lame Deer, John (Fire), and Richard Erdoes. *Lame Deer: Seeker of Visions.* New York: Simon and Schuster, 1972.

Larsen, Janet. "Dead Zones Increasing in World's Coastal Waters." Earth Policy Institute, June 16, 2004. http://www.earth-policy.org/Updates/Update41.htm (accessed June 20, 2004).

Lean, Geoffrey. "Why Antarctica Will Soon Be the *Only* Place to Live—Literally." *Independent,* May 2, 2004. http://news.independent.co.uk/world/environment/story.jsp?story=517321 (accessed May 6, 2004).

Ledeen, Michael. "Creative Destruction: How to Wage a Revolutionary War." *National Review Online,* September 20, 2001. http://www.nationalreview.com/contributors/ledeen092001.shtml (accessed May 17, 2003).

———. "Faster, Please." *National Review Online,* February 7, 2005. http://www.nationalreview.com/ledeen/ledeen200502070850.asp (accessed March 11, 2006).

———. "The Heart of Darkness: The Mullahs Make Terror Possible." *National Review Online,* December 12, 2002. At the Benador Associates website, http://www.benadorassociates.com/article/161 (accessed May 18, 2003).

———. "The Iranian Comedy Hour: In the U.S., the Silence Continues." *National Review Online,* October 23, 2002. At the Benador Associates website, http://www.benadorassociates.com/article/112 (accessed May 18, 2003).

———. "The Lincoln Speech." *National Review Online,* May 2, 2003. http://www.nationalreview.com/ledeen/ledeen050203.asp (accessed May 18, 2003).

———. "Machiavelli on Our War: Some Advice for Our Leaders." *National Review Online,* September 25, 2001. http://www.nationalreview.com/contributors/ledeen092501.shtml (accessed May 18, 2003).

———. "Scowcroft Strikes Out: A Familiar Cry." *National Review Online,* August 18, 2002. At the Benador Associates website, http://www.benadorassociates.com/article/71 (accessed May 18, 2003).

———. "The Temperature Rises: We Should Liberate Iran First—Now." *National Review Online,* November 12, 2002. At the Benador Associates website, http://www.benadorassociates.com/article/130 (accessed May 18, 2003).

———. "The Willful Blindness of Those Who Will Not See." *National Review Online,* February 18, 2003. http://www.nationalreview.com/ledeen/ledeen021803.asp (accessed May 18, 2003).

Leggett, Jeremy. *The Carbon War.* New York: Routledge, 2001.

LeGuin, Ursula. "Women/Wildness." In *Healing the Wounds.* Edited by Judith Plant. Philadelphia: New Society, 1989.

Liddell Hart, B. H., ed. *The Rommel Papers.* Translated by Paul Findlay. New York: Harcourt, Brace, and Company, 1953.

"Living in Reality: Indigenous and Campesino Resistance." *Green Anarchy,* no. 19, Spring 2005.

Livingston, John A. *The Fallacy of Wildlife Conservation.* Toronto: McClelland & Stewart, 1981.

Llanos, Miguel. "Study: Big Ocean Fish Nearly Gone." *MSNBC News,* May 14, 2003. http://www.msnbc.com/news/913074.asp?ocl=cR#BODY (accessed May 31, 2003).

Locke, John. *The Second Treatise on Government.* Edited with an introduction by J. W. Gough. New York: The MacMillan Company, 1956.

Lorde, Audre. "The Master's Tools Will Never Dismantle the Master's House." In *Sister/Outsider.* Trumansburg: The Crossing Press, 1984.

Losure, Mary. "Powerline Blues." *Minnesota Public Radio,* December 9, 2002. http://news.mpr .org/features/200212/08_losurem_powerline/ (accessed July 2, 2003).

Lyons, Dana. *Turn of the Wrench* (CD). Bellingham, WA: Reigning Records.

Malakoff, David. "Faulty Towers." *Audubon,* October 2001. http://magazine.audubon.org/ features0109/faulty_towers.html (accessed June 12, 2003).

Mallat, Chibli. "New Ways Out of the Arbitration Deadlock." *Daily Star,* December 19, 1996. http://www.soas.ac.uk/Centres/IslamicLaw/DS19-12-96EuroArabChib.html (accessed October 8, 2004).

Mann, Charles C. "1491." *Atlantic Monthly,* March 2002, 41–53. http://www.theatlantic.com/ issues/2002/03/mann.htm (accessed May 39, 2004).

Marcos, Subcomandante. *Our Word Is Our Weapon: Selected Writings of Subcomandante Insurgente Marcos.* New York: Seven Stories, 2001.

Martin, Brian. "Sabotage." Chap. 8 in *Nonviolence Versus Capitalism.* London: War Resisters' International, 2001. http://www.uow.edu.au/arts/sts/bmartin/pubs/01nvc/nvc08.html (accessed August 27, 2004).

Martin, Glen. "Battle of Battle Creek: Which Way to Save Salmon?" *San Francisco Chronicle,* March 15, 2004, A1, A11.

Martin, Harry V., with research assistance from David Caul. *FEMA: The Secret Government.* 1995. http://www.globalresearch.ca/articles/MAR402B.html (accessed March 12, 2006). This is all over the internet.

Marufu, L. T., B. F. Taubman, B. Bloomer, C. A. Piety, B. G. Doddridge, J. W. Stehr, and R. R. Dickerson. "The 2003 North American Electrical Blackout: An Accidental Experiment in Atmospheric Chemistry." *Geophysical Research Letters,* vol. 31, L13106,doi:10.1029/2004 GL019771, 2004. http://www.agu.org/pubs/crossref/2004/2004GL019771.shtml (accessed September 22, 2004).

Mason Jr., Herbert Molloy. *To Kill the Devil: The Attempts on the Life of Adolf Hitler.* New York: W. W. Norton & Company, 1978.

Matus, Victorino. "Big Bombs Are Best." *The Weekly Standard,* November 9, 2001. http://www.weeklystandard.com/Content/Public/Articles/000/000/000/514obizp.asp (accessed November 19, 2001).

McCarthy, Michael. "Disaster at Sea: Global Warming Hits UK Birds." *Independent*, 30 July 2004. http://news.independent.co.uk/uk/environment/story.jsp?story=546138 (accessed August 2, 2004).

———. "Greenhouse Gas 'Threatens Marine Life.'" *Independent*, February 4, 2005. http://news.independent.co.uk/world/environment/story.jsp?story=607579 (accessed February 9, 2005).

McConnell, Howard. "Remove the Dams on the Klamath River." *Eureka Times-Standard*, July 25, 2004. http://www.times-standard.com/Stories/0,1413,127~2906~2294032,00.html (accessed July 25, 2004).

McIntosh, Alistair. *Soil and Soul.* London: Aurum Press, 2002.

"Media March to War." *Fairness and Accuracy in Reporting*, September 17, 2001. http://www.fair.org/index.php?page=1853 (accessed March 11, 2006).

Melançon, Benjamin Maurice, with Vladimir Costés. "Landless Movement Regional Leader Jailed." *Narcosphere*, August 12, 2004. http://narcosphere.narconews.com/story/2004/8/13/0229/42902 (accessed October 8, 2004).

Merriam-Webster's Collegiate Dictionary, electronic ed., vers. 1.1, 1994–1995.

Mersereau, Adam. "Why Is Our Military Not Being Rebuilt? The Case for a Total War." *National Review Online*, May 24, 2002. http://www.nationalreview.com/comment/comment-mersereau052402.asp (accessed May 18, 2003).

Mies, Maria. *Patriarchy and Accumulation on a World Scale.* London: Zed Books, 1999.

Miller, Arthur. "Why I Wrote *The Crucible.*" *The New Yorker*, October 21, 1996, 158–64. http://www.newyorker.com/archive/content/?020422fr_archive02 (accessed December 1, 2004).

Ming Zhen Shakya. "What Is Zen Buddhism, Part II—Samsara and Nirvana." http://www.hsuyun.org/Dharma/zbohy/Literature/essays/mzs/whatzen2.html (accessed July 14, 2003).

Mitford, Jessica. *The American Prison Business.* New York: Penguin Books, 1977.

"MK84." FAS Military Analysis Network. http://www.fas.org/man/dod-101/sys/dumb/mk84.htm (accessed November 19, 2001).

Mokhiber, Russell. *Corporate Crime and Violence.* San Francisco: Sierra Club Books, 1988.

Mokhiber, Russell, and Robert Weissman. "Stossel Tries to Scam His Public." Essential Information, April 7, 2004. http://lists.essential.org/pipermail/corp-focus/2004/000177.html (accessed April 8, 2004).

Montgomery, David R. *King of Fish: The Thousand-Year Run of Salmon.* Boulder, CO: Westview, 2003.

Moodie, Donald. *The Record, or a Series of Official Papers Relative to the Condition and Treatment of the Native Tribes of South Africa.* Amsterdam: A. A. Balkema, 1960.

Moore, John Bassett. *A Digest of International Law.* Vol. 7. Washington, DC: Government Printing Office, 1906.

Moore, Kathleen Dean, and Jonathan W. Moore. "The Gift of Salmon." *Discover,* May 2003, 45–49.

Morgan, Edmund S. *American Slavery, American Freedom: The Ordeal of Colonial Virginia.* New York: W. W. Norton & Company, 1975.

Morgan, Jay. "Monks Always Get the Coolest Lines." Ordinary-Life.net. http://www.ordinary -life.net/blog/archives/002058.php (accessed July 29, 2003).

Mowat, Farley. *Sea of Slaughter.* Toronto: Seal, 1989.

Mulholland, Virginia. "The Plot to Assassinate Hitler." *Strategy & Tactics,* November/December 1976, 4–15.

Mullan, Bob, and Garry Marvin. *Zoo Culture: The Book About Watching People Watch Animals.* 2nd ed. Chicago: University of Illinois Press, 1999.

Mumford, Lewis. *The City in History: Its Origins, Its Transformations, and Its Prospects.* New York: Harcourt, Brace & World, 1961.

———. *The Myth of the Machine: Technics and Human Development.* New York: Harcourt Brace Jovanovich, 1966.

———. *The Myth of the Machine: The Pentagon of Power.* New York: Harcourt Brace Jovanovich, 1970.

Murray, Andrew. "Hostages of the Empire." *Guardian Unlimited,* Special Report: Iraq, July 1, 2003. http://www.guardian.co.uk/Iraq/Story/0,2763,988418,00.html (accessed October 10, 2004).

Letter from the National Science Foundation to the Center for Biological Diversity, October 16, 2002. http://www.biologicaldiversity.org/swcbd/species/beaked/NSFResponse.pdf (accessed October 26, 2002).

Neeteson, Kees. "The Dutch Low-Threshold [*sic*] Drugs Approach." http://people.zeelandnet .nl/scribeson/DutchApproach.html (accessed September 18, 2004).

New Columbia Encyclopedia. 4th ed. New York: Columbia University Press, 1975.

"New Iraq Abuse Pictures Surface." *Aljazeera.net,* May 6, 2004. http://english.aljazeera. net/NR/exeres/901052D2-7E43-49C3-A3F9-B0C3690CF59F.htm (accessed May 6, 2004).

"New World Vistas." Air Force Scientific Advisory Board, 1996, ancillary vol. 15.

"NMFS Refuses to Protect Habitat for World's Most Imperiled Whale: Despite Six Years of Continuous Sightings in SE Bering Sea, NMFS Claims It Can't Determine Critical Habitat for Right Whale." Center for Biological Diversity, February 20, 2002. http://www.biological diversity.org/swcbd/press/right2-20-02.html (accessed March 20, 2002).

NoFluoride. 2002. http://www.nofluoride.com/ (accessed January 21, 2002).

Nopper, Tamara Kil Ja Kim. "Yuri Kochiyama: On War, Imperialism, Osama bin Laden, and Black-Asian Politics." *AWOL Magazine,* 2003. http://awol.objector.org/yuri.html (accessed October 13, 2004).

Notes from Nowhere, eds. *We Are Everywhere: The Irresistible Rise of Global Anticapitalism.* New York: Verso, 2003.

Online Etymology Dictionary. http://www.etymonline.com/index.html (accessed August 14, 2004).

Oregon State Senate Bill 742, 72nd Legislative Assembly.

Orwell, George. *1984.* New York: New American Library, 1961.

Oxborrow, Judy. *The Oregonian,* January 20, 2002, F1.

Oxford English Dictionary, compact ed. Oxford: Oxford University Press, Oxford, 1985.

Patton Boggs, "Profile," http://www.pattonboggs.com/AboutUs/index.html (accessed July 22, 2004).

Paulson, Michael. "Deal Clears Way to Buy Elwha Dams: Dicks, Gorton and Babbitt Agree on Planning for Their Demolition." *Seattle PI,* October 20, 1999. http://seattlepi.nwsource.com/local/elwh20.shtml (accessed July 10, 2004).

Paz, Octavio. *The Labyrinth of Solitude.* New York: Grove Press, 1985.

Pearce, Joseph Chilton. *Magical Child.* New York: Plume, 1992.

Perlman, David. "Decline in Oceans' Phytoplankton Alarms Scientists: Experts Pondering Whether Reduction of Marine Plant Life Is Linked to Warming of the Seas." *San Francisco Chronicle,* October 6, 2003, A-6. http://sfgate.com/cgi-bin/article.cgi?f=/c/a/2003/10/06/MN31432.DTL&type=science (accessed October 28, 2003).

Peter, Laurence J. *Peter's Quotations: Ideas for Our Time.* New York: William Morrow and Company, 1977.

Pitt, William Rivers. "Kenny-Boy and George." *Truthout,* July 7, 2004. http://www.truthout.org/docs_04/070804A.shtml (accessed July 20, 2004).

Planck, Max. *Scientific Autobiography and Other Papers.* Translated by Frank Gaynor. New York: Philosophical Library, 1949.

"Population Increases and Democracy." http://www.eeeee.net/sd03048.htm (accessed September 23, 2002).

Priest, Dana, and Barton Gellman. "U.S. Uses Torture on Captive Terrorists: CIA Doesn't Spare the Rod in Interrogations." *San Francisco Chronicle,* December 26, 2002, A1, A21.

Project for the New American Century. "About PNAC." http://www.newamericancentury.org/aboutpnac.htm (accessed June 1, 2003).

PR Watch, Center for Media and Democracy. http://www.prwatch.org/cmd/ (accessed July 22, 2004).

Ramsland, Katherine. "Dr. Robert Hare: Expert on the Psychopath." Chap. 5, "The Psychopath Defined." *Court TV's Crime Library: Criminal Minds and Methods.* http://www.crimelibrary.com/criminal_mind/psychology/robert_hare/5.html?sect=19 (accessed August 6, 2004).

Rand, Ayn. Talk given at the United States Military Academy at West Point, NY, March 6, 1974. Nita Crabb, who has been of invaluable assistance in tracking down sources, was able to obtain a CD of this talk with the help of some extraordinary librarians at West Point. The sound quality is poor, but where you can hear what she says, it really is quite appalling.

"Rebuilding America's Defenses: Strategy, Forces and Resources for a New Century." A report of the Project for the New American Century, September 2000. http://www.newamerican century.org/RebuildingAmericasDefenses.pdf.

Reckard, E. Scott. "FBI Shift Crimps White-Collar Crime Probes: With More Agents Moved to Anti-Terrorism Duty, Corporate Fraud Cases are Routinely Put on Hold, Prosecutors Say." *Los Angeles Times*, August 30, 2004.

Regular, Arnon. "'Road Map Is a Life Saver for Us,' PM Abbas Tells Hamas." *Ha'aretz*, June 27, 2003. Also at *Unknown News* (and many other sites, though of course not in US capitalist newspapers), http://www.unknownnews.net/insanity7.html#quote (accessed June 30, 2003).

Reich, Wilhelm. *The Murder of Christ: The Emotional Plague of Mankind*. New York: Farrar, Strauss and Giroux, 1953.

Reitlinger, Gerald. *The Final Solution: The Attempt to Exterminate the Jews of Europe, 1939–1945*. 2nd ed. New York: Thomas Yoseloff, 1961.

Remedy. "Mattole Activists Assaulted, Arrested after Serving Subpoena for Pepper Spray Trial." *Treesit Blog*, August 27, 2004. http://www.contrast.org/treesit/ (accessed August 27, 2004).

"Report on the School of the Americas." Federation of American Scientists, March 6, 1997. http://www.fas.org/irp/congress/1997_rpt/soarpt.htm (accessed May 12, 2003).

"Report to the Seattle City Council WTO Accountability Committee by the Citizens' Panel on WTO Operations." Citizens' Panel on WTO Operations, September 7, 2000. http://www.cityofseattle.net/wtocommittee/panel3final.pdf (accessed March 17, 2002).

"Reviving the World's Rivers: Dam Removal." Part 4, Technical Challenges. International Rivers Network. http://www.irn.org/revival/decom/brochure/rrpt5.html (accessed July 11, 2004).

Revkin, Andrew C. "Bad News (and Good) on Arctic Warming." *New York Times*, October 30, 2004. The article is also at http://www.iht.com/bin/print_ipub.php?file=/articles/2004/10/29/news/arctic.html (accessed October 30, 2004).

Richardson, Paul. "Hojojutsu—The Art of Tying." Sukisha Ko Ryu: Bringing Together All the Elements of the Ninjutsu & Samuraijutsu Takamatsu-den Traditions. http://homepages .paradise.net.nz/sukisha/hojojutsu.html (accessed June 4, 2003).

"Rivers Reborn: Removing Dams and Restoring Rivers in California." Friends of the River. http://www.friendsoftheriver.org/Publications/RiversReborn/main3.html (accessed July 11, 2004).

Robbins, Tom. *Still Life with Woodpecker*. New York: Bantam, 1980.

Rogers, Lois. "Science Turns Monkeys into Drones—Humans Are Next, Genetic Experts Say." *Ottawa Citizen*, October 17, 2004. http://www.canada.com/ottawa/ottawacitizen/news/story.html?id=14314591-ee96-440f-8c83-11a9822d3d42 (accessed October 22, 2004).

Root, Deborah. *Cannibal Culture: Art, Appropriation, & the Commodification of Difference*. Boulder, CO: Westview Press, 1996.

Roycroft, Douglas. "Getting Well in Albion." *Anderson Valley Advertiser*, March 6, 2002, 8.

Russell, Diana E. H. *The Secret Trauma: Incest in the Lives of Girls and Women*. New York: Basic Books, 1986.

———. *Sexual Exploitation: Rape, Child Sexual Abuse, and Sexual Harassment*. Beverly Hills, 1984.

Rutten, Tim. "Cheney's History Needs a Revise." *Los Angeles Times*, November 26, 2005. http://fairuse.1accesshost.com/news3/latimes163.html (accessed November 28, 2005).

"Sabotage Blamed for Power Outage: Bolts Removed from 80-Foot Wisconsin Tower." *CNN.com*. http://www.cnn.com/2004/US/10/11/wisconsin.blackout.ap/ (accessed October 15, 2004).

Sadovi, Carlos. "Cell Phone Technology Killing Songbirds, Too." *Chicago Sun-Times*, November 30, 1999. Also at http://www.rense.com/politics5/songbirds.htm (accessed July 5, 2003).

Safire, William. "You Are a Suspect." *New York Times*, November 14, 2002.

Sale, Kirkpatrick. "An Illusion of Progress." *Ecologist*, June 2003. http://www.theecologist .org/archive_article.html?article=430&category=45 (accessed October 13, 2004).

San Francisco Chronicle. http://www.sfgate.com .

"Sardar Kartar Singh Saraba." Gateway to Sikhism. http://allaboutsikhs.com/martyrs/ sarabha.htm (accessed December 29, 2003). Citing Jagdev Singh Santokh., *Sikh Martyrs*. Birmingham, England: Sikh Missionary Resource Centre, 1995.

Savinar, Matt. "Life After the Oil Crash: Deal with Reality, or Reality Will Deal with You." http://www.lifeaftertheoilcrash.net/PageOne.html (accessed September 21, 2004).

Scheffer, Marten, Steve Carpenter, Jonathan A. Foley, Carl Folke, and Brian Walker. "Catastrophic Shifts in Ecosystems." *Nature*, October 11, 2001, 591–96.

Scherer, Glenn. "Religious Wrong: A Higher Power Informs the Republican Assault on the Environment." *E Magazine*, May 5, 2003, 35–39. Also available online under the title "Why Ecocide Is 'Good News' for the GOP." http://www.mindfully.org/Reform/2003/Ecocide -Is-Good-News5may03.htm (accessed May 21, 2003).

Schmitt, Diana. "Weapons in the War for Human Kindness: Why David Budbill Sits on a Mountaintop and Writes Poems." *The Sun*, March 2004.

Schor, Juliet B. *The Overworked American: The Unexpected Decline of Leisure*. New York: Basic Books, 1991.

Schuld, Andreas. "Dangers Associated with Fluoride." EcoMall: A Place to Save the Earth. http://www.ecomall.com/greenshopping/fluoride2.htm (accessed January 21, 2002).

Seekers of the Red Mist. http://www.seekersoftheredmist.com/ (accessed July 10, 2003). This particular comment was posted April 25, 2003, at 8:31 am.

Severn, David. "Vine Watch." *Anderson Valley Advertiser*, April 2, 2003, 8.

Shirer, William. *The Rise and Fall of the Third Reich: A History of Nazi Germany*. Greenwich: Fawcett Crest, 1970.

Shulman, Alix Kates. "Dances with Feminists." The Emma Goldman Papers, University of California, Berkely. http://sunsite.berkeley.edu/Goldman/Features/dances_shulman.html (accessed October 8, 2004). First published in the *Women's Review of Books* 9, no. 3 (December 1991).

"Signs to Look for in a Battering Personality." Projects for Victims of Family Violence, Inc. http://www.angelfire.com/ca6/soupandsalad/content13.htm (accessed November 17, 2002).

Silko, Leslie Marmon. "Tribal Councils: Puppets of the U.S. Government." In *Yellow Woman and a Beauty of the Spirit*. New York: Touchstone Books, 1997.

Sluka, Jeff. "National Liberation Movements in Global Context." Tamilnation.org. http://www.tamilnation.org/selfdetermination/fourthworld/jeffsluka.htm (accessed October 10, 2004).

Socially Responsible Shopping Guide. Global Exchange. http://www.globalexchange.org/economy/corporations/sweatshops/ftguide.html (accessed March 16, 2002).

Spretnak, Charlene. *States of Grace*. San Francisco: HarperSanFrancisco, 1991.

Stannard, David. *American Holocaust: Columbus and the Conquest of the New World*. Oxford: Oxford University Press, 1992.

Stark, Lisa, and Michelle Stark. "$100 Million Wasted: While Some Soldiers Paid Their Own Way, Thousands of Pentagon Airline Tickets Went Unused." *ABC News*. http://abcnews.go.com/sections/WNT/YourMoney/wasted_airline_tickets_040608-1.html (accessed June 9, 2004).

Star Wars. http://www.starwars.com/databank/location/deathstar/ (accessed April 23, 2004).

Star Wars 2. http://www.starwars.com/databank/location/deathstar/?id=eu (accessed April 24, 2004).

"States Get $16 Million for Endangered Species." *Environmental News Network*, September 28, 2001. http://www.enn.com/news/enn-stories/2001/09/09282001/s_45096.asp (accessed January 16, 2002).

St. Clair, Jeffrey. "Santorum: That's Latin for Asshole." *Anderson Valley Advertiser*, April 30, 2003, 8.

St. Clair, Jeffrey, and Alexander Cockburn. "Born Under a Bad Sky." *Anderson Valley Advertiser*, September 18, 2002, 5.

Steele, Jonathan. "Bombers' Justification: Russians Are Killing Our Children, So We Are Here to Kill Yours: Chechen Website Quotes Bible to Claim Carnage as Act of Legitimate Revenge." *Guardian Unlimited*, September 6, 2004. http://www.guardian.co.uk/russia/article/0,2763,1298075,00.html (accessed September 6, 2004).

"A Study of Assassination." This can be found at innumerable Web sites (well, a Google search shows 138). One version complete with drawings is at http://www.gwu.edu/~nsarchiv/NSAEBB/NSAEBB4/ciaguat2.html (accessed July 7, 2003).

"Study Says Five Percent of Greenhouse Gas Came from Exxon." *Planet Ark*, January 30, 2004. http://www.planetark.com/dailynewsstory.cfm/newsid/23638/story.htm (accessed June 22, 2004).

The Sun. http://thesunmagazine.org/.

Sunderland, Larry T. B. "California Indian Pre-Historic Demographics." Four Directions Institute. http://www.fourdir.com/california_indian_prehistoric_demographics.htm. See also the map "Native American Cultures Populations Per Square Mile at Time of European Contact." http://www.fourdir.com/aboriginal_population_per_sqmi.htm (accessed June 4, 2004).

Sweating for Nothing. Global Exchange, *Global Economy.* http://www.globalexchange.org/economy/corporations/ (accessed March 16, 2002).

Taber, Robert. *The War of the Flea.* 1st paperbound ed. New York: The Citadel Press, 1970.

Tebbel, John, and Keith Jennison. *The American Indian Wars.* Edison, NJ: Castle Books, 2003.

"Third National Incidence Study of Child Abuse and Neglect." Centers for Disease Control.

Thobi, Nizza. "Chanah Senesh." http://www.nizza-thobi.com/Senesh_engl.html (accessed December 3, 2004).

Thomas, Emory M. *The Confederate State of Richmond: A Biography of the Capital.* Austin: University of Texas Press, 1971.

Thomson, Bruce. "The Oil Crash and You." Great Change. http://greatchange.org/ov-thomson,convince_sheet.html (accessed September 28, 2004).

Thompson, Don. "Klamath Salmon Plight Worsens: State: Fish Kill May Be Double Previous Estimates." *The Daily Triplicate* (Crescent City, CA), July 31, 2004, A1, A10.

Tillich, Paul. *Systematic Theology.* Vol. 3, *Life and the Spirit; History and the Kingdom of God.* Chicago: University of Chicago Press, 1963.

Tokyo War Crimes Trial Decision. The International Military Tribunal for the Far East, May 3, 1946, to November 12, 1948.

Tomlinson, Chris. "Evidence of U.S. Bombs Killing Villagers." *San Francisco Chronicle,* December 6, 2001, A10.

"Too Hot for Uncle John's Bathroom Reader." Triviahalloffame. http://www.triviahalloffame.com/gandhi.htm (accessed August 8, 2004).

Towerkill. http://www.towerkill.com/ (accessed July 5, 2003).

Trial of the Major War Criminals before the International Military Tribunal, Nuremberg, 14 November 1945–1 October 1946. Nuremberg, 1947–1949.

Trudell, John. *Green Anarchy,* Fall 2003, 15.

Turner, Frederick. *Beyond Geography: The Western Spirit Against the Wilderness.* New Brunswick, NJ: Rutgers, 1992.

Udall, Stewart, Charles Conconi, and David Osterhout. *The Energy Balloon.* New York: McGraw-Hill, 1974.

U.S. Congress. *Congressional Record.* 56th Cong., 1st sess., 1900. Vol. 33, 704, 711–12.

U.S. et al. v. Goering et al. [The Nuremberg Trial]. Extra Lexis 1, 120, 6 F.R.D. 69 (1947) (accessed December 16, 2005).

"U.S. Military Spending to Exceed Rest of World Combined!" *Nexus*, September/October 2005, 9.

"The Victims: The Fight Against Terrorism." *The Oregonian*, January 16, 2002, A2.

Vidal, Gore. *The Decline and Fall of the American Empire*. Berkeley: Odonian Press, 1995.

Wacquant, Loïc. "Ghetto, Banlieue, Favela: Tools for Rethinking Urban Marginality." http://sociology.berkeley.edu/faculty/wacquant/condpref.pdf (accessed March 16, 2002).

Walker, Paul F., and Eric Stambler. ". . . And the Dirty Little Weapons: Cluster Bombs, Fuel-Air Explosives, and 'Daisy Cutters,' not Laser-Guided Weapons, Dominated the Gulf War." *Bulletin of the Atomic Scientists* 47, no. 4 (May 1991): 21–24. Also available at http://www.bullatomsci.org/issues/1991/may91/may91walker.html (accessed November 19, 2001).

Watson, Paul. "Report from the Galapagos." *Earth First! Journal*, Samhain/Yule 2003, 38.

"Weapons of American Terrorism: Torture." http://free.freespeech.org/americanstateterrorism/weapons/US-Torture.html (accessed May 12, 2003).

Weber, Max. *Max Weber: The Theory of Social and Economic Organization*. Translated by A. M. Henderson and Talcott Parsons. Edited with an introduction by Talcott Parsons. Oxford: Oxford University Press, 1947.

Webster's New Twentieth Century Dictionary of the English Language, 2nd ed. New York: Simon and Schuster, 1979.

Weiss, Rick. "Major Species Annihilated by Fishing, Says Study." *San Francisco Chronicle*, May 15, 2003, A13.

Weizenbaum, Joseph. *Computer Power and Human Reason: From Judgment to Calculation*. San Francisco: W. H. Freeman, 1976.

Whales. http://www.biologicaldiversity.org/swcbd/press/beaked10-15-2002.html, http://action network.org/campaign/whales/explanation, sites visited October 26, 2002. Also, http://www.faultline.org/news/2002/10/beaked.html, site visited October 27, 2002.

"What Is Depleted Uranium." http://www.web-light.nl/VISIE/depleted_uranium1.html (accessed January 23, 2002).

"What's the Dam Problem?" *The Why Files: The Science Behind the News*, January 16, 2003. http://whyfiles.org/169dam_remove/index.html (accessed July 11, 2004).

Wheatley, Margaret. *Turning to One Another: Simple Conversations to Restore Hope to the Future*. San Francisco: Berrett-Koehler Publishers, 2002.

White, Chris. "Why I Oppose the U.S. War on Terror: An Ex-Marine Sergeant Speaks Out." *The Grass Root* 3, no. 1 (Spring 2003). *The Grass Root* is the paper of the Kansas Greens (Kansas Green Party, Box 1482, Lawrence, KS 66044).

"Why Is Everybody Always Pickin' on Me?" Dispatches, *Outside Online*, July 1998. http://web.outsideonline.com/magazine/0798/9807disprod.html (accessed July 10, 2003).

Wikle, Thomas A. "Cellular Tower Proliferation in the United States." *The Geographical Review* 92, no. 1 (January 2002): 45–62.

Wilkinson, Bob. "Trained Killers." *Anderson Valley Advertiser*, April 30, 2003, 3.

Williams, Martyn. "UN Study: Think Upgrade Before Buying a New PC: New Report Finds 1.8 Tons of Material Are Used to Manufacture Desktop PC and Monitor." *Infoworld*, March 7, 2004. http://www.infoworld.com/article/04/03/07/hnunstudy_1.html (accessed March 12, 2004).

Wilson, Jim. "E-Bomb: In the Blink of an Eye, Electromagnetic Bombs Could Throw Civilization Back 200 Years. And Terrorists Can Build Them for $400." *Popular Mechanics*, September 2001. http://popularmechanics.com/science/military/2001/9/e-bomb/print.phtml (accessed August 22, 2003).

Wingate, Steve. "The OMEGA File—Concentration Camps: Federal Emergency Management Agency." http://www.posse-comitatus.org/govt/FEMA-Camp.html (accessed July 21, 2004).

"Witch Hunting and Population Policy." http://www.geocities.com/iconoclastes.geo/witches.html (accessed September 23, 2002).

The World Factbook, s.v. "Afghanistan." CIA. http://www.odci.gov/cia/publication/factbook/geos/af/html (accessed November 19, 2001).

Wyss, Jim. "Ecuador Free-for-All Threatens Tribes, Trees: Weak Government Lets Loggers Prevail." *San Francisco Chronicle*, September 3, 2004, W1.

Yergin, Daniel. *The Prize: The Epic Quest for Oil, Money & Power.* New York: Simon & Schuster, 1991.

Z Magazine, July/August 2000, 62.

About the Author

DERRICK JENSEN is the acclaimed author of *A Language Older Than Words* and *The Culture of Make Believe*, among many others. Author, teacher, activist, small farmer, and leading voice of uncompromising dissent, he regularly stirs auditoriums across the country with revolutionary spirit. Jensen lives in Northern California and organizes on issues surrounding deforestation, dam removal, restoration of habitat for salmon and other fish, habitat improvement for amphibians, the promotion of organic farms, and the preservation of family farms. He holds a degree in Creative Writing from Eastern Washington University, a degree in Mineral Engineering Physics from the Colorado School of Mines, and has taught at Eastern Washington University and Pelican Bay State Prison.